CAMBRIDGE STUDIES IN ADVANCED MATHEMATICS 19

Cellular structures in topology

Already published

1 W.M.L. Holcombe *Algebraic automata theory*
2 Karl Petersen *Ergodic theory*
3 Peter T. Johnstone *Stone spaces*
4 W.H. Schikhof *Ultrametric calculus*
5 J.-P. Kahane *Some random series of functions*
6 H. Cohn *Introduction to the construction of class fields*
7 J. Lambek & P. Scott *Introduction to higher-order categorical logic*
8 H. Matsumura *Commutative ring theory*
9 C.B. Thomas *Characteristic classes and the cohomology of finite groups*
10 M. Aschbacher *Finite group theory*
11 J.L. Alperin *Local representation theory*
12 Paul Koosis *The logarithmic integral: I*
13 A. Pietsch *Eigenvalues and s-numbers*
14 S.J. Patterson *Introduction to the theory of the Riemann zeta-function*
15 H.-J. Baues *Algebraic homotopy*
16 V.S. Varadarajan *Introduction to harmonic analysis on semisimple Lie groups*
17 M. Dunwoody & W. Dicks *Groups acting on graphs*
18 L. Corwin & F. Greenleaf *Representations of nilpotent Lie groups and their applications*
19 R. Fritsch & R. Piccinini *Cellular structures in topology*
20 H. Klingen *Introduction to modular functions*
21 M. Collins *Representations and characters of finite groups*

Cellular structures in topology

RUDOLF FRITSCH
Ludwig-Maximilians-Universität, München, Germany

RENZO A. PICCININI
Memorial University of Newfoundland, St John's, Canada

The right of the University of Cambridge to print and sell all manner of books was granted by Henry VIII in 1534. The University has printed and published continuously since 1584.

CAMBRIDGE UNIVERSITY PRESS
Cambridge
New York Port Chester
Melbourne Sydney

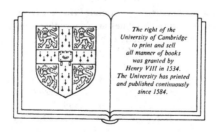

CAMBRIDGE UNIVERSITY PRESS
Cambridge, New York, Melbourne, Madrid, Cape Town, Singapore, São Paulo

Cambridge University Press
The Edinburgh Building, Cambridge CB2 8RU, UK

Published in the United States of America by Cambridge University Press, New York

www.cambridge.org
Information on this title: www.cambridge.org/9780521327848

First published 1990
This digitally printed version 2008

A catalogue record for this publication is available from the British Library

Library of Congress Cataloguing in Publication data

Fritsch, Rudolf,
Cellular structures in topology / Rudolf Fritsch, Renzo A. Piccinini.
p. cm.–(Cambridge studies in advanced mathematics)
Bibliography: p.
Includes index.
ISBN 0 521 32784 9
1. CW complexes. 2. Complexes. 3. *k*-spaces. I. Piccinini,
Renzo A., 1933– . II. Series.
QA611.35.F75 1990
514′.223–dc20 89–35772 CIP

ISBN 978-0-521-32784-8 hardback
ISBN 978-0-521-06387-6 paperback

To our wives Nair Piccinini and Gerda Fritsch
for their continuous and steadfast support

Contents

Preface

Felix, qui potuit rerum
cognoscere causas!

P. Vergilius Maro, Georgica 2,490

Cellular structures play an essential role in topology, analysis and geometry; they appear in the form of CW-complexes, simplicial sets and so on. The idea of this book is to give a unified treatment of their fundamental geometric and topological (in the sense of general topology) properties. As a common basis for their representation we have chosen the CW-complexes.

CW-complexes were formally introduced in the literature in 1949 by the great English mathematician John H.C. Whitehead. To appreciate better the depth and perception of Whitehead's ideas, it is worth looking back into the development of algebraic topology; on this trip through history we take Solomon Lefschetz as our Virgil. In his beautiful history of the early development of algebraic topology (see Lefschetz, 1970), Lefschetz shows us how homology was defined by Henri Poincaré – whom he calls the 'Founder' of algebraic topology – using spaces with a combinatorial structure; Lefschetz then points out the next stage in the development of the subject, namely the definition of homology for topological spaces and the introduction of the homotopy groups of spaces. What Whitehead did was to impose again a combinatorial structure on the spaces and to show how this leads to a much deeper insight into their homotopy groups. This and other particularly interesting properties of CW-complexes explain why their presence is felt throughout many branches of mathematics. The first two chapters of this book are devoted to the theory of CW-complexes.

Chapters 3 and 4 deal with the theory of simplicial complexes and simplicial sets; we feel that the existence of a very large body of research in that area and the importance of combinatorial structures in topology amply justify the relatively large size of these two chapters.

In the fifth chapter we study the category of spaces having the homotopy type of CW-complexes. We end the book with an appendix containing the results of homotopy theory, topology and dimension theory necessary to the development of the book. Normally we do not prove the results presented in the appendix but we indicate where the proofs can be found. The appendix should be read using the index, as sometimes the definitions are not written in order but, rather, following the flow of each section.

Because we emphasize geometric and combinatorial structures (and the arguments related to them), the material we borrowed from algebraic topology is mostly related to the theory of homotopy groups with only a minimal contribution from homology; in our minds we view homotopy groups as more intimately related to our geometric intuition than homology groups. As a consequence, the results about cellular structures that are heavily dependent on homology theory (e.g., cellular homology, obstruction theory, Wall obstruction to finiteness, classifying spaces, etc.) are not discussed in the book. However, we lay down the ground work needed for the development of these areas.

Although most of the exercises can be worked out easily using the material in the text, there are some exercises which require the reader to consult the references given in each case; the problems of this latter type have been inserted in the book in order to draw the reader's attention to interesting results which, however, could not be incorporated in the text without enlarging it to unmanageable dimensions. We apologise to their authors for presenting their work as exercises, possibly giving the impression that we do not consider it as important enough to be in the text; indeed, the contrary is true: in spite of the obvious lack of space, we did not just pass by and overlook these results!

With regard to the historical notes we wish to say that we have not done specific research to trace back carefully all the definitions and results presented in the book. We just give hints to our sources and apologise to all concerned if we have unintentionally given incorrect credits.

The reader is assumed to be familiar with the standard facts of general topology and category theory; as basic sources of information on these areas one can take, respectively, the classical books by John L. Kelley (1956) and Saunders MacLane (1971).

A few remarks about the notation used in the book: with the exception of Section A.1, the symbol *Top* denotes the category of weak Hausdorff k-spaces and continuous functions, explained just in that section. The word *map* always indicates continuity; a non-necessary continuous assignment between points of spaces of simply called a *function*. Finally, the symbol $\times$ between k-spaces always denotes the product in the category of k-spaces.

Many persons and institutions have given us a lot of support, encouragement and suggestions along the way; in particular, we wish to give our heartfelt thanks to: Professors Tammo tom Dieck, Philip Heath, Peter Hilton, Dana May Latch, Dieter Puppe; Drs Thomas Bartsch and Georg Peschke; Universität Konstanz, Ludwig-Maximilians-Universität, Memorial University of Newfoundland; DFG (Deutsche Forschungsgemeinschaft) and NSERC (Natural Sciences and Engineering

Research Council of Canada). Special thanks are due to the Max-Planck-Institut für Biochemie, in Martinsried near München, and in particular, to Dr Wolfgang Steigemann, Director of its Computer Centre, who introduced us to the wonders of 'computer text editing'; in this field we were also greatly helped by Professors Herb Gaskill, Edgar Goodaire and P.P. Narayanaswami. Last, but not least, we wish to thank Mr David Tranah, our friendly Mathematics Editor at Cambridge University Press, for his continuous assistance and support.

RUDOLF FRITSCH RENZO A. PICCININI

1

The fundamental properties of CW-complexes

Balls and balloons are the standard models for the cells used in the theory of CW-complexes; thus, the chapter starts by 'playing' a bit with such toys. Next, it continues with a discussion of the problem of attaching n-cells to a space and with the actual construction of CW-complexes, followed by a detailed study of the fundamental properties of such spaces.

The unusual number given to the first section of this chapter, namely 1.0, stems from the fact that the material discussed therein is really very elementary.

1.0 Balls, spheres and projective spaces

The *ball* in the Euclidean space $\mathbf{R}^{n+1}$ is the space

$$B^{n+1} = \{s = (s_0, s_1, \ldots, s_n) : |s| \leqslant 1\};$$

its topological boundary is the *sphere*

$$\delta B^{n+1} = S^n = \{s \in B^{n+1} : |s| = 1\}$$

and the difference

$$\mathring{B}^{n+1} = B^{n+1} \backslash S^n$$

is the interior of the ball B^{n+1}, namely, the *open ball.* Observe that the ball $B^1 = [-1, 1]$ does not coincide with the unit interval $I = [0, 1]$ (in the sequel, the boundary of I will be denoted by $\dot{I}$).

Intuitively, one may view a sphere as the skin of a ball (i.e., a balloon). To blow up a balloon, there must be an opening, a 'base point'; thus, set the point $e_0 = (1, 0, \ldots, 0)$ as the base point of both B^{n+1} and S^n.

Spheres do not appear only as boundaries of balls; in addition to the inclusions

$$i^n : S^n \to B^{n+1},$$

it will be necessary to discuss several standard maps relating spheres and balls. The list of such maps described in this section is actually longer than that needed to develop the material herein. The primary two reasons are:

these maps could be used to fill in the details for the material sketched in the appendix;

some of the maps discussed could be used in the homology of cellular structures (e.g., the Hurewicz isomorphism theorem). Although homology is beyond the scope of this volume, it is a natural continuation for the theory here developed.

It is often convenient to view all balls B^{n+1} and all spheres S^n as contained in the space $\mathbf{R}^\infty$ of all sequences which vanish almost everywhere, via the embeddings $s \mapsto (s, 0, 0, \ldots)$; the topology of $\mathbf{R}^\infty$ is determined by the family of all Euclidean subspaces $\mathbf{R}^n$ (see Section A.2). Within this framework, consider the origin of $\mathbf{R}^\infty$ as the 0-*ball*

$$B^0 = \{0\},$$

whose boundary is the 'sphere'

$$\delta B^0 = S^{-1} = \varnothing,$$

and which coincides with its interior

$$\mathring{B}^0 = B^0.$$

In contrast with these 'minimal' models B^0 and S^{-1}, one has the *infinite ball* $B^\infty = \bigcup_{n \geqslant 0} B^n$ and the *infinite sphere* $S^\infty = \bigcup_{n \geqslant 0} S^n$ as subspaces of R^∞. Notice that these two infinite models are determined by the corresponding families of finite models (see Corollary A.2.3).

The ball B^n is embedded into the ball B^{n+1} as a strong deformation retract; a suitable retraction is the map

$$j^n : B^{n+1} \to B^n,$$

given by

$$j^n(s) = (s_0, \ldots, s_{n-1}).$$

Define the 'eggs of Columbus' using the map j^n, i.e. the inclusions

$$j_+, j_- : B^{n+1} \rightarrowtail B^{n+1}$$

given by

$$j_+(s) = (s_0, \ldots, s_{n-1}, \tfrac{1}{2}(s_n + \sqrt{1 - |j^n(s)|^2}))$$

and

$$j_-(s) = (s_0, \ldots, s_{n-1}, \tfrac{1}{2}(s_n - \sqrt{1 - |j^n(s)|^2})).$$

The function j_+ (resp. j_-) maps the upper (resp. lower) hemisphere onto itself and the lower (resp. upper) hemisphere onto the equatorial ball B^n (see Figure 1).

The deformation

$$d^n : (B^n \times B^n) \times I \to B^n \times B^n$$

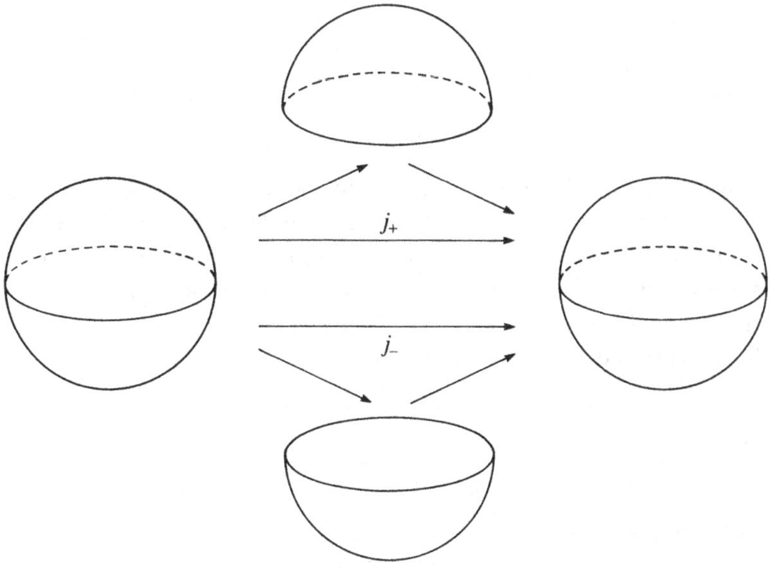

Figure 1

defined by

$$d^n((s, s'), t) = ((1-t)s + t(\tfrac{1}{2}(s+s')), (1-t)s' + t(\tfrac{1}{2}(s+s'))),$$

for every $(s, s') \in B^n \times B^n$ and every $t \in I$, shows that the diagonal subspace $\Delta B^n \subset B^n \times B^n$ is a strong deformation retract of $B^n \times B^n$; thus, balls are LEC spaces (see Section A.4, page 253).

The sphere S^{n-1} is included into the sphere S^n as its equator, and this inclusion, in turn, extends to embeddings

$$i_-, i_+ : B^n \rightarrowtail S^n$$

of the ball B^n into the southern, respectively northern hemisphere of S^n, given by

$$i_-(s) = (s, -\sqrt{1-|s|^2})$$

and

$$i_+(s) = (s, \sqrt{1-|s|^2}),$$

respectively, having $j^n | S^n$ as common left inverse.

The maps i_-, i_+ are homotopic only in a very curious way; in fact, a homotopy can be constructed by observing that both maps are homotopic to the constant map onto the base point, but there is no homotopy between them relative to the boundary (see the end of this paragraph). Viewed as maps into B^{n+1} the maps i_-, i_+ are homotopic in a neat manner namely, rel. S^{n-1} via the map

$$h^n : B^n \times I \to B^{n+1},$$

given by

$$h^n(s,t) = (s, (2t-1)\sqrt{1-|s|^2}).$$

The importance of this map h^n resides in the fact that every homotopy rel. S^{n-1} given between two maps defined on B^n factors through h^n. In particular this shows: If i_-, i_+ were homotopic rel. S^{n-1}, any corresponding homotopy factored through h^n would yield a retraction of B^{n+1} onto S^n, contradicting Brouwer theorem (see TheoremA.9.4).

Next, recall that the map (Figure 2)

$$c^n : S^n \times I \to B^{n+1}$$

given by

$$c^n(s,t) = (1-t)e_0 + ts$$

induces a homeomorphism

$$S^n \wedge I \to B^{n+1}$$

where the symbol $\wedge$ denotes the usual *smash product*

$$S^n \wedge I = S^n \times I / S^n \times \{0\} \cup \{e_0\} \times I.$$

The formation of the smash product with one factor equal to I is also known as the *reduced cone construction.* The *reduced suspension* of a based space (X, x_0) is one step further away; this construction is given on the based space (X, x_0) by

$$\Sigma.X = X \times I / X \times \dot{I} \cup x_0 \times I$$

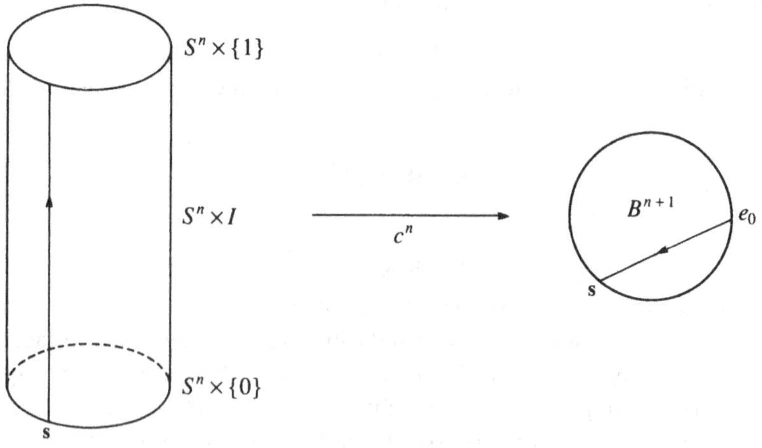

Figure 2

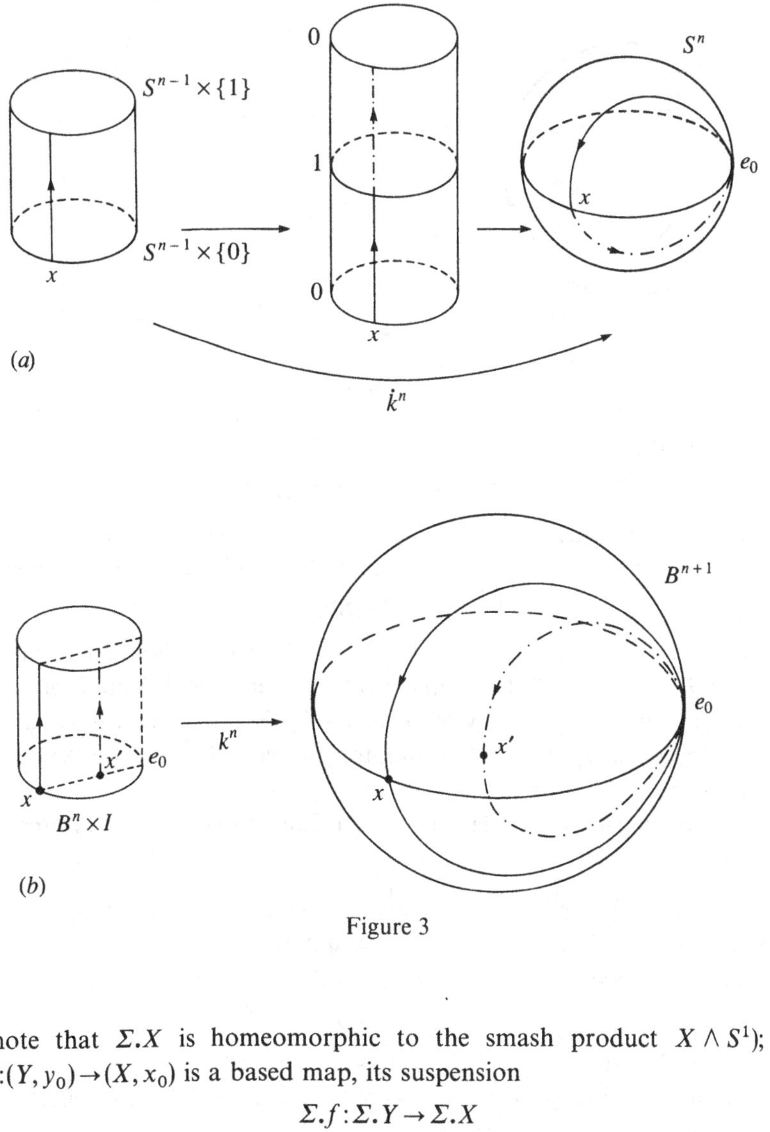

Figure 3

(note that $\Sigma.X$ is homeomorphic to the smash product $X\wedge S^1$); if $f:(Y,y_0)\to(X,x_0)$ is a based map, its suspension

$$\Sigma.f:\Sigma.Y\to\Sigma.X$$

is the map induced by $f\times 1: Y\times I\to X\times I$.

For $n\geqslant 1$, define $\dot{k}^n: S^{n-1}\times I\to S^n$ (see Figure 3(*a*)) by

$$\dot{k}^n(s,t)=\begin{cases} i_+c^{n-1}(s,2t), & 0\leqslant t\leqslant\frac{1}{2}\\ i_-c^{n-1}(s,2-2t), & \frac{1}{2}\leqslant t\leqslant 1;\end{cases}$$

the map $\dot{k}^n$ takes $S^{n-1}\times\dot{I}\cup e_0\times I$ into e_0 and is bijective outside that space; thus, it induces a homeomorphism $\Sigma.S^{n+1}\to S^n$. Moreover, the map $\dot{k}^n$ can be extended to a map $k^n: B^n\times I\to B^{n+1}$ (see Figure 3(*b*))

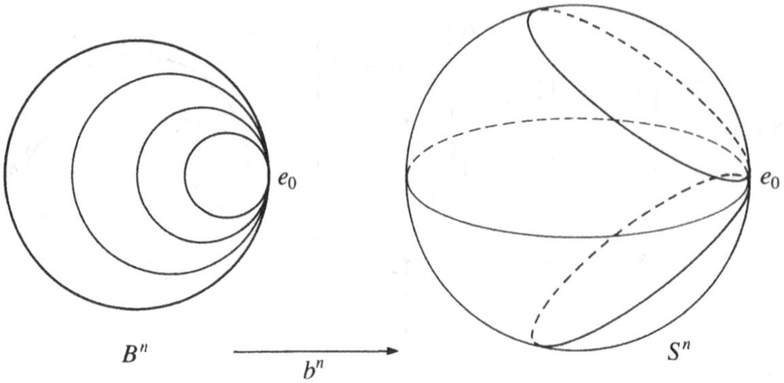

Figure 4

simply by taking

$$k^n(c^{n-1}(s, t'), t) = c^n(\dot{k}^n(s, t), t');$$

this latter map induces a homeomorphism $\Sigma.B^n \to B^{n+1}$. Finally, notice that the map $\dot{k}^n$ factors through the map c^{n-1}, and thus induces a map

$$b^n : B^n \to S^n;$$

formally, $b^n \circ c^{n-1} = \dot{k}^n$. In turn, the map b^n gives a homeomorphism between B^n/S^{n-1} and S^n. It is convenient to extend the definition of b^n to include $b^0 : B^0 \to S^0$ given by $b^0(B^0) = \{-1\}$. Figure 4 indicates that b^n is homotopic rel. $\{e_0\}$ to i_+ via a homotopy moving S^{n-1} only in the lower hemisphere.

The following maps are relevant to the definition of homotopy groups:

(i) the *units*

$$u^n : B^{n+1} \to B^{n+1}, \qquad \dot{u}^n : S^n \to S^n$$

defined for all $n \in \mathbf{N}$ as the constant-based maps;

(ii) the *inversions*

$$l^n : B^{n+1} \to B^{n+1},$$

defined by $l^n(k^n(s, t)) = k^n(s, 1-t)$ for every $(s, t) \in B^n \times I$; this inversion on B^{n+1} induces an inversion $\dot{l}^n : S^n \to S^n$ on S^n; notice that l^n, $\dot{l}^n$ are reflections about the hyperplane $\mathbf{R}^n \subset \mathbf{R}^{n+1}$:

$$l^n(s_0, \ldots, s_n, s_{n+1}) = (s_0, \ldots, s_n, -s_{n+1});$$

(iii) for $n \geqslant 1$, the *pinchings* (see Figure 5)

$$p^n : B^{n+1} \to B^{n+1} \vee B^{n+1}$$

given by

$$p^n(k^n(s, t)) = \begin{cases} (k^n(s, 2t), e_0), & 0 \leqslant t \leqslant \frac{1}{2} \\ (e_0, k^n(s, 2t-1)), & \frac{1}{2} \leqslant t \leqslant 1; \end{cases}$$

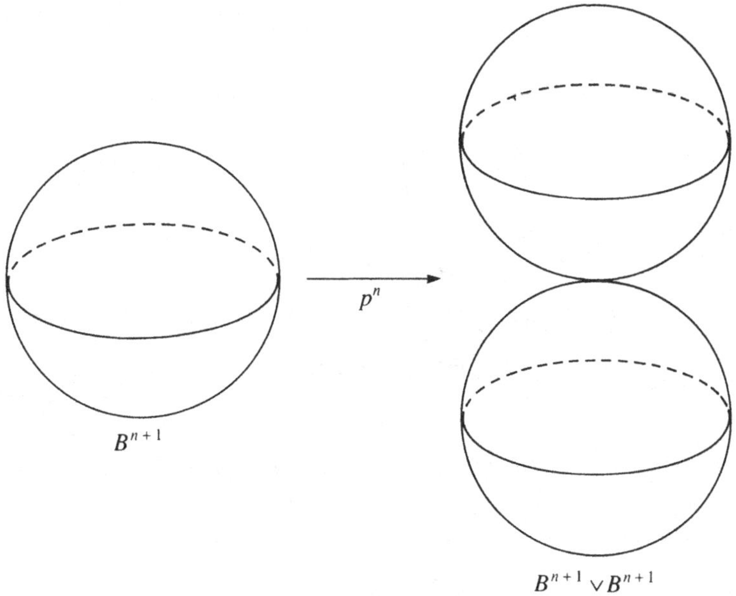

Figure 5

this means that the points with last coordinate equal to zero are mapped into the wedge point (e_0, e_0).

The maps p^n induce the *pinching of the spheres*

$$\dot{p}^n : S^n \to S^n \vee S^n.$$

(The symbol $\vee$ denotes the usual *wedge product*: for any pair of based spaces, say (X, x_0), (Y, y_0), the space $X \vee Y$ is defined to be $X \times \{y_0\} \cup \{x_0\} \times Y$, regarded as a subspace of $X \times Y$.)

An inaccurate but graphic description of the pinching is provided by cell division, a basic process in biology.

For $n \geqslant 2$, there is another useful type of *pinching*:

$$\hat{p} : B^{n+1} \to B^{n+1} \vee B^{n+1}$$

given by

$$\hat{p}^n(k^n(k^{n-1}(s,u),t)) = \begin{cases} (k^n(k^{n-1}(s,2u),t), e_0), & 0 \leqslant u < \frac{1}{2}, \\ (e_0, k^n(k^{n-1}(s,2u-1),t)), & \frac{1}{2} \leqslant u \leqslant 1. \end{cases}$$

This means that the points with penultimate coordinate equal to zero are mapped into the wedge point (e_0, e_0).

Next, consider the map obtained by projecting $B^{n+1} \times I$ onto $B^{n+1} \times \{0\} \cup S^n \times I$ from $(0, 2)$ in $\mathbf{R}^{n+1} \times \mathbf{R}$ (see Figure 6):

$$r^{n+1} : B^{n+1} \times I \to B^{n+1} \times \{0\} \cup S^n \times I$$

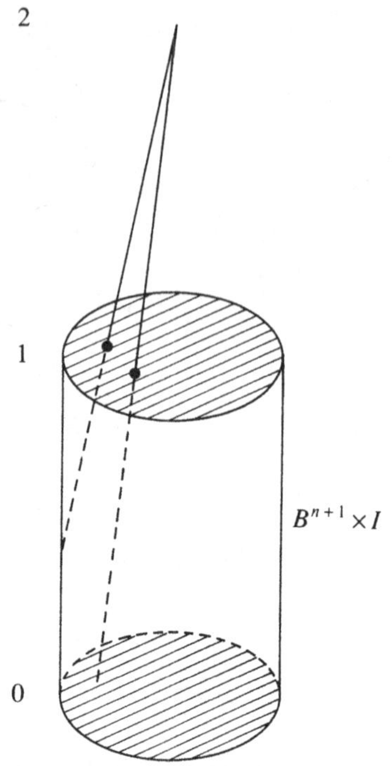

Figure 6

$$r^{n+1}(s,t)=\begin{cases}\dfrac{2}{2-t}(s,0), & 0\leqslant t\leqslant 2(1-|s|),\\[2ex] \dfrac{1}{|s|}(s,2|s|+t-2), & 2(1-|s|)\leqslant t\leqslant 1, |s|\neq 0.\end{cases}$$

Notice that the restriction of r^{n+1} to $B^{n+1}\times\{0\}\cup S^n\times I$ is the identity and that the composition of r^{n+1} with the inclusion of the latter space into $B^{n+1}\times I$ is homotopic rel. $B^{n+1}\times\{0\}\cup S^n\times I$ to the identity map, via the homotopy

$$R^{n+1}: B^{n+1}\times I\times I\to B^{n+1}\times I$$

given by

$$R^{n+1}(s,t,u)=u(s,t)+(1-u)r^{n+1}(s,t);$$

thus, $B^{n+1}\times\{0\}\cup S^n\times I$ is a strong deformation retract of $B^{n+1}\times I$. This means that the inclusion of S^n in B^{n+1} is a closed cofibration (see Example 1, Section A.4).

The restriction of the homotopy R^n to $B^n\times\{1\}\times I\simeq B^n\times I$ factors through the map h^n, thereby inducing a homeomorphism (see Figure 7)

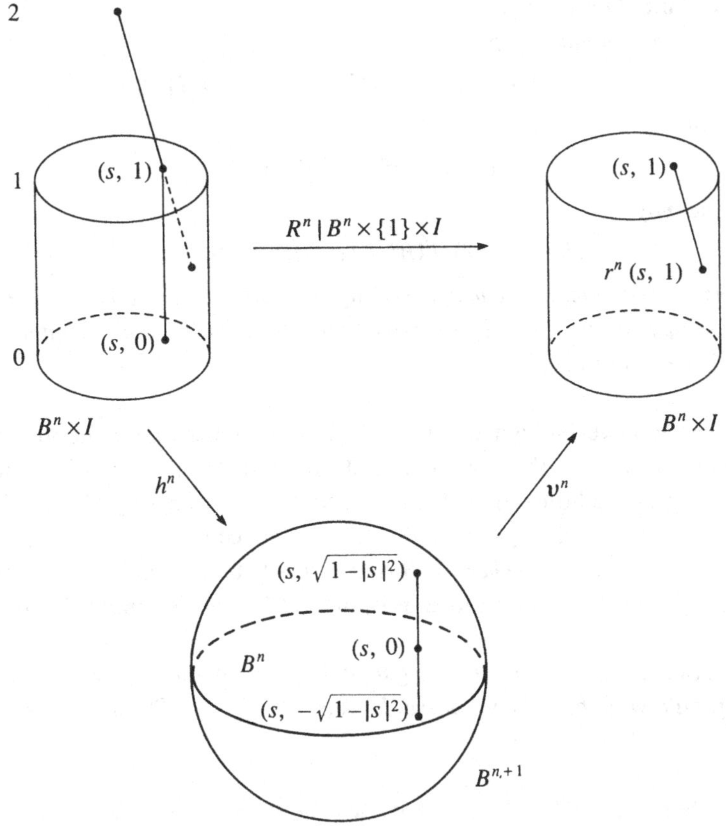

Figure 7

$$v^n : B^{n+1} \to B^n \times I;$$

one should notice that, regarding i_+, i_- as inclusions of B^n into B^{n+1}, then

$$v^n \circ i_- = r^n | B^n \times \{1\} \quad \text{and} \quad v^n \circ i_+ = \text{inclusion}.$$

The homeomorphism v^n, interesting in its own right, can be used to interchange the components $B^n \times \{0\} \cup S^{n-1} \times I$ and $B^n \times \{1\}$ of the boundary of $B^n \times I$: to see this, first note that v^n maps the upper hemisphere of S^{n+1} onto $B^n \times \{1\}$ and its lower hemisphere onto $B^n \times \{0\} \cup S^{n-1} \times I$; the actual interchange is then effected by the composite function $w^n = v^n \circ l^n \circ (v^n)^{-1}$. Two more remarks about the map v^n are called for: firstly, v^n induces a homeomorphism

$$\dot{v}^n : S^n \to B^n \times \dot{I} \cup S^{n-1} \times I;$$

secondly, v^n combines with the two pinchings p^n and $\hat{p}^n$ to yield an interesting commutative property:

Lemma 1.0.1 *For every $n \geqslant 1$,*

(i) *there is a unique map*

$$q^n : (B^n \vee B^n) \times I \to (B^n \times I) \vee (B^n \times I)$$

such that

$$q^n \circ (p^n \times 1) \circ v^n = (v^n \vee v^n) \circ \hat{p}^n;$$

(ii) *the map*

$$\hat{q}^n : (B^n \times I) \vee (B^n \times I) \to (B^n \vee B^n) \times I$$

induced by the obvious inclusions is a left homotopy inverse to q^n: there is a homotopy $\hat{q}^n \circ q^n \simeq 1$ rel. $((e_0, e_0), 1)$ and transforms the boundary of $(B^n \vee B^n) \times I$ into itself. □

In order to have enough fun in this game of balls and balloons, one actually needs more than one ball and one balloon in every dimension. Thus every space homeomorphic to the ball B^n (respectively, $\mathring{B}^n$) is called an *n-ball* (respectively, *open n-ball*) and every space homeomorphic to the sphere S^n is called an *n-sphere*. If B is any $(n+1)$-ball, its boundary sphere i.e., the image of S^n under a homeomorphism $B^{n+1} \to B$, is denoted by δB.

Proposition 1.0.2 *For any non-negative integers p and q, $B^p \times B^q$ is a $(p+q)$-ball with boundary sphere $B^p \times S^{q-1} \cup S^{p-1} \times B^q$; moreover, for every $n > 0$, $(B^1)^n$ is an n-ball.*

Proof Define $\Phi : B^p \times B^q \to B^{p+q}$ by setting, for every $(s, s') \in B^p \times B^q$,

$$\Phi(s, s') = \{\max(|s|, |s'|)/\sqrt{|s|^2 + |s'|^2}\}(s, s'),$$

if $(s, s') \neq (0, 0)$ and

$$\Phi(0, 0) = 0.$$

The continuity of Φ is not difficult to prove. Its inverse is obtained as follows. Let $s = (s_1, \ldots, s_p, \ldots, s_{p+q}) \in B^{p+q}$ be given. Set $s' = (s_1, \ldots, s_p)$ and $s'' = (s_{p+1}, \ldots, s_{p+q})$; then, define

$$\Phi^{-1}(s) = \{|s|/\max(|s'|, |s''|)\}(s', s'').$$

The restriction of Φ to $\delta(B^p \times B^q)$ gives the second homeomorphism announced in the statement. The third homeomorphism is obtained by induction on n. □

Projective spaces

From the topological point of view, projective spaces are intimately connected to spheres. However, before exhibiting this connection, one must give the definition of 'projective space' over a field.

Let **F** be a (not necessarily commutative) field. The *n-dimensional projective space over* **F**, denoted by $\mathbf{F}P^n$, is defined as the set of all 1-dimensional (left) vector subspaces of the $(n+1)$-dimensional (left) vector space $\mathbf{F}^{n+1}$. The space $\mathbf{F}P^n$ can be identified with the set $(\mathbf{F}^{n+1}\backslash\{0\})/\sim$, where $\sim$ is the equivalence relation defined by: $s \sim s'$ iff there is a scalar $t \in \mathbf{F}$ with $s' = ts$.

If **F** is a topological field the projective space $\mathbf{F}P^n$ is given the identification topology induced by the projection $\mathbf{F}^{n+1}\backslash\{0\} \to \mathbf{F}P^n$. In this book, **F** represents the field **R** of real numbers, the field **C** of complex numbers or the skew-field **H** of quaternions. Then the space $\mathbf{F}^{n+1}$ can be identified with one of the Euclidean spaces $\mathbf{R}^{n+1}$, $\mathbf{R}^{2n+2}$ or $\mathbf{R}^{4n+4}$. Note that one can find, for every point in the projective space, a representative of length 1 in the corresponding Euclidean space, i.e., a point in the spheres S^n, S^{2n+1} or S^{4n+3}. These identifications yield, respectively, the identification maps

$$q_{\mathbf{R}}{}^n : S^n \to \mathbf{R}P^n,$$
$$q_{\mathbf{C}}{}^n : S^{2n+1} \to \mathbf{C}P^n,$$
$$q_{\mathbf{H}}{}^n : S^{4n+3} \to \mathbf{H}P^n.$$

The inverse image of a point in the projective space is a pair of antipodal points in S^n for $\mathbf{F} = \mathbf{R}$, a circle (= 1-sphere) in S^{2n+1} for $\mathbf{F} = \mathbf{C}$ and a 3-sphere in S^{4n+3} for $\mathbf{F} = \mathbf{H}$.

1.1 Adjunction of *n*-cells

The reader should always bear in mind that all the work in this book is done within the context of the category of weak Hausdorff *k*-spaces, denoted simply by *Top* (except in Section A.1, where it is denoted by *wHk*(*Top*)).

Intuitively, a CW-complex is a space which can be considered as a union of disjoint 'open cells'. For instance, the ball B^{n+1} can be considered as the union of an $(n+1)$-cell, namely the open ball $\mathring{B}^{n+1}$, an n-cell, namely the punctured sphere $S^n\backslash\{e_0\}$, and the 0-cell $\{e_0\}$:

$$B^{n+1} = \mathring{B}^{n+1} \cup (S^n\backslash\{e_0\}) \cup \{e_0\}.$$

In this book the term 'cell' will often be preceded by the adjectives 'open', 'closed', 'regular', or the combination 'closed regular'. The following list is intended to make matters clear. A subspace e of a space X is said to be

- an *open n-cell* in X ($n \in \mathbf{N}$), if it is an open n-ball (recall that an open n-ball is a space homeomorphic to the open ball $\mathring{B}^n$);
- a *closed n-cell* in X, if it is the closure (in X) of an open n-cell;
- a *regular n-cell* in X, if it is an open n-cell whose closure is an n-ball and whose boundary in the closure is an $(n-1)$-sphere;
- a *closed regular n-cell* in X, if it is the closure of a regular n-cell.

Observe that an n-cell does not have to be regular: the punctured sphere $S^n \backslash \{e_0\}, n > 0$, as a subspace of the sphere S^n, is an example of this fact. For open, regular or closed regular n-cells e, the natural number n is the dimension of e: $\dim e = n$ (see Section A.9). By abuse of language, one also assigns to a closed n-cell the dimension n, although, outside the theory of CW-complexes, this does not necessarily coincide with the covering dimension of the space under consideration (see Example 5). But, if a space X contains an n-cell of any type, then $\dim X \geqslant n$, because inside each open n-cell there are closed n-balls (see Corollary A.9.2).

The ball B^{n+1} was decomposed into a union of open cells at the beginning of this section. In what follows, one should have this sort of cellular decomposition in mind. For the sake of simplicity, the formal constructions and proofs will often proceed in a slightly different manner.

A pair (X, A) is an *adjunction of n-cells*, $n \in \mathbb{N}$, if X can be viewed as an adjunction space (see Section A.4)

$$X = A \sqcup_f Y$$

where Y is a topological sum of n-balls and the domain of f consists of the boundary spheres of the balls forming Y; in other words, if X is given by a pushout of the form

$$\begin{array}{ccc} \bigsqcup_\lambda B_\lambda = Y & \longrightarrow & X \\ \uparrow & & \uparrow \\ \bigsqcup_\lambda S_\lambda & \longrightarrow & A, \end{array}$$

with B_λ an n-ball and $S_\lambda = \delta B_\lambda$, for all indices λ in an arbitrary index set Λ. If $n = 0$ the definition means simply that X is a topological sum of A and a discrete space. If (X, A) is an adjunction of n-cells, any path-component of $X \backslash A$ is an open n-cell in X, called an *n-cell* of (X, A). Each induced map $B_\lambda \to X$ is called a *characteristic map* for the λth cell; each induced map $S_\lambda \to A$ is an *attaching map* for the λth cell. If A is a based space and every map $S_\lambda \to A$ is based, the pair (X, A) is said to be a *based* adjunction of n-cells.

Proposition 1.1.1 *If (A, a_0) is path-connected and (X, A) is an adjunction of n-cells, $n > 0$, there exists a based adjunction of n-cells (X', A), such that X' is homotopically equivalent to X via homotopies rel. A.*

Proof Suppose that $X = A \sqcup_f (\sqcup_\lambda B_\lambda)$. Let $f_\lambda : S_\lambda \to A$ be the attaching map for the λth cell and let $\omega_\lambda : I \to A$ be a path such that $\omega_\lambda(0) = f_\lambda(e_0)$,

$\omega_\lambda(1) = a_0$; choose a representative for $(\omega_\lambda)_n^{-1}([f_\lambda])$ (see page 287 in the appendix), for every index λ. The maps f'_λ together define a based adjunction of n-cells (X', A) with the properties required (see Proposition A.4.15).

□

Example 1 For every $n > 0$, the pair (B^n, S^{n-1}) is an adjunction of just one regular n-cell; one can take the identity of S^{n-1} as an attaching map and the identity of B^n as a characteristic map.

Example 2 For every $n \in \mathbf{N}$, the pair $(S^n, \{e_0\})$ is an adjunction of just one non-regular n-cell. If $n > 0$, the map $b^n: B^n \to S^n$ (see page 6) can be used as a characteristic map; here there is no choice for the attaching map: it has to be the constant map.

Example 3 For every $n > 0$, the pair (S^n, S^{n-1}) is an adjunction of two regular n-cells. Take as components of the characteristic map the embeddings i_+, i_- (see page 3) of the ball B^n as the upper, respectively the lower, hemisphere into the sphere S^n.

The next example is not so trivial.

Example 4 For every $n \in \mathbf{N}$, the pair $(B^{n+1} \cup S^{n+1}, B^n \cup S^n)$ is an adjunction of exactly four regular $(n+1)$-cells. To prove this assertion, first observe that

$$B^{n+1} \cup S^{n+1} = B^{n+1} \sqcup_{B^n \cup S^n} (B^n \cup S^{n+1});$$

then note that because of the addition law (L3), it is enough to show that each of the pairs $(B^{n+1}, B^n \cup S^n)$ and $(B^n \cup S^{n+1}, B^n \cup S^n)$ is an adjunction of just two $(n+1)$-cells. Example 3 and the horizontal composition law (L1) are used to show that the pair $(B^n \cup S^{n+1}, B^n \cup S^n)$ is an adjunction of two $(n+1)$-cells. To prove that the pair $(B^{n+1}, B^n \cup S^n)$ is an adjunction of just two $(n+1)$-cells, construct the appropriate pushout using the 'eggs of Columbus' (see page 2) as components of the characteristic map.

Example 5 Let $f : B^1 \to B^n, n > 2$, be a *Peano curve*, i.e., a map from B^1 *onto* B^n. Then, the composition $f \circ j^1 | S^1$ defines a partial map $g : B^2 -/\to B^n$ (for the definition of the map j^1, see page 2). The pair $(B^n \sqcup_g B^2, B^n)$ is an adjunction of just one 2-cell. The corresponding closed 2-cell has covering dimension $n > 2$!

Example 6 Let $\mathbf{F}$ be one of the fields $\mathbf{R}$, $\mathbf{C}$, $\mathbf{H}$, of the real, complex or quaternionic numbers, respectively; also, let d be the dimension of $\mathbf{F}$ as a vector space over $\mathbf{R}$. Then, for every $n > 0$, the pair $(\mathbf{F}P^n, \mathbf{F}P^{n-1})$ is an adjunction of just one non-regular dn-cell. The composition of the inclusion $i_+ : B^{dn} \rightarrowtail S^{dn}$ (see page 3), the embedding $S^{dn} \rightarrowtail S^{dn+d-1}$, and the projection $q^n_{\mathbf{F}} : S^{dn+d-1} \to \mathbf{F}P^n$ (see page 11) may serve as characteristic map for the adjunction; this characteristic map induces the attaching map $q^{n-1}_{\mathbf{F}} : S^{dn-1} \to \mathbf{F}P^{n-1}$.

Proposition 1.1.2 *Let (X, A) be an adjunction of n-cells, say*

$$X = A \bigsqcup_f \left(\bigsqcup_\lambda B_\lambda \right).$$

Then the following statements hold true:

(i) *the inclusion $A \rightarrowtail X$ is a closed cofibration;*

(ii) *the space X is (perfectly) normal, whenever the subspace A is (perfectly) normal;*

(iii) *the space X has dimension n, whenever the subspace A is a normal space of dimension $\leqslant n$ and the index set is not empty;*

(iv) *$X \backslash A$ is a topological sum of open n-cells, one for each index λ;*

(v) *for any map $f' : A \to A'$, the pair $(A' \bigsqcup_{f'} X, A')$ is an adjunction of n-cells.*

Proof The inclusion $dom\, f \rightarrowtail Y$ is a topological sum of closed cofibrations, and therefore is itself a closed cofibration. Thus (i) follows because the attaching process preserves cofibrations.

Since $\bigsqcup_\lambda B_\lambda$ is perfectly normal, the adjunction space X is (perfectly) normal if A is (perfectly) normal (see Proposition A.4.8 (iv)).

To prove (iii) note that under the first part of the condition given, the space X has dimension $\leqslant n$ (see Proposition A.4.8 (v)); if n-cells are really present, $\dim X \geqslant n$.

Part (iv) follows from the fact that $X \backslash A$ is homeomorphic to $Y \backslash dom\, f = \sqcup (B_\lambda \backslash S_\lambda)$.

Finally, (v) follows from the law of horizontal composition of Section A.4. □

Remark According to (iv), the index set for $\sqcup B_\lambda$ can be viewed as the set $\pi(X \backslash A)$ of path-components of $X \backslash A$. Give the discrete topology to the set $\pi(X \backslash A)$; then the space $B^n \times \pi(X \backslash A)$ can be viewed as the domain of the characteristic map for the adjunction of n-cells (X, A), and the space $S^{n-1} \times \pi(X \backslash A)$ can be viewed as the domain of the attaching map for the same adjunction. □

The bridge between the point of view of considering globally all the cells used in the adjunction, and that of considering successive attachings of single n-cells, is given by the following result.

Proposition 1.1.3 *The pair* (X, A) *is an adjunction of n-cells, iff*

(i) *for every path-component* e *of* $X\backslash A$ *the pair* $(A\cup e, A)$ *is an adjunction of just one n-cell* ($A\cup e$ *is considered as a subspace of* X) *and*

(ii) *the space* X *is determined by the family* $\{A\}\cup\{\bar{e} : e\in\pi(X\backslash A)\}$.

Proof '⇒': (i) Let e be an n-cell of (X, A) with attaching map f_e and characteristic map $\bar{f}_e$.

To prove the equality

$$A\cup e = A\bigsqcup\nolimits_{f_e} B_e,$$

observe first that $A\cup e = A\cup(B_e\backslash S_e)$, as sets. It remains to show that the subspace topology of $A\cup e$ is the same topology as that of the adjunction space $A\bigsqcup_{f_e} B_e$. Notice that by the universal property of the adjunction, the space $A\bigsqcup_{f_e} B_e$ has a finer topology than $A\cup e$. Next, let $V\subset A\cup e$ be such that $V\cap A$ is closed in A, and $\bar{f}_e^{-1}(V)$ is closed in B_e. Because X is a weak Hausdorff k-space, $V\cap\bar{f}_e(B_e)=\bar{f}_e(\bar{f}_e^{-1}(V))$ is closed in X (see Lemma A.1.1), and, hence, in $A\cup e$; this, together with the fact that $V\cap A$ is closed in A, implies that V is closed in X.

(ii) Let $U\subset X$ be such that $U\cap A$ and $U\cap\bar{e}$ are closed respectively in A and $\bar{e}$, for each $e\in\pi(X\backslash A)$. Then, if $\bar{f}$ is the characteristic map of the adjunction, $\bar{f}^{-1}(U)=\sqcup\bar{f}_e^{-1}(U\cap\bar{e})$ is closed.

'⇐': For every $e\in\pi(X\backslash A)$, let $f_e : S_e\to A$ denote an attaching map generating the adjunction space $A\cup e = A\bigsqcup_{f_e} B_e$. Let $f:\square S_e\to A$ be the map defined by the maps f_e, and let $\hat{X}$ be the adjunction space $A\bigsqcup_f(\bigsqcup B_e)$ with a fixed characteristic map $\bar{f}$. The pair $(\hat{X}, A)$ is an adjunction of n-cells, and thus it suffices to show that the spaces X and $\hat{X}$ coincide (up to canonical homeomorphism). The universal property of the adjunction space $\hat{X}$ gives rise to a bijective map $\hat{X}\to X$; thus, assume that the spaces $\hat{X}$ and X have the same underlying sets, and the topology of $\hat{X}$ is finer than that of X.

Notice that $\bar{f}(B_e)=\bar{e}$ because X is a weak Hausdorff k-space. Let $V\subset X$ be such that $V\cap A$ and $\bar{f}^{-1}(V)$ are closed in A and $\sqcup B_e$, respectively. Hence $\bar{f}^{-1}(V)\cap B_e$ is closed in B_e, for every $e\in\pi(X\backslash A)$; because $\bar{f}(\bar{f}^{-1}(V)\cap B_e)=V\cap\bar{e}$, it follows that $V\cap\bar{e}$ is closed in $\bar{e}$, for every $e\in\pi(X\backslash A)$. Condition (ii) implies that the set V is also closed in X. □

The following result, which is actually contained in the previous proof, has some interest in its own right.

Lemma 1.1.4 *Let (X, A) be an adjunction of n-cells and let e be an n-cell of (X, A). Then,*

$$\bar{e} = \bar{f}(B),$$

where $\bar{f}$ denotes a characteristic map for e and B denotes an n-ball in the domain of $\bar{f}$. □

An advantage of looking at the adjunction of just one cell at a time lies in the fact that this process can be characterized without the explicit construction of a pushout diagram.

Lemma 1.1.5 *The pair (X, A) is an adjunction of just one n-cell iff*
(i) *A is closed in X*
and
(ii) *there is a map $B^n \to X$ inducing a homeomorphism $\mathring{B}^n \to X \backslash A$.*

Proof '⇒': clear from the definition.

'⇐': Let $\bar{f} : B^n \to X$ be a map as described in condition (ii). First, prove that $\bar{f}$ takes the boundary S^{n-1} of the ball B^n into the space A. To this end, assume the existence of a point $s \in S^{n-1}$ such that $\bar{f}(s) \in X \backslash A$. Then there is a unique point $s' \in \mathring{B}^n$ such that $\bar{f}(s) = \bar{f}(s')$; furthermore, the inverse image of every neighbourhood of $\bar{f}(s')$ contains points close to s, contradicting the assumption that $\bar{f}$ induces a homeomorphism $\mathring{B}^n \to X \backslash A$.

Denote by $f : S^{n-1} \to A$ the map induced by $\bar{f}$, and form the commutative square

$$\begin{array}{ccc} B_n & \xrightarrow{\bar{f}} & X \\ \uparrow & & \uparrow \\ S^{n-1} & \xrightarrow[f]{} & A \end{array}$$

It remains to prove that X has the final topology with respect to $\bar{f}$ and the inclusion $A \subset X$. To show this, first observe that the subspace A is closed in X, by (i), and the subspace $\bar{f}(B^n)$ is closed in X, because X is weak Hausdorff. Since the space X is the union of these two closed subspaces, a function with domain X is continuous iff its restrictions to the subspaces A and $\bar{f}(B^n)$ are continuous. □

The condition (i) in this lemma is necessary, as one can deduce from the following.

Example 7 The pair $(B^2, B^2 \backslash \mathring{B}^1)$ satisfies condition (ii) for $n = 1$, but fails to be an adjunction of a 1-cell.

Because a space with a finite closed covering is determined by that covering, condition (ii) in Proposition 1.1.3 is superfluous if one deals with adjunctions of only finitely many cells. The following is an example showing that this condition is unavoidable in the general case.

Example 8 Let $\{B_\lambda : \lambda \in \mathbf{N}\setminus\{0\}\}$ be a countable set of copies of the ball B^1. For every index λ, let $\bar{f}_\lambda : B_\lambda \to I$ denote an embedding whose image is the interval $[1/(\lambda+1), 1/\lambda]$ and define $\bar{f} : \sqcup B_\lambda \to I$ by taking $\bar{f}|B_\lambda = \bar{f}_\lambda$. Since $\bar{f}(\sqcup S_\lambda)$ is contained in

$$A = \{0\} \cup \left\{ \frac{1}{\lambda} : \lambda \in \mathbf{N}\setminus\{0\} \right\},$$

$\bar{f}$ induces a map $f : \sqcup S_\lambda \to A$ and a commutative square

$$\begin{array}{ccc} \sqcup B_\lambda & \xrightarrow{\bar{f}} & I \\ \uparrow & & \uparrow \\ \sqcup S_\lambda & \xrightarrow[f]{} & A \end{array}$$

Now, for every index λ, define the 1-cell $e_\lambda = \bar{f}(\mathring{B}_\lambda)$ in I; then

$$A \cup e_\lambda = A \cup \left\{ t \in \mathbf{R} : \frac{1}{\lambda+1} \leqslant t \leqslant \frac{1}{\lambda} \right\}$$

and the pair $(A \cup e_\lambda, A)$ is an adjunction of just one 1-cell (see Lemma 1.1.5). But I is not determined by the family $\{A \cup e_\lambda\}$! To see this, consider the sequence $\{(2\lambda+1)/2\lambda(\lambda+1)\}$. This sequence meets every space $A \cup e_\lambda$ in just one point, thus it is closed in the topology determined by $\{A \cup e_\lambda\}$; however, it converges to 0 in the usual topology of the unit interval I. (This situation may also serve as a counterexample in general topology: it is easy to see that, with respect to the topology of I determined by $\{A \cup e_\lambda\}$, 0 is a cluster point of $\bar{f}(\sqcup \mathring{B}_\lambda)$, but no sequence in $\bar{f}(\sqcup \mathring{B}_\lambda)$ converges to 0; this means that the resulting space is not a Fréchet space. Similar ideas will be used in Example 13 of the next section.)

There are two more relevant examples of pairs which are adjunctions of infinitely many n-cells.

Example 9 The concept of the wedge of two spheres $S^n \vee S^n$ was briefly discussed in Section 1.0; this concept has the following generalization. Let Γ be any set; for every $\gamma \in \Gamma$ take a copy of the n-sphere S^n with its base point e_0, i.e., $(S^n{}_\gamma, e_0) = (S^n, e_0)$. The *wedge product* of the family of based spaces

$$\{(S^n_\gamma, e_0) : \gamma \in \Gamma\}$$

(also called a *bouquet of n-spheres*) is the based space $(\vee_\Gamma S^n_\gamma, *)$, given by the set

$$\bigvee_\Gamma S^n_\gamma = \left\{ (s_\gamma) \in \prod_{\gamma \in \Gamma} S^n_\gamma : s_\gamma \neq e_0, \text{ for at most one } \gamma \in \Gamma \right\},$$

endowed with the final topology with respect to the canonical map

$$p : \bigsqcup_{\gamma \in \Gamma} S^n_\gamma \to \bigvee_{\gamma \in \Gamma} S^n_\gamma,$$

and the point $*$ taken to be the element (e_0). Note that if Γ is finite, this topology coincides with the subspace topology induced by $\prod_{\gamma \in \Gamma} S^n_\gamma$.

The pair $(\vee_\Gamma S^n, *)$ is an adjunction of n-cells; notice that there are as many n-cells as there are elements in Γ. A characteristic map for this adjunction is given by the map

$$\bar{f} : \bigsqcup_{\gamma \in \Gamma} B^n \cong B^n \times \Gamma \to \bigvee_\Gamma S^n$$

(here Γ is given the discrete topology) defined by $\bar{f}(s, \gamma) = (s_\gamma)$, where $s_\gamma = b^n(s)$.

Example 10 Let π be an abelian group and let n be a natural number > 1. Let $FA(\pi)$ be the free abelian group generated by the elements of π, and let Γ be a basis of the kernel of the canonical homomorphism $FA(\pi) \to \pi$. Let $\varphi : FA(\pi) \to \pi_n(\vee_\pi S^n, *)$ denote the homomorphism which assigns to a generator α of $FA(\pi)$ the homotopy class of the inclusion of S^n into the π-fold wedge $\vee_\pi S^n$ of S^n as the αth factor. Next, for each $\gamma \in \Gamma$, choose a representative $f_\gamma : S^n \to \vee_\pi S^n$ of the homotopy class $\varphi(\gamma)$. The maps f_γ define a partial map $f : B^{n+1} \times \Gamma \rightharpoonup \vee_\pi S^n$, whose resulting adjunction space $M(\pi, n)$ is called a *Moore space* of type (π, n). The construction of $M(\pi, n)$ shows that the pair $(M(\pi, n), \vee_\pi S^n)$ is an adjunction of $(n+1)$-cells.

What follows is more than just an example.

Theorem 1.1.6 (i) *Let (X, A) be an adjunction of n-cells and let $p : \tilde{X} \to X$ be a covering projection. Then, the pair $(\tilde{X}, \tilde{A})$ with $\tilde{A} = p^{-1}(A)$, is also an adjunction of n-cells.*

(ii) *Let (X, A) be an adjunction of n-cells, $n > 2$, and let $p : \tilde{A} \to A$ be a covering projection. Then, there are an adjunction of n-cells $(\tilde{X}, \tilde{A})$ and a* *covering projection* $q:\tilde{X}\to X$, *such that p is induced from q by the inclusion* $A\to X$. *In particular, if p is a universal covering projection, so is q.*

Proof (i) Consider $\Lambda=\pi(X\backslash A)$ as the index set for the n-cells of the adjunction (X,A). For every $\lambda\in\Lambda$, choose a characteristic map $\bar{c}_\lambda : B^n\to X$ for the cell e_λ. Then, take $\tilde{\Lambda}=\{(z,\lambda)\in\tilde{X}\times\Lambda : p(z)=\bar{c}_\lambda(e_0)\}$ and let $\bar{c}_{\tilde{\lambda}}$ denote the unique lifting of $\bar{c}_\lambda$ with $\bar{c}_{\tilde{\lambda}}(e_0)=z$, for any $\tilde{\lambda}=(z,\lambda)\in\tilde{\Lambda}$ (see Theorem A.8.5). Next, define $\bar{f}:B^n\times\tilde{\Lambda}\to\tilde{X}$ by $(s,\tilde{\lambda})\mapsto\bar{c}_{\tilde{\lambda}}(s)$. The restriction $\bar{f}|S^{n-1}\times\tilde{\Lambda}$ factors through $\tilde{A}$ therefore, inducing a map $f:S^{n-1}\times\tilde{\Lambda}\to\tilde{A}$. It will be shown that $\tilde{X}$ may be viewed as being obtained from $\tilde{A}$ by adjoining $B^n\times\tilde{\Lambda}$ via f.

First, prove that every point $\tilde{x}\in\tilde{X}\backslash\tilde{A}$ corresponds to a unique point in $\mathring{B}^n\times\tilde{\Lambda}$. To this end, notice that $p(\tilde{x})\notin A$, and so $p(\tilde{x})=\bar{c}_\lambda(s)$, for a unique $\lambda\in\Lambda$ and a unique $s\in\mathring{B}^n$. Now, let W denote the line segment in B^n connecting s to e_0 and let $\omega : W\to\tilde{X}$ denote the unique lifting of $\bar{c}_\lambda|W$, with $\omega(s)=\tilde{x}$. Then, $\tilde{x}=\bar{c}_{\tilde{\lambda}}(s)$, with $\tilde{\lambda}=(\omega(e_0),\lambda)$.

Second, $\tilde{X}$ has the right topology. It will be shown that a subset $U\subset\tilde{X}$ is open if $U\cap\tilde{A}$ is open in $\tilde{A}$ and $\bar{c}_{\tilde{\lambda}}^{-1}(U)$ is open in B^n, for every $\tilde{\lambda}\in\tilde{\Lambda}$. Because p is a covering projection, there is an open cover $\{V_\gamma : \gamma\in\Gamma\}$ of $\tilde{X}$ such that the induced map $V_\gamma\to p(V_\gamma)$ is a homeomorphism and $p(V_\gamma)$ is open in X, for every $\gamma\in\Gamma$. Since U is open in $\tilde{X}$ iff $U\cap V_\gamma$ is open, for every γ, it suffices to assume $U\subset V_\gamma$, for some γ. But then, U is open iff $p(U)$ is open in X. Now $p(U)\cap A=p(U\cap\tilde{A})$ is open in A and $\bar{c}_\lambda^{-1}(p(U))=\bigcup_z\bar{c}_{\tilde{\lambda}}^{-1}(U)$ where the union is taken over all z's such that $(z,\lambda)\in\tilde{\Lambda}$, is open in B^n, for every $\lambda\in\Lambda$; thus, because X has the final topology with respect to the inclusion of A and the characteristic maps $\bar{c}_\lambda$, the set $p(U)$ is open in X.

(ii) According to the condition on n, each attaching map for an n-cell of (X,A) has a simply connected domain, and so it has liftings to $\tilde{A}$. Use each of these liftings to attach an n-cell to $\tilde{A}$. The result is a space $\tilde{X}$ and the universal property of the attachings determines the covering projection q. □

Collaring

Whenever dealing with pairs (X,A) which are adjunctions of n-cells, $n>0$, sometimes it is necessary to enlarge open sets of the subspace A to appropriate open sets of X. This can be done by the technique of 'collaring', which is described next.

Let $\bar{f} : \sqcup B_\lambda\to X$ be a characteristic map, and let $f : \sqcup S_\lambda\to A$ be the corresponding attaching map. Assume that every ball B_λ is just a copy of

B^n; thus, one can multiply any $s \in B_\lambda$ (viewed as a vector of $\mathbf{R}^n$) by a scalar $t \in I$; the product ts is still a point of B_λ. The $\bar{f}$-*collar* of a set $V \subset A$ is defined to be the subset

$$C_{\bar{f}}(V) = V \cup \bar{f}(\{ts : s \in f^{-1}(V), \quad \tfrac{1}{2} < t \leqslant 1\}).$$

The following is an immediate consequence of the definition.

Lemma 1.1.7 *Let (X, A) be an adjunction of n-cells, let $\bar{f}$ be a characteristic map for the adjunction, and let V be a subset of A. Then*

(i) $C_{\bar{f}}(V) \cap A = V$;

(ii) $\bar{f}^{-1}(C_{\bar{f}}(V)) = \{ts : s \in f^{-1}(V), \quad \tfrac{1}{2} < t \leqslant 1\}$;

(iii) $C_{\bar{f}}(V)$ *is open in X iff V is open in A;*

(iv) *if V is a closed subset of A, the closure of the $\bar{f}$-collar of V is the set*

$$\overline{C_{\bar{f}}(V)} = V \cup \bar{f}(\{ts : s \in f^{-1}(V), \quad \tfrac{1}{2} \leqslant t \leqslant 1\});$$

(v) *if e is an n-cell of (X, A), then $e \cap \overline{C_{\bar{f}}(V)} \neq \varnothing$ iff $e \cap C_{\bar{f}}(V) \neq \varnothing$ iff $\bar{e} \cap V \neq \varnothing$;*

(vi) $C_{\bar{f}}(V)$ *contains V as a strong deformation retract;*

Moreover, if (V_γ) is a locally finite family of subsets of A (respectively, a family of pairwise disjoint subsets of A), then

(vii) $(C_{\bar{f}}(V_\gamma))$ *is a locally finite family of subsets of X (respectively, a family of pairwise disjoint subsets of X).* □

The next result requires a little work.

Lemma 1.1.8 *Let (X, A) be an adjunction of n-cells and $\bar{f}$ be a characteristic map for the adjunction. If $V \subset A$ is open or closed in A, then*

$$C_{\bar{f}}(V) = V \bigsqcup_g (f^{-1}(V) \times (\tfrac{1}{2}, 1])$$

where f is the attaching map corresponding to $\bar{f}$ and $g : f^{-1}(V) \to V$ is the map induced by f.

Proof Assume first that V is open in A. Then, because of Lemma 1.1.7 (iii), $C_{\bar{f}}(V)$ is open in X; the stated result now follows by application of the restriction law (L4) of adjunction spaces, and parts (i) and (ii) of the previous lemma. In particular, notice that

$$C_{\bar{f}}(A) = A \bigsqcup_f (f^{-1}(A) \times (\tfrac{1}{2}, 1]).$$

Now if V is closed in A, then $C_{\bar{f}}(V)$ is closed in $C_{\bar{f}}(A)$, and so the statement follows again by (L4). □

Corollary 1.1.9 *If V is an open or closed subspace of A, then the inclusion $V \rightarrowtail C_{\bar{f}}(V)$ is a closed cofibration.* □

The fact that the characteristic maps are not unique might be quite advantageous; indeed, it permits the choice of the 'right coordinates' for a variety of purposes, as proved by the next proposition.

Proposition 1.1.10 *Let (X, A) be an adjunction of n-cells, V be a closed set of A and U be an open subset of X containing V. Then there is a characteristic map $\bar{f}$ for the adjunction such that the closure $\overline{C_{\bar{f}}(V)}$ is still contained in U.*

Proof Choose arbitrarily a characteristic map $\tilde{f} : \sqcup B_e \to X$, where the index e runs through all the n-cells of the adjunction. The map $\tilde{f}$ determines an attaching map $f : S_e \to A$ whose restriction to a sphere S_e will be denoted by f_e. The objective is to construct cellwise a 'transformation of coordinates', which keeps the attaching map f invariant. Notice that $\tilde{f}$ must be modified only for cells e, such that

$$(*) \quad \tilde{f}(\{ts{:}s \in S_e, f(s) \in V, \tfrac{1}{2} \leqslant t \leqslant 1\}) \not\subset U.$$

Let e be such a cell. Then $V_e = f_e^{-1}(V)$ is non-empty and $\tilde{f}(B_e)$ is not completely contained in U. Hence the set $U_e = B_e \backslash \tilde{f}^{-1}(U)$ is a non-empty closed subset of B_e which does not meet the closed set V_e. The distance δ_e between the closed sets U_e and V_e is defined and different from 0, because these two closed sets are both compact subsets of a metric space. It is easy to conclude from $(*)$ that $\delta_e \leqslant \frac{1}{2}$. Next, select a homeomorphism $h_e : B_e \to B_e$ which coincides with the identity map on the boundary of B_e and shrinks the ball $\{s \in B_e : |s| \leqslant 1 - \delta_e\}$ radially into the ball $\{s \in B_e : |s| \leqslant \frac{1}{3}\}$; then define

$$\bar{f} | B_e = \tilde{f} | B_e \circ h_e^{-1}.$$

This completes the construction of the desired characteristic map $\bar{f}$. □

Exercises

1. Let (Y, D) be an adjunction of n-cells and let A be a contractible space. Show that any map $f : D \to A$ can be extended over Y.
2. Let $M(\pi, n)$ be a Moore space of type (π, n). Show that $M(\pi, n)$ is up to homotopy independent of the choice of the basis Γ selected for the kernel of the canonical homomorphism $FA(\pi) \to \pi$; show also that $M(\pi, n)$ does not depend on the choice of the representatives $f_\gamma{:}S^n \to \vee_\pi S^n$ (see Example 10).

3. Show by a counterexample that, under the hypotheses of Proposition 1.1.10, without the assumption that V is closed, there might be no collar of V contained in U.

1.2 CW-complexes

One of the main objectives of this book is the study of CW-complexes; these can now be defined.

A *filtration* of a space X is a finite or infinite sequence $\{X^n : n=0,1,\ldots\}$ of closed subspaces of X which is a covering of X and such that X^{n-1} is a subspace of X^n for $n=1,2,\ldots$.

A *CW-structure* for a space X is a filtration of X such that

(0) X^0 is a discrete space,
(1) for every $n>0$ the pair (X^n, X^{n-1}) is an adjunction of n-cells, and
(2) X is determined by the family of subspaces $\{X^n : n \in \mathbf{N}\}$.

There are occasions when it is convenient to start the filtration with X^{-1}. If $X^{-1}=\varnothing$, the introduction of this extra space does not change matters; otherwise, one is led to the notion of *relative CW-complex*, whose basic properties are introduced at the end of this section.

If the filtration is finite, condition (2) above is superfluous. If the filtration is infinite, condition (1) shows that $\{X^n : n=0,1,\ldots\}$ is an expanding sequence; in this case, condition (2) makes sure that X is its union space (see Section A.5).

A *CW-complex* is a space endowed with a CW-structure. If one wishes to be perfectly clear, it is convenient to use the notation

$$\{X; X^n : n=0,1,\ldots\}$$

or

$$\{X; X^0, X^1, \ldots, X^m\}$$

to describe the CW-complex consisting of the *global* space X and the corresponding filtration; otherwise, if the filtration is clearly understood, just write X instead of the previous lengthy expressions. By abuse of language, a space X is said to be a CW-complex if a CW-structure is implicitly given. Conversely, whenever referring to topological properties of a CW-complex $\{X; X^n : n=0,1,\ldots\}$ e.g., dimension, it is understood that these are properties of the space X.

Proposition 1.2.1 *Any CW-complex is a perfectly normal k-space.*

Proof (1) A discrete space is perfectly normal; (2) adjunction preserves perfect normality (Proposition A.4.8 (iv)); (3) the union space of an expanding sequence of perfectly normal spaces is perfectly normal (Proposition A.5.1 (iv)). □

If $\{X; X^n : n=0,1,\ldots\}$ is a CW-complex, then the closed subspace $X^n, n=0,1,\ldots,$ has dimension $\leqslant n$ (see Proposition 1.1.2 (iii)) and is called the *n-skeleton* of X; the space X^n inherits, in the obvious manner, a CW-structure, and thus it can be considered a CW-complex.

Attention: whenever a letter X represents a CW-complex, X^n will denote the n-skeleton of X and not its n-fold product.

Any (open) n-cell e of the adjunction (X^n, X^{n-1}) is called an *n-cell* of $X, n>0$; the points of the discrete space X^0 are called *0-cells*. A CW-complex (CW-structure) is *regular* if all its cells are regular cells.

If e is an n-cell of a CW-complex X, any map $\bar{c}_e : B^n \to X$ inducing a homeomorphism $\mathring{B}^n \to e$ is called a *characteristic map* for e; the compact image of $\bar{c}_e$ in X, which, incidentally, is just the closure $\bar{e}$ of the open cell e (see Lemma 1.1.4), is called a *closed cell* of X; in contrast to Example 5 of Section 1.1, the closed n-cells of a CW-complex have covering dimension n as subspaces of the n-skeleton (see Corollary A.9.2). For $n>0$, any characteristic map $\bar{c}_e$ induces an *attaching map* $c_e : S^{n-1} = \delta B^n \to X^{n-1}$ for the n-cell e. Clearly, there are characteristic maps for every cell; indeed, any characteristic map for the adjunction (X^n, X^{n-1}) gives rise to a family of characteristic maps for all the n-cells of X.

Notice again that from the set-theoretical point of view, a CW-complex is the disjoint union of its open cells. Observe that the open cells of a CW-complex X, in general, are not open sets of X; for instance, the only open n-cell of the $(n+1)$-ball referred to in the beginning of Section 1.1 is not open in B^{n+1}. The unique cell e containing a given point $x \in X$ is called the *carrier* of x.

The decomposition of a CW-complex into cells suggests the possibility of defining another topology, which, however, is equivalent to the topology determined by its skeleta.

Proposition 1.2.2 *A CW-complex is determined by the family of its closed cells.*

Proof Let $\{X; X^n : n=0,1,\ldots\}$ be a CW-complex. If $U \subset X$ is closed in X then, clearly $U \cap \bar{e}$ is closed in $\bar{e}$, for all cells e of X. Conversely, suppose that $U \subset X$ intersects every closed cell of X in a closed set. It is clear that $U \cap X^0$ is closed in the discrete space X^0. Assume, by induction, that $U \cap X^{n-1}$ is closed in X^{n-1}; since the n-skeleton X^n is determined by X^{n-1} and all the closed n-cells $\bar{e}$ (see Proposition 1.1.3 (ii)), the set $U \cap X^n$

is closed in X. It follows that $U \cap X^n$ is closed for all $n \in \mathbf{N}$, and since X is determined by the family of its own skeleta, U is closed in X. □

Remark This proposition says that a CW-complex can also be viewed as the codomain of an identification map from a coproduct of balls. Indeed, take characteristic maps for all cells; then, form the coproduct of their domains and the induced map to the CW-complex.

Example 1 The ball B^{n+1} has a *canonical* CW-structure

$$\{\{e_0\},\ldots,\{e_0\},S^n,B^{n+1}\}$$

(see the beginning of Section 1.1).

As pointed out in that section, the resulting CW-complex B^{n+1} is not regular, if $n>0$. Since S^n is here the n-skeleton of B^{n+1}, it inherits a non-regular CW-structure. There are, however, regular CW-structures for balls and spheres, as we can see in the next example.

Example 2 In view of Example 3 of Section 1.1,

$$\{S^0,S^1,\ldots,S^m,\ldots,S^n,B^{n+1}\}$$

is a regular CW-structure for B^{n+1}. This also proves that S^n, as the n-skeleton of this CW-structure, can be viewed as a regular CW-complex.

Example 3 Example 4 of Section 1.1 gives another regular CW-structure for the ball B^{n+1}, namely

$$\{B^0 \cup S^0,\ldots,B^n \cup S^n,B^{n+1}\}.$$

Example 4 The sequence $\{S^n : n=0,1,\ldots\}$ is an expanding sequence with union space S^∞, and, again, in view of Example 3 of Section 1.1, it is a CW-structure for S^∞.

Example 5 Unfortunately the sequence $\{B^n : n=0,1,\ldots\}$ fails to be a CW-structure for the infinite ball B^∞, because the pairs (B^n,B^{n-1}) are not adjunctions of n-cells. Nevertheless, there is a CW-structure for B^∞, namely

$$\{B^0 \cup S^0,\ldots,B^n \cup S^n,\ldots\}.$$

The union space of this expanding sequence and B^∞ clearly coincide as sets. To prove that they have the same topology, consider the expanding sequence

$$\{B^0,B^0 \cup S^0,B^1,\ldots,B^n,B^n \cup S^n,B^{n+1},B^{n+1} \cup S^{n+1},\ldots\},$$

whose union space again concides, as a set, with the set B^∞. But now, if f is a function with domain B^∞ and values in a given space, all the restrictions $f|B^n \cup S^n$ are continuous iff all the restrictions $f|B^n$ are continuous; this gives the desired result.

Example 6 Example 6 of Section 1.1 proves that

$$\{\mathbf{R}P^0, \ldots, \mathbf{R}P^m, \ldots, \mathbf{R}P^n\}$$

is a CW-structure for the real projective space $\mathbf{R}P^n, n \in \mathbf{N}$.

Example 7 The previous example proves that the sequence $\{\mathbf{R}P^n : n = 0, 1, \ldots\}$ is a CW-structure for the real infinite projective space $\mathbf{R}P^\infty$; this CW-structure for $\mathbf{R}P^\infty$ has exactly one cell in each dimension.

Example 8 For every $n \in \mathbf{N}$, define

$$X^{2n} = X^{2n+1} = \mathbf{C}P^n.$$

Then $\{X^0, X^1, \ldots, X^{2n}\}$ is a CW-structure for the complex projective space $\mathbf{C}P^n$, and the infinite sequence $\{X^n : n \in \mathbf{N}\}$ is a CW-structure for the infinite complex projective space $\mathbf{C}P^\infty$; this CW-structure for $\mathbf{C}P^\infty$ has exactly one cell in each even dimension.

Example 9 For every $n \in \mathbf{N}$, define

$$X^{4n} = X^{4n+1} = X^{4n+2} = X^{4n+3} = \mathbf{H}P^n.$$

Then $\{X^0, X^1, \ldots, X^{4n+3}\}$ is a CW-structure for the quaternionic projective space $\mathbf{H}P^n$, and the infinite sequence $\{X^n : n \in \mathbf{N}\}$ is a CW-structure for the infinite quaternionic projective space $\mathbf{H}P^\infty$; as in the previous case, this latter structure has exactly one cell in the dimensions $4n, n \in \mathbf{N}$.

The next example is of a totally different nature.

Example 10 The 2-term sequence $\{\mathbf{Z}, \mathbf{R}\}$ is a CW-structure for the real line $\mathbf{R}$.

Example 11 The sequence

$$\{\mathbf{Z} \times \mathbf{Z}, \mathbf{R} \times \mathbf{Z} \cup \mathbf{Z} \times \mathbf{R}, \mathbf{R}^2\},$$

can be taken as a CW-structure for $\mathbf{R}^2$.

Example 12 For an abelian group π and a natural number $n > 0$,

$$\{\{*\}, \ldots, \{*\}, \bigvee_\pi S^n, M(\pi, n)\}$$

is a CW-structure for the Moore space $M(\pi, n)$ (see Section 1.1, Example 10).

All the examples given up to now discuss the CW-structure of well-behaved spaces; the example bearing the unlucky (or lucky, according to one's point of view) number 13 presents a space with a bad property. Actually, part of the scope in describing the CW-complex of this example is to show that a CW-complex is not necessarily a Fréchet space.

Example 13 Let X be the CW-complex having $\{0\}$ and $\{1\}$ as 0-cells, the open interval $(0, 1)$ as the only 1-cell and for every natural number $\lambda \neq 0$, a 2-cell e_λ, with $\bar{e}_\lambda \backslash e_\lambda = \{1/\lambda\}$. Notice that $0 \in \overline{X \backslash X^1}$, but no sequence in $X \backslash X^1$ coverges to 0: if (x_λ) is a sequence in $X \backslash X^1$ which converges to a point in X^1, then at least one 2-cell $e_{\lambda'}$ has to contain a subsequence (y_λ), otherwise the sequence would be closed in X and could not converge to a point outside $X \backslash X^1$. But then, $1 \backslash \lambda' = \lim_{\lambda \to \infty} y_\lambda = \lim_{\lambda \to \infty} x_\lambda \neq 0$.

The concept of *relative CW-complex* is useful for several purposes. A *relative CW-structure* for a space X is a filtration $\{X^{(n)} : n = -1, 0, 1, \ldots\}$ of X such that:

(1) for every $n \geqslant 0$, $(X^{(n)}, X^{(n-1)})$ is an adjunction of n-cells;
(2) X is determined by the spaces $X^{(n)}$.

Notice that there are no conditions attached to the space $X^{(-1)}$. If $X^{(-1)} = \varnothing$, then X is an honest CW-complex with $X^n = X^{(n)}$, for $n \geqslant 0$; moreover, if X is any relative CW-complex, then $X/X^{(-1)}$ is a CW-complex.

Let X be a relative CW-complex; if $X^{(-1)}$ is a singleton space, say $X^{(-1)} = \{x_0\}$, one obtains a *based* CW-complex; in this case, $X^0 = X^{(0)}$ and $X^n = X^{(n)}$, for every $n > 0$, define an ordinary CW-structure for X. If X is a based CW-complex, it will be considered as a based space with the only point in $X^{(-1)}$ as base point.

Given a CW-complex X, to choose a base point means to construct a based CW-complex by defining $X^{(-1)} = \{x_0\}$ for the $x_0 \in X^0$ selected, and also, $X^{(n)} = X^n$, for all $n \geqslant 0$.

Example 14 An adjunction of n-cells (X, A) is a relative CW-complex, with $X^{(k)} = A$ for every $k < n$ and $X^{(k)} = X$ for $k \geqslant n$.

Proposition 1.2.3 *Let X be a relative CW-complex. Then,*

(i) *if $X^{(-1)}$ is a (perfectly) normal space, so is X;*
(ii) *X is determined by the family consisting of its closed cells and of $X^{(-1)}$;*
(iii) *the inclusion of any $X^{(n)}$ into X is a closed cofibration.*

Proof (i) Compare with the proof of Proposition 1.2.1.
(ii) The proof is similar to that of Proposition 1.2.2.
(iii) follows from Proposition A.5.1(iii) □

Exercises

1. Prove that every base $\{\mathbf{b}_1, \mathbf{b}_2, \ldots, \mathbf{b}_n\}$ of the vector space $\mathbf{R}^n$ defines a CW-structure for that space by taking the k-skeleton of $\mathbf{R}^n$ to be the set
$$(\mathbf{R}^n)^k = \left\{ \sum_{i=1}^{n} s_i \boldsymbol{b}_i : s_i \in \mathbf{R}, s_i \in \mathbf{Z} \text{ for at least } n-k \text{ indices } i \right\}.$$
2. A space X is said to be a *sequential space* if a subset A of X is closed iff together with any sequence it contains all its cluster points. Show that any CW-complex is a sequential space. (Note: Any Fréchet space is a sequential space but not vice versa – see Example 13.)
3. Let X be a CW-complex. Consider the identification map described in the Remark following Proposition 1.2.2. Show that this map is closed and that the inverse image of any point of X is compact iff X is metrizable. (Hint: see Morita & Hanai, 1956, Theorem 1, or Stone, 1956, cf. Proposition 1.5.11.)

1.3 Some topological properties

It is not very easy to describe the open sets of a CW-complex. The notion of *collaring* defined previously is used in this section to construct open sets of a CW-complex.

Let $\{X; X^n : n = 0, 1, \ldots\}$ be a CW-complex, and let $\{\bar{f}^n : n \in \mathbf{N}\}$ be a sequence of characteristic maps for the adjunctions (X^n, X^{n-1}). If V_m is a subset of a given m-skeleton X^m, the *infinite collar* $C_\infty(V_m)$ is defined as follows. For every $n \geqslant m$, define
$$V_{n+1} = C_{\bar{f}^{n+1}}(V_n)$$
(the $\bar{f}^{n+1}$-collar defined in Section 1.1), and then take
$$C_\infty(V_m) = \bigcup_{n \geq m} V_n$$
as a subspace of X. Clearly, the space $C_\infty(V_m)$ depends on the sequence $\{\bar{f}^n : n \in \mathbf{N}\}$ of characteristic maps.

The main properties of the infinite collaring are collected in the following.

Proposition 1.3.1 *Let $\{X; X^n : n \in \mathbf{N}\}$ be a CW-complex, let $\{\bar{f}^n : n \in \mathbf{N}\}$ be a sequence of characteristic maps for the adjunctions (X^n, X^{n-1}) and let V*

be an open or closed subset of an m-skeleton X^m. *Then the infinite collar* $C_\infty(V)$ *of* V

(i) *intersects* X^m *in* V;

(ii) *is open in* X *iff* V *is open in* X^m;

(iii) *has as closure the union in* X *of the closures of the intermediate collars*;

(iv) *is the union space of the expanding sequence of the intermediate collars*;

(v) *contains* V *as a strong deformation retract.*

Moreover, if (V_γ) *is a locally finite family of subsets of* X^m *(respectively, a family of pairwise disjoint subsets of* X^m*), then*

(vi) *the family of their infinite collars is again locally finite (respectively, is a family of pairwise disjoint subsets of* X*).*

Proof Parts (i), (ii), and (vi) are trivial; the normality of X (see Proposition 1.2.1) ensures that $C_\infty(V)$ has properties (iii) and (iv) (see Proposition A.5.4). Statement (v) follows from the fact that each intermediate collar V_n is a strong deformation retract of its successor (see Corollary A.5.8). □

The next result is a first application of the technique of infinite collaring; it shows that CW-complexes are locally contractible in a strong sense. In general, a space X is said to be *locally contractible* if for each point $x \in X$ and each neighbourhood U of x there is a smaller neighbourhood V of x such that the inclusion $V \to U$ is homotopic to a constant map. For CW-complexes the following result holds true.

Theorem 1.3.2 *Let* X *be a CW-complex, let* x_0 *be a point of* X, *and let* U *be an open neighbourhood of* x_0 *in* X. *Then there is a contractible open neighbourhood* V *of* x_0 *whose closure* $\bar{V}$ *is still contained in* U.

Proof Let the m-cell e be the carrier of x_0 in X and let $\bar{c} : B^m \to X$ be a characteristic map for e. Notice that the point $y = \bar{c}^{-1}(x_0)$ lies in the interior of B^m, and $\bar{c}^{-1}(U)$ is an open neighbourhood of y in B^m. Now choose a small m-ball B in $\mathring{B}^m$ such that $y \in B \subset \bar{c}^{-1}(U)$. Since the interior $\mathring{B} = B \setminus \delta B$ is a contractible neighbourhood of y in $\mathring{B}^m$, the set $V_m = \bar{c}(\mathring{B})$ is a contractible open neighbourhood of x_0 in X^m, whose closure $\bar{V}_m$ is contained in U. Selecting inductively the 'right coordinates' (see Proposition 1.1.10), one obtains an infinite collar $V = C_\infty(V_m)$ whose closure is still contained in U. By statement (ii) of the previous proposition, V is an open neighbourhood of x_0 in X; because of statement (v) there, V contracts to V_m, and hence to x_0. □

Recall that a space X is said to be *locally path connected at* x if, for every neighbourhood U of x, there is a path-connected neighbourhood V of x contained in U. Since a contractible set is path connected, Theorem 1.3.2 has the following consequence.

Corollary 1.3.3 *A CW-complex is locally path connected, and hence locally connected.* □

Corollary 1.3.4 *Let U be an open subset of a CW-complex X. Then U is connected iff U is path connected. In particular, a CW-complex is connected iff path connected.*

Proof In a locally path-connected space, connected and path-connected components coincide.[†] □

Another application of collaring is the proof of the paracompactness of CW-complexes.

Theorem 1.3.5 *A CW-complex is paracompact.*

Proof[‡] Let X be a CW-complex and let $\{U_\lambda : \lambda \in \Lambda\}$ be an open covering of X. The objective is to construct inductively a graded index set

$$\Gamma = \bigsqcup_{n=0}^{\infty} \Gamma_n$$

and subsets $V_{\gamma,n}$, for every $\gamma \in \Gamma$ and every $n \in \mathbf{N}$, such that the family $\{V_\gamma : \gamma \in \Gamma\}$ with

$$V_\gamma = \bigcup_{n=0}^{\infty} V_{\gamma,n}$$

is an open, locally finite refinement of the covering $\{U_\lambda\}$. Moreover, it will be shown that for a fixed $m \in \mathbf{N}$, the family $\{V_{\gamma,m}\}$ is an open, locally finite refinement of the covering $\{X^m \cap U_\lambda\}$ for the m-skeleton X^m.

An index γ is said to have *degree n* (notation: $\deg \gamma = n$), whenever $\gamma \in \Gamma_n$. As soon as an index γ is constructed, another index $\lambda = \lambda(\gamma)$ will be selected and the constructions will be done in such a way that $V_{\gamma,n}$ is contained in $U_{\lambda(\gamma)}$. Furthermore, the set $V_{\gamma,m}$ will be taken as a non-empty subset of $X^m \backslash X^{m-1}$ for $m = \deg \gamma$, $V_{\gamma,m} = \varnothing$ for $m < \deg \gamma$ and $V_{\gamma,m} \supset V_{\gamma,\deg\gamma}$ for $m > \deg \gamma$.

[†] For another proof of the second part see Corollary 1.4.12.

[‡] For an alternative proof see Section 1.5, Exercise 1.

Start the construction of Γ by taking $\Gamma_0 = X^0$; then, for every $\gamma \in \Gamma_0$, select an index $\lambda(\gamma)$ such that $\gamma \in X^0 \cap U_{\lambda(\gamma)}$, and, for these γ, define

$$V_{\gamma,0} = \{\gamma\}.$$

Now assume that the sets Γ_n are constructed up to and including the index $m-1$, together with the corresponding sets $V_{\gamma,n}, \gamma \in \Gamma_n$, and the corresponding indices $\lambda(\gamma)$.

In order to take the induction step, first define $V_{\gamma,m}, \gamma \in \Gamma_n, n = 0, 1, \ldots, m-1$. Choose an arbitrary characteristic map $\bar{f}$ for the adjunction (X^m, X^{m-1}) of m-cells, and, for any γ as before, define

$$V_{\gamma,m} = C_{\bar{f}}(V_{\gamma,m-1}) \cap U_{\lambda(\gamma)}.$$

Because of the induction hypothesis, the family $(V_{\gamma,m-1})$ is open and locally finite, and so is the family $(V_{\gamma,m})$ (see Lemma 1.1.7(iii) & (vii)). Note that $V_m = \bigcup_\gamma V_{\gamma,m}$ (where γ runs over the sets $\Gamma_n, n = 0, 1, \ldots, m-1$) is an open set in X^m which contains X^{m-1}. With a suitable choice of the 'right coordinates' (see Proposition 1.1.10), one can arrange matters so that the closure of the collar of X^{m-1} is still contained in V_m.

Let e be an m-cell and let $\bar{f}_e: B^m \to X$ be its corresponding characteristic map. The family $\{U'_\lambda = \bar{f}^{-1}(U_\lambda)\}$ covers the m-ball $B' = \{s \in B^m : |s| \leqslant \frac{3}{4}\}$; since B' is compact, finitely many members of this family, say $U'_{\lambda_1}, \ldots, U'_{\lambda_k}$, suffice to cover B'. With this in mind, define

$$\Gamma_e = \{\lambda_1, \ldots, \lambda_k\};$$

moreover, for every $\gamma \in \Gamma_e$, define

$$\lambda(\gamma) = \gamma$$

and

$$V_{\gamma,m} = \bar{f}(\mathring{B}' \cap U'_\gamma).$$

Finally, define

$$\Gamma_m = \bigsqcup_e \Gamma_e,$$

where e runs over the set of all m-cells of X.

One must prove now that the indices γ and the sets $V_{\gamma,m}$ actually perform the tasks they are supposed to. First, prove that the family $\{V_{\gamma,m} : \deg \gamma \leqslant m\}$ is locally finite. Let x be a point of X whose carrier is an m-cell e. On the one hand, it is already known that x has a neighbourhood U which meets only finitely many sets $V_{\gamma,m}$, with $\deg \gamma < m$, and, on the other hand, e meets only finitely many sets $V_{\gamma,m}$ with $\deg \gamma = m$, namely those with $\gamma \in \Gamma_e$; then $U \cap e$ is a neighbourhood of x in X^m which meets only finitely many $V_{\gamma,m}$ with $\deg \gamma \leqslant m$. Suppose now that $x \in X^{m-1}$. The idea now is to prove that any open neighbourhood U of x in X^{m-1} meeting only finitely many $V_{\gamma,m-1}$, with $\deg \gamma < m$, can be enlarged to a neighbourhood

U' of x in X^m, meeting only finitely many $V_{\gamma,m}$, with $\deg\gamma \leqslant m$. This will be done by analysing carefully the various collaring processes. Let $\bar{f}$ be one of the characteristic maps used in the construction of the set V_m; then, according to Lemma 1.1.7 (vii), the collar $C_{\bar{f}}(U)$ interesects only finitely many $V_{\gamma,m}$, with $\deg\gamma < m$. Notice that $\tilde{B} = \bigcup_e \bar{f}_e(\mathring{B}')$ is a closed subset of X^m which does not meet X^{m-1}. Therefore, $U' = C_{\bar{f}}(U)\backslash\tilde{B}$ is a neighbourhood of x in X^m having the desired property.

At this point, one should notice a property which will be needed in the sequel: U' intersects $V_{\gamma,m}$ iff U intersects $V_{\gamma,m-1}$.

The final step of the proof can now be taken. For every index γ,

$$V_\gamma \cap X^m = V_{\gamma,m},$$

and therefore V_γ is open in X. Moreover, the inclusions

$$V_\gamma \subset U_{\lambda(\gamma)}$$

show that the family (V_γ) is a refinement of the family (U_λ). For an arbitrary point $x \in X$, take a non-negative integer m such that $x \in X^m$ and choose a neighbourhood U_m of x in X^m which intersects only a finite number of sets V_γ. As proved before, the neighbourhood U_m of x can be enlarged, inductively, to neighbourhoods U_n of x in X^n, all of which intersect only a finite number of sets V_γ, which already intersect U_m. Thus $\bigcup_{n=m}^{\infty} U_n$ is a neighbourhood of x in X intersecting only finitely many sets V_γ; this proves that the family (V_γ) is locally finite. □

Remark Theorem 1.3.5 actually follows from the fact that the formation of adjunction spaces and union spaces of expanding sequences preserves paracompactness (see Exercise 5 of the Appendix and Proposition A.5.1 (v)); the text, however, offers an independent direct proof; one reason for selecting this procedure is to pinpoint the use of the axiom of choice. The reader will have noted that several choices were made during the proof of Theorem 1.3.5. Then, one may ask if the axiom of choice is really necessary to prove the paracompactness of CW-complexes. Couldn't one have the same situation as in the Tychonoff theorem (the Cartesian product of a collection of compact spaces is compact relative to the product topology) which is known to be equivalent to the axiom of choice? A systematic examination of the proof of Theorem 1.3.5 shows that the following two weak forms of the axiom of choice would be enough to yield the theorem.

(i) Axiom of countable choice: it is possible to select just one element from any member of a countable family of sets. This is used to choose the characteristic maps $\bar{f}$ needed to construct the sets V_m.

(ii) Axiom of multiple choice: it is possible to select a finite set of elements from any member of an arbitrary family of sets. With this axiom, instead of choosing one $\lambda(\gamma)$ for each γ, one chooses a finite set L_γ of indices λ, so that the sets $U_\lambda, \lambda \in L_\gamma$ contain the set $V_{\gamma,m}$ in question. Then one can proceed by assuming that the subsequent sets $V_{\gamma,n}$ are all contained in the open set $\bigcap_{\lambda \in L_\lambda} U_\lambda$.

The choice of a finite subfamily of the covering (U'_λ) of the ball B' is not really necessary. One can get the Lebesgue number ε of this covering without using any axiom of choice, and then it is possible to construct explicitly open sets with diameter ε. The selection of the 'right coordinates' does not require a choice.

The previous considerations show that one is not really faced with a situation analogous to that of Tychonoff's theorem. For compact Hausdorff spaces, it is known that the theorem of Tychonoff is equivalent to the ultrafilter theorem, a weaker form of the axiom of choice, but in another direction. The ultrafilter theorem is not equivalent to the axiom of multiple choice. This question leads to more sophisticated problems in set theory.

Theorem 1.3.6 *Every CW-complex is an LEC space.*

Proof Let X be a CW-complex. Because coproducts of balls are LEC spaces, Corollary A.4.14 shows, by induction, that all the skeleta X^n are LEC and so is their union space X (see Corollary A.5.6). □

Remark LEC spaces are locally contractible (see Proposition A.4.5); thus, the previous theorem proves again that CW-complexes are locally contractible. However, the previous proof of this fact (see Theorem 1.3.2) is not wasted, since it indeed furnishes a stronger result.

Corollary 1.3.7 *Let X be a CW-complex and x be an arbitrary point of X. Then, the inclusion map $\{x\} \to X$ is a closed cofibration.*

Proof Because X is LEC, there is a neighbourhood U of ΔX in $X \times X$ which is deformable to ΔX in $X \times X$ rel. ΔX; moreover, there is a map $\alpha : X \times X \to I$ such that $\alpha^{-1}(0) = \Delta X$ and $\alpha|(X \times X \setminus U) = 1$ (see Proposition A.4.1 (iv)). Take

$$\alpha_x : X \to I$$

to be $\alpha_x(y) = \alpha(y, x)$, for every $y \in X$, and $U_x = \alpha_x{}^{-1}([0,1))$. Then, $\alpha_x{}^{-1}(0) = \{x\}$ and $\alpha_x|(X \setminus U_x) = 1$. Now, if $H : U \times I \to X \times X$ is a suitable deformation, take $H_x : U_x \times I \to X$ to be

$$H_x(y,t) = \begin{cases} pr_1 \circ H((y,x), 2t), & 0 \leqslant t \leqslant \frac{1}{2} \\ pr_2 \circ H((y,x), 2-2t), & \frac{1}{2} \leqslant t \leqslant 1 \end{cases},$$

for every $y \in U_x, t \in I$. Notice that H_x deforms the neighbourhood U_x of x into $\{x\}$ rel. $\{x\}$. Thus, the inclusion of $\{x\}$ into X is a closed cofibration (see again Proposition A.4.1 (iv)). □

Exercises

1 A space is *hereditarily paracompact* if every subspace is paracompact. Show that it suffices to require this condition for open subspaces and that CW-complexes are hereditarily paracompact.

2 Prove that CW-complexes are stratifiable. (Borges, 1966)

3 Let X be a CW-complex and C be a compact Hausdorff space. Show that the function space X^C is stratifiable. (Cauty, 1976)

Later on, it is proved that X^C has the type of a CW-complex (see Corollary 5.3.6).

4 Generalize the statement of Exercise 3 to spaces C for which there is a sequence $\{C_n : n \in \mathbf{N}\}$ of compact Hausdorff subspaces C_n such that every compact Hausdorff subspace of C is contained in some C_n. (Thus, C may be any locally compact CW-complex or any locally compact metric space satisfying the second axiom of countability.)

1.4 Subcomplexes

Let X be a CW-complex. A CW-complex A is a *subcomplex* of X, if

(1) A is a subspace of X and

(2) for all $n \in \mathbf{N}$, the n-skeleton of A is the intersection of A with the n-skeleton of X:

$$A^n = X^n \cap A.$$

Condition (2) implies that the CW-structure of A is determined by the space A and the CW-structure of X; thus, by abuse of language, one also says that a subspace A of the CW-complex X is a subcomplex of X, if the filtration

$$X^0 \cap A \subset X^1 \cap A \subset \cdots \subset X^n \cap A \subset \cdots$$

is a CW-structure for A.

Every skeleton X^n of X is a subcomplex of X. Moreover, a subcomplex of a subcomplex of X is again a subcomplex of X.

Lemma 1.4.1 *If A is a subcomplex of the CW-complex X then every cell of A is a cell of X.*

Proof Let e be an n-cell of A; as a subset of $A^n \backslash A^{n-1}$, the set e is also a subset of $X^n \backslash X^{n-1}$, that is to say, e is contained in an open n-cell e' of X. By the theorem of invariance of domain, e is an open subset of e'. On the other hand, $\bar{e} \backslash e \subset A^{n-1} \subset X^{n-1}$, and so $\bar{e} \cap e' = e$. But $\bar{e}$ is compact, and hence is closed in X. This implies that e is also closed in e'. Being a non-empty open and closed subset of the connected set e', the set e must coincide with e'. □

Corollary 1.4.2 *Let A be a subcomplex of a CW-complex X and let e be a cell of X. Then, e is a cell of A iff $e \cap A$ is not empty.*

Proof Only the sufficiency will be proved, since the necessary condition is trivial. Let x be a point in $e \cap A$ and let e' be the carrier of x in A. The lemma implies that e' is also a cell of X, and, since $e \cap e'$ is not empty, it follows that the cells e and e' must coincide. □

The skeleta of a CW-complex and those of a subcomplex are related as follows.

Lemma 1.4.3 *Let A be a subcomplex of a CW-complex X. Then,*

(i) *the pairs $(X^{n-1} \cup A^n, X^{n-1})$ are adjunctions of n-cells;*
(ii) *the pairs $(X^n, X^{n-1} \cup A^n)$ are adjunctions of n-cells;*
(iii) *the inclusions of A^n into X^n are closed cofibrations.*

Proof Let $j : A \to X$ denote the inclusion of A into X. If e is an n-cell of A and $\bar{c}' : B^n \to A$ is a characteristic map for e, then the composition $j \circ \bar{c}' : B^n \to X$ is a characteristic map for e considered as an n-cell of X. Conversely, if e is an n-cell of A and $\bar{c} : B^n \to X$ is a characteristic map for e considered as a cell of X, then $\bar{c}$ factors through A, giving a characteristic map $\bar{c}' : B^n \to A$ for the cell e. Consider simultaneously the characteristic maps for all n-cells of X; then A^n is closed in X^n (by induction) and (i) follows from Proposition 1.1.2(v) while (ii) is a consequence of the addition law for adjunction spaces.

Finally, assume inductively that the inclusion of A^{n-1} into X^{n-1} is a closed cofibration. Then, by the horizontal composition law,

$$X^{n-1} \cup A^n = X^{n-1} \sqcup_{A^{n-1}} A^n,$$

and therefore the inclusion of A^n into $X^{n-1} \cup A^n$ is a closed cofibration. Composition of this inclusion with the closed cofibration $X^{n-1} \cup A^n \to X^n$ of (ii) completes the induction. □

As seen before, a subcomplex of a CW-complex is a union of cells of the CW-complex; one should then ask: which unions of cells of a CW-complex form a subcomplex? This question is taken up by the following result.

Proposition 1.4.4 *Let X be a CW-complex, let Ω be a set of cells of X and let A be the union of all cells of Ω. The following conditions are equivalent:*

(i) *A is a subcomplex of X;*

(ii) *A is a closed subspace of X;*

(iii) *for every cell $e \in \Omega$, the closure $\bar{e}$ of e is contained in A.*

Proof (i)$\Rightarrow$(ii): Since X has the final topology with respect to the X^n and since $X^n \cap A = A^n$, it follows that A is closed in X.

(ii)$\Rightarrow$(iii): Trivial.

(iii)$\Rightarrow$(i): Define $A^n = X^n \cap A$; then A is determined by the subspaces A^n, according to Proposition A.5.4 (ii). Thus, it remains to prove that every pair (A^n, A^{n-1}) is an adjunction of n-cells, for every $n \in \mathbb{N}$. This will be done by induction on n, establishing at the same time the fact that A^n is closed in X^n. Choose a characteristic map $\bar{c}_\lambda : B_\lambda \to X$ for every n-cell e_λ of Ω. Condition (iii) implies that the corresponding attaching maps c_λ factor through A^{n-1}. It is immediate that at the set theoretical level

$$A^n = A^{n-1} \sqcup \bigsqcup_\lambda (B_\lambda \backslash S_\lambda).$$

To show that A^n has the correct topology, take a subset U of A^n such that $U \cap A^{n-1}$ and $\bar{c}_\lambda^{-1}(U)$ are closed in A^{n-1} and B_λ, respectively, for every λ. Now, since A^{n-1} is closed in X^{n-1} (use the inductive hypothesis), it follows that U is closed in X^n; for the special case $U = A^n$ this already shows that A^n is closed in X^n. But if U is closed in X^n it is also closed in the subspace A^n of X^n. □

Corollary 1.4.5 *Arbitrary unions and intersections of subcomplexes of a CW-complex X are subcomplexes of X.* □

Proposition 1.4.6 *Let X be a CW-complex and A be a subcomplex. Then the filtration*

$$\{X^{(n)} = X^n \cup A : n = -1, 0, \ldots\}$$

gives X the structure of a relative CW-complex.

Proof The space $X^{(n)}$ is the union of the closed subspaces $X^{(n-1)}$ and X^n whose intersection is $X^{n-1} \cup A^n$. Since the inclusion $X^{n-1} \cup A^n \subset X^n$ is a closed cofibration (see Lemma 1.4.3(ii)) $X^{(n)}$ may be viewed as obtained from $X^{(n-1)}$ by attaching X^n via the inclusion $X^{n-1} \cup A_n \subset X^{(n-1)}$. But the pair $(X^n, X^{n-1} \cup A^n)$ is an adjunction of n-cells (see again Lemma 1.4.3 (ii)), and so the pair $(X^{(n)}, X^{(n-1)})$ is also an adjunction of n-cells (see the horizontal composition law (L1)).

Since X is determined by the family $\{X^n : n \in \mathbb{N}\}$, it is also determined by the family of the subspaces $X^{(n)}$ (see Proposition A.2.1). □

In the previous proof one uses the fact that a subcomplex is always a closed subspace; this is part of Proposition 1.4.4. Moreover, the following finer result holds true.

Corollary 1.4.7 *The inclusion of a subcomplex into a CW-complex is a closed cofibration.*

Proof See Proposition 1.2.3 (iii). □

Example 1 For any subset L of a CW-complex X, the intersection of all subcomplexes of X containing L is a subcomplex $X(L)$ of X. Moreover, because any subcomplex is closed, it follows that

$$X(L) = X(\bar{L}).$$

In general, the closed cells of a CW-complex are not subcomplexes (see Example 13 in Section 1.2).

Example 2 For any subset L of a CW-complex X, the *star of* L is the subcomplex

$$\mathrm{St}(L) = \bigcup_{\bar{e} \cap L \neq \varnothing} X(e).$$

A pair consisting of a CW-complex and one of its subcomplexes behaves nicely with respect to Serre fibrations.

Proposition 1.4.8 *Let D be a subcomplex of the CW-complex Y and let $p : Z \to X$ be a Serre fibration. Moreover, let a homotopy $H : Y \times I \to X$ and a map $\hat{H} : Y \times \{0\} \cup D \times I \to Z$ with $p \circ \hat{H} = H | Y \times \{0\} \cup D \times I$ be given. Then, there exists a homotopy $\tilde{H} : Y \times I \to Z$ such that $\tilde{H} | Y \times \{0\} \cup D \times I = \hat{H}$ and $p \circ \tilde{H} = H$.*

Proof The proof is done using the 'method of the least criminal'. Using Zorn's lemma, find a subcomplex A of Y containing D and which is maximal with respect to the property: there is a homotopy $G : A \times I \to Z$ such that $G | A \times \{0\} \cup D \times I = \hat{H} | A \times \{0\} \cup D \times I$ and $p \circ G = H | A \times I$. Assume A to be different from Y and take a cell $e \subset Y \backslash A$ of lowest dimension, say $\dim e = n$. Let $\bar{c} : B^n \to Y$ be a characteristic map for the cell e; its associated attaching map factors through a unique map $c : S^{n-1} \to A$. The space $A' = A \cup e$, obtained by attaching B^n to A via c, is a subcomplex of Y strictly bigger than A. It will be shown that the homotopy G can be extended suitably over $A' \times I$, contradicting the maximality of A.

To simplify the notation, assume that $A = D$ and that $A' = Y$. Note that in this case, $Y \times I$ is obtained from $Y \times \{0\} \cup D \times I$ by attaching $B^n \times I$ via the restriction $f : B^n \times \{0\} \cup S^{n-1} \times I \to Y \times \{0\} \cup D \times I$ of $\bar{c} \times 1_I$; moreover, $p \circ \hat{H} \circ f = H \circ (\bar{c} \times 1_I) | B^n \times \{0\} \cup S^{n-1} \times I$. Next, take a homeomorphism

$$h : B^n \times I \to B^n \times I$$

which induces a homeomorphism

$$\hat{h} : B^n \times \{0\} \to B^n \times \{0\} \cup S^{n-1} \times I$$

(this can be obtained from the homeomorphism w^n defined on page 9 by 'turning top to bottom'). Now apply the defining property of Serre fibrations in order to obtain a homotopy $C : B^n \times I \to Z$, such that $C | B^n \times \{0\} = \hat{H} \circ f \circ \hat{h}$ and $p \circ C = H \circ \bar{c} \times 1_I \circ h$. Since $C \circ h^{-1} | B^n \times \{0\} \cup S^{n-1} \times I = \hat{H} \circ f$, the universal property of attaching yields a homotopy $C' : Y \times I \to Z$ with the desired properties. □

Corollary 1.4.9 *Let D be a subcomplex of the CW-complex Y which is a strong deformation retract of Y and let $p : Z \to X$ be a Serre fibration. Moreover, let maps $f : Y \to X$ and $\bar{f} : D \to Z$ with $p \circ \bar{f} = f | D$ be given. Then, there is a map $\tilde{f} : Y \to Z$ such that $\tilde{f} | D = \bar{f}$ and $p \circ \tilde{f} = f$.*

Proof Let $H : Y \times I \to Y$ be a deformation of Y into D, i.e., a homotopy rel. D from a map which factors through a retraction $r : Y \to D$ to 1_Y. The restriction of $f \circ H$ to $Y \times \{0\} \cup D \times I$ can be decomposed in the form $p \circ \hat{H}$; the proposition proves the existence of a homotopy $\tilde{H} : Y \times I \to Z$ with $\tilde{H} | Y \times \{0\} \cup D \times I = \hat{H}$ and $p \circ \tilde{H} = H$. Now take $\tilde{f} : Y \to Z, y \mapsto \tilde{H}(y, 1)$. □

Theorem 1.4.10 *If X is a regular CW-complex, the closure of any cell e of X is a subcomplex, i.e., $\bar{e} = X(e)$.*

Proof Clear, if $\dim e = 0$. Assume the theorem true for all cells of dimension strictly less than n. Let e be an n-cell of X. It is enough to prove that $\bar{e} \backslash e$ (which is an $(n-1)$-dimensional sphere, by the regularity assumption) is a subcomplex. Let e' be an open $(n-1)$-cell meeting $\bar{e} \backslash e$. Clearly, $e' \cap (\bar{e} \backslash e)$ is closed in e' and open in $\bar{e} \backslash e$, and thus it is also open in e' (see the theorem of invariance of domain). It follows that $e' \cap (\bar{e} \backslash e)$ is a non-empty component of e', and, hence, $e' \subset \bar{e} \backslash e$. Now let Ω' denote the set of all $(n-1)$-cells of X meeting $\bar{e} \backslash e$; by the induction hypothesis, the closures of all these $(n-1)$-cells e' are subcomplexes of X and so is $A = \bigcup_{e' \in \Omega'} \bar{e}'$.

It is clear that $A \subset \bar{e} \setminus e$; however, the opposite inclusion also holds true, as can be seen via the following decreasing induction:

$$\bar{e} \setminus e \subset A \cup X^q \Rightarrow \bar{e} \setminus e \subset A \cup X^{q-1}.$$

It is already known that $\bar{e} \setminus e \subset A \cup X^{n-2}$. For the inductive step, take a q-cell e'' with $e'' \cap A = \emptyset$; this implies that e'' is open in $A \cup X^q$, and thus $e'' \cap (\bar{e} \setminus e)$ is open in $\bar{e} \setminus e$. If $e'' \cap (\bar{e} \setminus e) \neq \emptyset$, then one would have an open $(n-1)$-ball contained in the open q-ball e'' and this, because $q \leqslant n-2$, is a contradiction to the theorem of invariance of domain. Therefore, $e'' \cap (\bar{e} \setminus e) = \emptyset$, implying that $\bar{e} \setminus e \subset A \cup X^{q-1}$. □

Proposition 1.4.4 also has some consequences about the connectivity of a CW-complex.

Proposition 1.4.11 *A connected component (respectively, a path-component) of a CW-complex X is a subcomplex of X.*

Proof Let A be a connected component (respectively, a path-component) of X. Any cell (respectively, any closed cell) meeting A is completely contained in A. Thus, A is a union of cells and condition (iii) of Proposition 1.4.4 holds true. □

The previous ideas permit one to re-prove one of the conclusions of Corollary 1.3.4:

Corollary 1.4.12 *Any connected CW-complex is path-connected.*

Proof Let A be a path-component of a connected CW-complex X. Both A and $X \setminus A$ are subcomplexes of X, the latter as the union of the remaining path-components of X. Thus, A is open and closed in X and therefore it is equal to X. □

A similar argument proves the following fact.

Corollary 1.4.13 *Any CW-complex is the topological sum of its (path-) components.* □

In particular, this means that a path-component of a CW-complex is open. This property is shared by all spaces which are dominated by CW-complexes:

Proposition 1.4.14 *If a space Y is dominated by a CW-complex X, then the path-components of Y are open.*

Proof Let $f: Y \to X$ and $g: X \to Y$ be maps such that $g \circ f$ is homotopic to the identity of Y and let U be a path-component of Y. Let y be an arbitrary point in U and let V be a path-connected neighbourhood of $f(y)$ in X (see Corollary 1.3.3).

Any point $z \in f^{-1}(V)$ can be joined by a path to $g \circ f(z)$ by means of a homotopy $1_Y \simeq g \circ f$; since $f(z) \in V$, there is a path from $f(z)$ to $f(y)$ which is transformed into a path from $g \circ f(z)$ to $g \circ f(y)$ by the map g. Using a homotopy $g \circ f \simeq 1_Y$, there is a path from $g \circ f(y)$ to y. Altogether, this gives a path from z to y, implying that $f^{-1}(V) \subset U$; thus, U is a neighbourhood of y, and, since this fact holds true for all $y \in U$, the set U is open in Y. □

Remark The previous result is not as general as it seems, because, as will be seen in Chapter 5, spaces dominated by CW-complexes already have the type of CW-complexes (see Proposition 5.1.1). □

A trivial consequence of Proposition 1.4.4 is that any collection of 0-cells is a subcomplex. One should finally note that any non-empty CW-complex contains 0-cells.

Proposition 1.4.15 *If X is a CW-complex then, every path-component of X (respectively, X^m) contains at least one 0-cell.*

Proof By its very definition, a path-component of a space is non-empty. Thus, a given path-component of X must contain a cell e_0. Clearly, the result holds true if e_0 is a 0-cell. Otherwise, $\bar{e}_0 \backslash e_0 \neq \varnothing$, and so there exists a cell e_1 of lower dimension. If e_1 is not yet a 0-cell, the same argument shows the existence of a cell e_2 of even lower dimension. A 0-cell must be obtained after only finitely many steps. □

Exercises

1 If X is a contractible CW-complex and A is a contractible subcomplex, prove that A is a strong deformation retract of X.

2 Let A and Y be subcomplexes of a CW-complex X with $X = A \cup Y$ and $A, Y, A \cap Y$ contractible. Prove that X is contractible.

1.5 Finiteness and countability

In the theory of CW-complexes, one uses several kinds of countability assumptions: a CW-complex (CW-structure) X is

finite, if X has only finitely many cells;
locally finite, if every (open) cell of X meets only finitely many closed cells of X;
of finite type, if every skeleton X^n is a finite CW-complex;
countable, if X has countably many cells.

There is still another type of finiteness condition, satisfied by every CW-complex.

Theorem 1.5.1 *A CW-complex is closure finite, i.e., the closure of every cell meets only finitely many (open) cells.*

Warning: Be careful to distinguish between the notion of 'closure finiteness' given in 1.5.1 and that of 'local finiteness' described before. The latter is a very restrictive condition (see Propositions 1.5.10 and 1.5.17). A way to view this distinction is perhaps to observe that for any given cell e, closure finiteness deals with the *lower* dimensional cells which meet the closures $\bar{e}$ of e, while local finiteness has to do with the *higher* dimensional cells whose closure meet the open cell e. □

The proof of Theorem 1.5.1 is an immediate consequence of the following.

Proposition 1.5.2 *A compact subset of a CW-complex is contained in a finite union of (open) cells of the CW-complex.*

Proof Let X be a CW-complex and K be a compact subset of X. Let E denote the set of all cells of X which intersect K. Choose a point $x_e \in K$ in every cell $e \in E$. It will be shown inductively that the set $Z = \{x_e : e \in E\}$ intersects any skeleton in only finitely many points; thus, Z will be a discrete closed subset of X, and also of K. Recalling that any discrete closed subset of a compact space is finite, one obtains that Z is finite.

Clearly, $Z \cap X^0 = K \cap X^0$ is a discrete closed subset of the compact space K, and so it is finite. Assume now that $Z \cap X^{n-1}$ is finite. Since Z meets any closed n-cell in a finite number of points, and X^n is determined by the family consisting of X^{n-1} and all closed n-cells of X (see Proposition 1.1.3 (ii)), $Z \cap X^n$ is a discrete closed subset of X^n contained in the compact space K, and hence is finite. □

Proposition 1.5.2 has other interesting consequences.

Corollary 1.5.3 *If X is a CW-complex and e is any of its cells, then the subcomplex $X(e)$ is finite.*

Proof The proof is by induction on the dimension of the cells. If e is a 0-cell, then $X(e)$ coincides with e itself and therefore is finite. Suppose that $X(e)$ is finite for every cell e of dimension strictly less than n. Let e be an n-cell of X. Because $\bar{e}\backslash e$ is compact and contained in X^{n-1}, the proposition implies that $\bar{e}\backslash e$ is contained in the union of finitely many open cells

$$e_1, e_2, \ldots, e_k$$

of dimensions $< n$. By the induction hypothesis, the subcomplexes $X(e_i)$, $i = 1, \ldots, k$ are finite; moreover, the union

$$\tilde{X} = e \cup X(e_1) \cup \cdots \cup X(e_k)$$

is a (finite) subcomplex of X (Proposition 1.4.4 (iii)). The proof is completed by noticing that $X(e)$ is contained in the finite subcomplex $\tilde{X}$. □

Corollary 1.5.4 *If X is a CW-complex and L is a relatively compact subset of X then the subcomplex $X(L)$ is finite.*

Proof Since $X(L) = X(\bar{L})$, one may assume that L is compact. By Proposition 1.5.2, L is contained in a union of finitely many cells, say $e_1, e_2, \ldots, e_k$; but then

$$X(L) \subset X(e_1) \cup \cdots \cup X(e_k),$$

the latter being a finite subcomplex of X. □

This corollary has a very general application (but read the remark after Proposition 1.4.14).

Corollary 1.5.5 *A compact space dominated by a CW-complex is dominated by a finite CW-complex.*

Proof Let X be a compact space and let Y be a CW-complex which dominates X; let $f : X \to Y$ and $g : Y \to X$ be maps such that $g \circ f \simeq 1_X$. Since X is compact, $f(X)$ is compact, and hence contained in a finite subcomplex Y' of Y, which also dominates X. □

Another application of Proposition 1.5.2 concerns the notion of path-connectivity. A CW-complex X is said to be *cell path connected* if, for every two points $x, y \in X$, there is a sequence $\{x = x_0, \ldots, x_n = y\}$ of points of X such that, for every $i = 1, \ldots, n$, $\{x_{i-1}, x_i\}$ is contained in some closed cell of X.

Proposition 1.5.6 *A CW-complex is path connected iff it is cell path connected.*

Proof Clearly, a CW-complex which is cell path connected is also path connected. Assume that one is given a path-connected CW-complex X, two different points $x, y \in X$ and a path $\sigma : I \to X$ from x to y. The compactness of the path implies that it meets only finitely many cells of X (see Proposition 1.5.2). Perform an induction on the number of these cells. If the path is contained entirely in one cell, there is nothing to prove. Otherwise, from the set of cells under consideration whose closure contains x, select one cell e_1 having maximal dimension and take $x_1 = \sigma(t_1)$ to be the last point of the path that belongs to the closure of e_1. Notice that x_1 is a point on the boundary of e_1 and that $\sigma([t_1, 1]) \cap e_1 = \emptyset$. Now we apply the inductive hypothesis to the path $\sigma|[t_1, 1]$. □

Some subcomplexes of a CW-complex are always cell path connected, independently of the fact that the CW-complex is path connected or not. The following is an example.

Proposition 1.5.7 *Let e be a cell of a CW-complex X. Then $X(e) = X(\bar{e})$ is cell path connected.*

Proof By induction on the dimension of the cell e. The proposition is obvious if e is 0-dimensional; suppose that $\dim e = n$ and that the result is true for all cells of dimension $< n$.

As seen in the proof of Corollary 1.5.3, the CW-complex $X(e)$ can be written in the form

$$X(e) = X(\bar{e}) = e \cup X(e_1) \cup \cdots \cup X(e_p),$$

where $e_1, \ldots, e_p$ are the finitely many cells of X that intersect $\bar{e}$. Now take $x, y \in X(e)$. The result is clearly true if $x, y \in \bar{e}$. If $x \in \bar{e}$ and $y \in X(e_i)$, let $z \in \bar{e} \cap e_i \neq \emptyset$; by the induction hypothesis, there is a sequence $\{x'_0, \ldots, x'_q\}$ of points belonging to $X(e_i)$, and therefore to X, and cells $e'_1, \ldots, e'_q$ of $X(e_i)$ such that $x'_0 = z, x'_q = y$ and $\{x'_{j-1}, x'_j\} \subset \bar{e}'_j$. Now take the sequence $\{x_0, \ldots, x_{q+1}\}$ and the cells $e_1, \ldots, e_{q+1}$, so that $x_0 = x, x_1 = x'_0, \ldots, x_{q+1} = x'_q$ and $e_1 = e, e_2 = e'_1, \ldots, e_{q+1} = e'_q$. Finally, if $x \in X(e_i)$ and $y \in X(e_j)$, take $z \in \bar{e} \cap e_i, w \in \bar{e} \cap e_j$ and reduce the argument to the previous case. □

A property P of a CW-complex X is said to be *topologically invariant* if P depends only on the space X and not on the specific CW-structure chosen for X. As a consequence of Proposition 1.5.2, the property of 'finiteness' is topologically invariant.

Proposition 1.5.8 *A CW-complex is finite iff it is compact.* □

Finite-dimensional balls, spheres and projective spaces with the CW-structures described in Examples 1, 2, 3, 6, 8 and 9 of Section 1.2 are examples of finite CW-complexes. Clearly, all finite CW-complexes are locally finite; the CW-structures of the spaces $\mathbf{R}^1$ and $\mathbf{R}^2$ described in Examples 10 and 11 of Section 1.2 respectively are locally finite but not finite (the same will be true for the CW-structures of the higher-dimensional Euclidean spaces given in Example 1 of Section 2.2). Local finiteness is another topologically invariant property of CW-complexes.

Lemma 1.5.9 *If X is a locally finite CW-complex and e is any of its cells, then the star* $\operatorname{St}(e)$ *of e is a compact neighbourhood of e.*

Proof Because X is locally finite, e meets only finitely many closed cells of X, and thus

$$\operatorname{St}(e) = \bigcup_{\bar{e}' \cap e \neq \varnothing} X(e')$$

is a finite subcomplex of X, and so is a compact space.

Now one has to show that $\operatorname{St}(e)$ is a neighbourhood of e. Let Ω be the finite set of all closed cells of X which meet e; the union

$$W = \bigcup_{\bar{e}'' \in \Omega} \bar{e}''$$

is a subset of $\operatorname{St}(e)$. The idea is to prove that W is already a neighbourhood of e. To this end, let Ω' be the set of all closed sets of X which meet W but not e. Closure finiteness (see Theorem 1.5.1) and the hypothesis imply that Ω' too is finite; thus the union

$$C = \bigcup_{\bar{e}' \in \Omega' \setminus \Omega} \bar{e}'$$

is a compact, in particular, a closed subset of X disjoint to e. Next, choose an infinite collar $V = C_\infty(e)$ of e. Since V is open (see Proposition 1.3.1 (ii)), the difference set $V \setminus C$ is an open neighbourhood of e. It will be shown by induction that, for every $n \geqslant m = \dim e$, the set $V_n = X^n \cap (V \setminus C)$ is contained in W, implying that $V \setminus C \subset W$ as desired. Clearly, $V_m = e \subset W$. For the induction step, consider a point $y \in V_{n+1} \setminus V_n$, with carrier e_y. Since y belongs to a collar of V_n, the closed cell $\bar{e}_y$ meets V_n (see Lemma 1.1.7 (v)), and thus W, by the induction hypothesis. Now, if $\bar{e}_y$ would not belong to the set Ω, it would be an element of Ω' and, consequently, a subset of C, contradicting the assumption that $y \in \bar{e}_y \cap (V \setminus C)$. But $\bar{e}_y \in \Omega$ implies that $y \in \bar{e}_y \subset W$, as desired. □

One may think that the assumption of local finiteness on X in Lemma 1.5.9 is only necessary to prove the compactness of St(e) and that St(e) is always a neighbourhood of e, for any cell e of a CW-complex X. To see that this is not the case, take the CW-complex X constructed in Example 13 of Section 1.2 and observe that the St$(\{0\}) = X^1$ is not a neighbourhood of $\{0\}$.

Proposition 1.5.10 *A CW-complex is locally finite iff it is locally compact.*

Proof '$\Rightarrow$': See Lemma 1.5.9.
'$\Leftarrow$': Let X be a locally compact CW-complex and let e be a cell of X. Every point of the closed cell $\bar{e}$ has a compact neighbourhood; since $\bar{e}$ is compact, e is covered by finitely many of these compact neighbourhoods, and therefore e has a compact neighbourhood V in X. Now observe that, on the one hand, e does not intersect the closure of any cell of X contained in $X \setminus V$, because V is a neighbourhood of e, and, on the other, V intersects only finitely many cells of X, V being compact (see Proposition 1.5.2). These observations prove that the cell e intersects only finitely many closed cells of X. □

Corollary 1.5.11 *A CW-complex X is locally finite iff its closed cells form a locally finite (closed) covering of X.*

Proof '$\Rightarrow$': Let $x \in X$ be an arbitrary point of X and let K be a compact neighbourhood of x. According to Proposition 1.5.2, K is contained in a finite union of open cells of X. Now, local finiteness implies that each of the open cells meeting K interesects only finitely many closed cells.
'$\Leftarrow$': Let e be an open cell of X. For each point $x \in \bar{e}$, choose a neighbourhood U_x of x with the property that U_x meets only finitely many closed cells of X. Since $\bar{e}$ is compact, finitely many neighbourhoods $U_{x_0}, \ldots, U_{x_r}$ are enough to cover the cell e. Then each closed cell of X which encounters e must meet one such set U_{x_i}, $i = 0, \ldots, r$. Since each U_{x_i} intersects only finitely many closed cells of X, the open cell e meets only finitely many closed cells of X. □

Remark There are still some other topological characterizations of local finiteness whose proofs depend on results not yet stated; for this reason, they are postponed to Proposition 1.5.17. □

The discussion about locally finite CW-complexes is continued with the following result.

Proposition 1.5.12 *A locally finite and connected CW-complex X is countable.*

Proof Let Ω be the set of all cells of X and let e_0 be a fixed cell of X; for each $n \in \mathbf{N}$, let

$$A_n = \{(e_0, e_1, \ldots, e_n) : e_i \in \Omega, \bar{e}_{i-1} \cap \bar{e}_i \neq \emptyset, i = 1, \ldots, n\}.$$

Claim: the sets $A_n, n \in \mathbf{N}$ are all finite. In fact, every component e_i of any element $(e_0, e_1, \ldots, e_n) \in A_n$ is a cell of $\mathrm{St}(\bar{e}_{i-1}), i = 1, \ldots, n$; then, use the fact that each $\mathrm{St}(e_{i-1})$ is finite (this follows from the local and closure finiteness properties of X).

Now take the countable set

$$A = \bigcup_{n \in \mathbf{N}} A_n$$

and define the function

$$\alpha : A \to \Omega$$

whose restriction to A_n is the function taking $(e_0, \ldots, e_n)$ into e_n. Because of cell path-connectivity (see Proposition 1.5.6) α is an epimorphism and therefore Ω is also countable. □

Proposition 1.5.13 *A locally finite and countable CW-complex is the union space of an expanding sequence of finite subcomplexes X_n such that, for every n, X_n is contained in the interior of X_{n+1} (the interior taken with respect to the topology of X).*

Proof Let $e_0, e_1, \ldots, e_j, \ldots$ be an enumeration of the cells of X. Define X_0 as the empty set and assume that X_n has been constructed. Consider the integer i defined by

$$i = \min\{j : e_j \not\subset X_n\},$$

and also the finite set Ω of all cells contained in X_n; then define the subcomplex X_{n+1} by

$$X_{n+1} = \mathrm{St}(e_i) \cup \bigcup_{e \in \Omega} \mathrm{St}(e).$$

Because of Lemma 1.5.9, X_{n+1} is a finite subcomplex of X, and, moreover, is a neighbourhood of X_n. By construction, each cell e_j belongs to some X_n; thus,

$$X = \bigcup_{n \in \mathbf{N}} X_n$$

(see Proposition A.5.3). □

Another topologically invariant property for CW-complexes is the dimension. The finite case can be described easily.

Proposition 1.5.14 *A CW-complex X is finite-dimensional iff it coincides with one of its skeleta, i.e., if the CW-structure can be described by a finite filtration. If this is the case, the smallest number m such that $X = X^m$ is the dimension of X:*

$$\dim X = \min\{m \in \mathbf{N} : X = X^m\}.$$

Proof If $\dim X = n$ then the space X cannot contain m-cells with $m > n$ (see Corollary A.9.2); thus $X = X^n$. Conversely, if X coincides with one of its skeleta it has the finite dimension of that skeleton.

Clearly, $X = X^m$ implies $\dim X \leqslant m$; thus the dimension of X is smaller than or equal to the minimum described. If this minimum is n, then $X \neq X^{n-1}$ implies that there must be m-cells with $m \geqslant n$ i.e., $\dim X \geqslant n$. □

The questions coming up now have to do with embeddings of CW-complexes into Euclidean spaces.

Theorem 1.5.15 *Every locally finite and countable CW-complex of dimension m can be embedded in $\mathbf{R}^{2m+1}$.*

Proof Let X be a locally finite, countable and m-dimensional CW-complex. The bulk of the proof consists in the explicit construction of an embedding $f : X \to \mathbf{R}^{k(m)}$, with $k(m) = \frac{1}{2}(m+1)(m+2)$. Once this is done, it follows that X is metrizable and satisfies the second axiom of countability; thus, as a space of dimension m, X can be embedded in $\mathbf{R}^{2m+1}$ (see Theorem A.9.7).

The construction of the map f will be done by induction on the skeleta of X. Start by enumerating the 0-cells of X and defining $f_0 : X^0 \to \mathbf{R}^1$ as the function which sends the only point of the jth 0-cell of X into $2j \in \mathbf{R}$. Suppose that $f_n : X^n \to \mathbf{R}^{k(n)}$ has been defined. Let

$$e_0, e_1, \ldots, e_j, \ldots$$

be an enumeration of the open $(n+1)$-cells of X, and for each $j \in \mathbf{N}$ let $\bar{c}_j$ (respectively, c_j) denote a characteristic map (respectively, the induced attaching map) for the cell e_j. Then, define the injection

$$f_{n+1}(x) = \begin{cases} (f_n(x), 0), & x \in X^n \\ 2j(1-t)e_{k(n)+1} + [tf_n(c_j(s)), (1-t)ts, 1-t], & x = \bar{c}_j(ts) \in e_j, \end{cases}$$

where $e_{k(n)+1} \in \mathbf{R}^{k(n+1)}$ is the unit vector with the $(k(n)+1)$th coordinate equal to 1. Finally, set $f = f_m$.

One now proves, using an induction procedure, that each $f_n, n=0,\ldots,m$, is an embedding. This is visibly so for $n=0$. Assume that f_n has been proved to be an embedding and take $f_{n+1} : X^{n+1} \to \mathbf{R}^{k(n+1)}$. Consider the given enumeration of the $(n+1)$-cells and observe that $f_{n+1}(e_j) = f(X^{n+1}) \cap V_j$, where

$$V_j = \{z = (z_1, \ldots, z_{k(n+1)}) \in \mathbf{R}^{k(n+1)} : (2j-1)z_{k(n+1)} < z_{k(n)+1} < (2j+1)z_{k(n+1)}\}$$

for each $j \in \mathbf{N}$; therefore, $f_{n+1}(e_j)$ is open in $f(X^{n+1})$. Finally, one has to show that f_{n+1} takes open sets of X^{n+1} into open sets of $f(X^{n+1})$. Let V be an arbitrary open set in X^{n+1} and take $x \in V$. Claim: $f_{n+1}(x)$ is an interior point of $f_{n+1}(V)$ with respect to $f_{n+1}(X^{n+1})$.

Case 1 Assume $x \in X^{n+1} \backslash X^n$ and let the $(n+1)$-cell e_j be its carrier. Because $f_{n+1}|\bar{e}_j$ is a homeomorphism, $f_{n+1}(V \cap e_j)$ is open in $f_{n+1}(\bar{e}_j)$; thus, $f_{n+1}(V \cap e_j)$ is open in $f_{n+1}(e_j)$ and therefore in $f_{n+1}(X^{n+1})$.
Case 2 Now suppose that $x \in X^n$. According to Corollary 1.5.11, assume that V meets only finitely many closed $(n+1)$-cells, say $\bar{e}_{j_0}, \ldots, \bar{e}_{j_r}$. It suffices to prove that no sequence in $f_{n+1}(X^{n+1}) \backslash f_{n+1}(V)$ converges to $f_{n+1}(x) = f_n(x)$. Assume the contrary i.e., suppose that there is sequence $\{x_i : i \in \mathbf{N}\}$ in $X^{n+1} \backslash V$ such that

$$\lim_{i \to \infty} f_{n+1}(x_i) = f_{n+1}(x) = f_n(x).$$

By the induction hypothesis, $f_{n+1}(V \cap X^n)$ is open in $f_{n+1}(X^n)$, and therefore the sequence $\{x_i\}$ cannot have a subsequence contained in X^n. Hence, one may assume that $\{x_i : i \in \mathbf{N}\} \subset X^{n+1} \backslash X^n$; this means that each x_i is of the form

$$x_i = \bar{c}_{j_{p(i)}}(t_i s_i)$$

for some $p(i) \in \mathbf{N}, t_i \in [0, 1)$ and $s_i \in S^n$. Considering that the last coordinate of $f_{n+1}(x_i)$ is $1 - t_i$ and that the last coordinate of $f_n(x)$ is 0, it follows that

$$\lim_{i \to \infty} (1 - t_i) = 0,$$

that is to say,

$$\lim_{i \to \infty} t_i = 1.$$

This implies that

$$f_n(x) = \lim_{i \to \infty} f_{n+1}(c_{j_{p(i)}}(s_i)) = f_n\left(\lim_{i \to \infty} (c_{j_{p(i)}}(s_i)) \right).$$

From the induction hypothesis one obtains that

$$x = \lim_{i \to \infty} (c_{j_{p(i)}}(s_i)),$$

so one can assume that

$$\{c_{j_{p(i)}}(s_i) : i \in \mathbf{N}\} \subset X^n \cap V;$$

hence, $\{j_{p(i)} : i \in \mathbf{N}\} \subset \{j_0, \ldots, j_r\}$. This implies that the sequence $\{p(i)\}$ must contain a constant sequence, i.e., there is a subsequence $\{y_k : k \in \mathbf{N}\}$ of the sequence $\{x_i\}$ which is contained in one open $(n+1)$-cell e_j, with $0 \leqslant s \leqslant r$. Finally, this shows that $x = \lim_{k \to \infty} y_k$, contradicting the fact that $\{y_k\} \subset \bar{e}_{j_s} \backslash V$. □

Disregarding the hypothesis on dimension in the previous theorem, one still obtains an interesting embedding theorem.

Theorem 1.5.16 *Every locally finite and countable CW-complex can be embedded in the Hilbert cube.*

Proof Let X be a locally finite and countable CW-complex. Take an expanding sequence $\{X_n : n \in \mathbf{N}\}$ as described in Proposition 1.5.13. Every X_n can be embedded into a Euclidean space (by the previous theorem). Now one can construct a countable basis of $X = \bigcup_{n=0}^{\infty} \mathring{X}_n$ using Cantor's diagonal procedure. Because X is normal, Urysohn's metrization theorem implies the metrizability of X. Thus, X is metrizable and satisfies the second axiom of countability; consequently, it can be embedded in the Hilbert cube (see Theorem A.9.8). □

Now the stage is ready for the presentation of the other topological characterizations of local finiteness announced earlier (see the Remark after Corollary 1.5.11).

Proposition 1.5.17 *For a CW-complex X, the following conditions are equivalent:*

(i) *X is locally finite;*
(ii) *X is metrizable;*
(iii) *X satisfies the First Axiom of Countability.*

Proof (i)⇒(ii): If X is locally finite, so is each one of its path-components. Thus the path-components, being locally finite and countable (see Proposition 1.5.12), are metrizable (see Theorem 1.5.16). Finally, X, being the topological sum of its path-components (see Corollary 1.4.13), is metrizable.

(ii)⇒(iii): Trivial.

(iii)⇒(i): Assume X not to be locally finite. Then, there is a cell e in X which meets the closure of infinitely many cells. Choose a sequence $\{e_j : j \in \mathbf{N}\}$ of pairwise distinct cells such that $e \cap \bar{e}_j \neq \emptyset$, and, for every

$j \in \mathbf{N}$, choose a point $x_j \in e \cap \bar{e}_j$. Because $\bar{e}$ is compact, the sequence $\{x_j : j \in \mathbf{N}\}$ contains a convergent subsequence; thus, one may assume, without loss of generality, that the sequence $\{x_j\}$ is itself convergent, say to a point x.

Now let $U_0 \supset U_1 \supset \cdots \supset U_n \cdots$ be an open basis of the neighbourhood system of x. Notice that each U_i meets infinitely many open cells e_j. Define a sequence of natural numbers $\{j_i : i \in \mathbf{N}\}$ by taking

$$j_0 = \min\{j : U_0 \cap e_j \neq \varnothing\},$$
$$j_{i+1} = \min\{j : j > j_i \text{ and } U_{i+1} \cap e_j \neq \varnothing\}.$$

Next, for every $i \in \mathbf{N}$, choose a point $z_i \in U_i \cap e_{j_i}$. On the one hand, the set $\{z_i : i \in \mathbf{N}\}$ is closed, because any open cell of X contains at most one element of this sequence, and thus, by closure finiteness, any closed cell contains at most finitely many points z_i. On the other hand, every neighbourhood U of x contains one U_n, and thus all the points z_i with $i \geqslant n$; this implies that $x = \lim_{i \to \infty} z_i$, contradicting the fact that $\{z_i : i \in \mathbf{N}\}$ is a discrete subset of X. □

Remark For a better understanding of the proof of this theorem, the reader should go back to Example 13 of Section 1.2; that is, to an example of a CW-complex which is not locally finite. □

The embedding theorems given before (Theorems 1.5.15 and 1.5.16) have a converse.

Theorem 1.5.18 *Let X be a CW-complex.*

(i) *If X is embeddable in the Hilbert cube, then X is locally finite and countable.*

(ii) *If X is embeddable in the Euclidean space $\mathbf{R}^m$, then X is locally finite, countable and has dimension $\leqslant m$.*

Proof (i) As a subspace of the Hilbert cube, X satisfies both axioms of countability. By the previous results, the first axiom implies that X is locally finite. Moreover, its path-components are locally finite and countable (see Proposition 1.5.12). If X is not countable, then it cannot have a countable number of path-components and therefore it cannot satisfy the second axiom.

(ii) By the theorem of invariance of domain, $\mathbf{R}^m$ cannot contain open cells of dimension $> m$. □

The concluding results of this section are based on a simple consequence of closure finiteness.

Lemma 1.5.19 *Let X be a CW-complex and let Z be a subset of X which meets every open cell of X in at most one point. Then Z forms a discrete closed subspace of X.*

Proof Because of closure finiteness, every subset of Z meets a closed cell of X in a finite and, therefore, closed subset. Since X is determined by its closed cells this implies that every subset of Z is closed in X and so, in Z, yielding the result. □

One application of this fact is the topological invariance of countability.

Proposition 1.5.20 *A CW-complex is countable iff it does not contain an uncountable discrete subset.*

Proof '$\Rightarrow$': Assume X to be a countable CW-complex with an uncountable discrete subset A. But X is a codomain of an identification map $f : B \to X$, where B is a coproduct of countably many balls. Since B satisfies the second axiom of countability, it cannot contain an uncountable discrete subset; on the other hand, by taking one point in the inverse image of each point in A, one obtains an uncountable discrete subset of B.

'$\Leftarrow$': See Lemma 1.5.19. □

Finally, there is an analogue to proposition 1.5.2 and Corollary 1.5.4 for the Lindelöf property. Recall that a Hausdorff space is called *Lindelöf* if every open covering contains a countable subcovering.

Proposition 1.5.21 *If K is a Lindelöf subspace of a CW-complex X, then $X(K)$ is a countable subcomplex of X.*

Proof Let E denote the set of all open cells of X which intersect K. The choice of a point $x_e \in K$ in each cell $e \in E$ produces a discrete closed subspace Z of X (see Lemma 1.5.19), and therefore of K. By the Lindelöf property, Z must be countable. This, together with closure finiteness, implies the result. □

Exercises

1 Prove that a CW-complex is topologically dominated by the family of its finite subcomplexes (this gives another proof for the paracompactness of CW-complexes – see Theorem A.2.5).

2 Give an example of a CW-complex which is not topologically dominated by the family of its closed cells.

3 A CW-complex X is locally countable if every (open) cell of X meets only countably many closed cells of X. Show that a CW-complex X is locally countable iff each point $x \in X$ has a neighbourhood meeting only countably many cells. (Lundell & Weingram, 1969, Proposition 2.3.6; compare also with Lemma 1.5.9 and Proposition 1.5.10 given before in this section.)

1.6 Whitehead complexes

The objective of this section is to prove that the definition of CW-complexes given in Section 1.2 coincides with that originally given by J. H. C. Whitehead.

First recall the definitions introduced by Whitehead. A *cell complex* is a Hausdorff space X, which is the union of disjoint open cells subject to the following condition. The closure $\bar{e}$ of each n-cell e of X is the image of a map $\bar{f} : B^n \to \bar{e}$ such that:

(1) $\bar{f}$ induces a homeomorphism $\mathring{B}^n \to e$;
(2) $\bar{f}(\delta B^n) = \bar{e} \backslash e \subset X^{n-1}$, where X^{n-1} – the $(n-1)$-*skeleton of* X – is the union of all m-cells of X with $m < n$. (Note: $X^{-1} = \varnothing$.)

The definitions given in Sections 1.1 and 1.2 show that any CW-complex is a cell complex; the converse is not true.

Example 1 Take $X = I$ and consider every point of X as a 0-cell.

A *cell subcomplex* A *of* X is a union of cells of X such that if e is a cell of A then $\bar{e} \subset A$. It is clear from this definition that the union and the intersection of a set (finite or infinite) of cell subcomplexes of X are cell subcomplexes of X. Moreover, from (2) above, for every non-negative integer n, X^n is a cell subcomplex of X. A cell subcomplex of X is *finite* if it is the union of finitely many cells of X. Notice that a finite cell subcomplex A of X is a closed subset of X, and, indeed, a compact subset of X, for A is a finite union of finitely many compact spaces $\bar{e}$, e in A. However, an arbitrary cell subcomplex of X need not be closed: in Example 1, every subset A of X is a subcomplex, even if A is not closed in X.

If L is a subset of X, analogously to the definition given in Section 1.4, define $X(L)$ to be the intersection of all cell subcomplexes of X which contain L; clearly $X(L)$ is a cell subcomplex of X. Note that $X(L)$ is the union of all $X(e)$, for all cells e of X which intersect L:

$$X(L) = \bigcup_{e \cap L \neq \varnothing} X(e).$$

In fact, the union on the right-hand side is a cell complex containing L, and, conversely, if $x \in e \cap L \neq \varnothing$, $X\{x\} = X(e) \subset X(L)$. Moreover, $X(e) = X(\bar{e})$, for every cell e of X.

As in Section 1.5, a cell complex is said to be *closure finite* if the closure of each one of its cells meets only finitely many cells.

Lemma 1.6.1 *A cell complex X is closure finite iff $X(e)$ is finite, for every cell e of X.*

Proof '⇒': The claim is analogous to the statement of Corollary 1.5.3 for CW-complexes. The proof given there remains valid in this more general situation.

'⇐': Suppose that $X(e)$ is finite for every cell e of X. Notice that for any given cell e_0 of X,

$$X(e_0) = X(\bar{e}_0) = \bigcup_{e' \cap \bar{e}_0 \neq \varnothing} X(e').$$

Since every cell e' of $\bigcup_{e' \cap \bar{e}_0 \neq \varnothing} X(e')$ appears also in $X(\bar{e}_0)$ and the latter subcomplex is finite by hypothesis, there are only finitely many cells e' meeting $\bar{e}_0$. □

The cell complex of Example 1 is closure finite.

Example 2 Let X be the space of the ball B^2 with a 0-cell for each point of S^1 and just one 2-cell, namely $B^2 \backslash S^1$. The cell complex obtained is not closure finite. In contrast to the cell complex of Example 1, this new space is determined by the family of the closures of all of its cells. □

The necessary definitions to formulate Whitehead's definition are now all in place: a *Whitehead complex* is a cell complex X such that

(1) X is closure finite,
(2) X is determined by the family consisting of the closures of all cells $e \in X$.

As stated by Whitehead in his paper, the name *CW-complex* is an abbreviation for **C***losure finite complex with the* **W***eak topology* (i.e., the topology determined by the family of the closure of all cells).

Condition (1) above implies that, for any cell e of a Whitehead complex X, the cell subcomplex $X(e) = X(\bar{e})$ is finite. Thus, every closed cell is contained in a finite subcomplex; this, together with condition (2) shows that:

(2′) X is determined by the family of all finite cell subcomplexes (see Proposition A.2.1).

In contrast with the situation in an ordinary cell complex, subcomplexes of Whitehead complexes behave as expected.

Lemma 1.6.2 *Let A be a cell subcomplex of a Whitehead complex X. Then A is a closed subset of X and is itself a Whitehead complex. In particular, the skeleta X^n are Whitehead complexes.*

Proof The intersection of A with any finite cell subcomplex of X is a finite cell subcomplex, and thus is closed.

In order to see that A is a Whitehead complex, it is only necessary to check that A is determined by the family consisting of the closures of its cells, since its closure finiteness is evident. Let L be a subset of A such that $\bar{e} \cap L$ is closed in $\bar{e}$, for all cells e of A. Take any closed cell $\bar{e}'$ of X. Because of closure finiteness, $\bar{e}' \cap A$ is contained in a finite union of finitely many cells $e_1, \ldots, e_k$ of A; thus

$$\bar{e}' \cap A = \bar{e}' \cap \bigcup_{i=1}^{k} \bar{e}_i = \bigcup_{i=1}^{k} \bar{e}' \cap \bar{e}_i,$$

and thus

$$\bar{e}' \cap L = \bar{e}' \cap L \cap A = \bigcup_{i=1}^{k} \bar{e}' \cap \bar{e}_i \cap L$$

is a finite union of closed sets and therefore is itself closed. Because X is determined by the family $\{\bar{e}'\}$, it follows that L is closed in X, and hence also in A. □

Theorem 1.6.3 *The skeletal filtration of a cell complex X is a CW structure iff X is a Whitehead complex.*

Proof '⇒': Proposition 1.2.2 shows that X is determined by the family of the closure of its cells. Theorem 1.5.1 shows closure finiteness.

'⇐': (1) X^0 is discrete: Let L be a subset of X^0. Since X is closure finite, every closed cell $\bar{e}$ meets only finitely many 0-cells, and hence contains only a finite subset of L; therefore $\bar{e} \cap L$ is closed. But X is determined by the closures of its cells; so L is closed in X and thus in X^0.

(2) For every $n > 1$, the pair (X^n, X^{n-1}) is an adjunction of n-cells: If e is an arbitrary n-cell of X, condition (1) of the definition of cell complex and Lemma 1.1.5 prove that $(X^{n-1} \cup e, X^{n-1})$ is an adjunction of just one n-cell. Moreover, X^n is determined by the family consisting of all closed cells of dimension at most n (see Lemma 1.6.2), and therefore is determined by the family $\{X^{n-1}\} \cup \{\bar{e} : e \in \pi(X^n \backslash X^{n-1})\}$, according to Proposition A.2.1. The claim is now proved (see Proposition 1.1.3).

(3) The covering of X by the closed cells refines the covering by the skeleta; thus X is determined by the subspaces X^n (see again Proposition A.2.1). □

Remark In the construction of CW-complexes given in Section 1.2, it is not necessary to worry about Hausdorffness; this is a trivial consequence of Proposition 1.2.1. However, in the approach presented by Whitehead, such a separation axiom is an intrinsic part of the definition. What follows is an example of a cell complex which is closure finite and is determined by the family of its closed cells, and yet fails to be Hausdorff. □

Example 3 Let X be the 'interval with a double point' i.e., the quotient space

$$I \times \{0,1\}/[(t,0) \sim (t,1) : 0 \leqslant t < 1].$$

The cellular structure of X is given as follows. Let $p : I \times \{0,1\} \to X$ be the identification map and take $p((0,0)) = p((0,1)), p((1,0)), p((1,1))$ as 0-cells and $p((I \backslash \{0,1\}) \times \{0\}) = p((I \backslash \{0,1\}) \times \{1\})$ as the only 1-cell.

Exercise

Given a (not necessarily Hausdorff) space X and a family $\{f_\lambda : \lambda \in \Lambda\}$ of maps

$$f_\lambda : B^{n_\lambda} \to X,$$

let X^n denote the union of all $e_\lambda = f_\lambda(\mathring{B}^{n_\lambda})$ with $n_\lambda \leqslant n$. Prove that the filtration $\{X^n : n \in \mathbf{N}\}$ of X is a CW-structure for X if the following conditions are satisfied:

(i) each e_λ is an open n_λ-cell of X via the map f_λ;
(ii) X is the disjoint union of the cells e_λ;
(iii) for every $\lambda \in \Lambda$, $f_\lambda(\delta B^{n_\lambda}) \subset X^{n_\lambda}$;
(iv) a subset of X^n is closed whenever its inverse image in each $B^{n_\lambda}, n_\lambda \leqslant n$ is closed. (Milnor, 1956).

Notes to Chapter 1

Balls, spheres, projective spaces and most of the maps relating them are due to John Folklore. The construction of suspension was introduced first in Freundenthal (1938), with the German name *Einhängung*.

The notion of 'CW-complex', the basic notion of the whole text, was introduced in Whitehead (1949a), where also many of the results given in Chapters 1 and 2 are proved, including the famous realizability theorem (see Section 2.5). The development of the theory of CW-complexes presented in the book, although leading to the same geometric structures invented by J. H. C. Whitehead (see Section 1.6), has a more categorical flavour; this allows us to streamline many of the proofs and to obtain new results. This approach is not really new, but was employed before, for instance, in Hu (1964), Brown (1968) and Piccinini (1973). In contrast to the more or less widespread use of the term 'weak topology', we elected to follow the advice of Ernest Michael and speak of 'topologies determined by...'; one reason for this is the fact that the 'weak' topology on a simplicial complex in general has more open sets than the 'strong' topology (see Section 3.3, Example 3). The basic processes in our presentation are: (1) attachings, and (2) formation of the union space of expanding sequences in which the attached spaces are certain metric spaces; if one allows the attaching of arbitrary metric spaces, one obtains the slightly larger category of *M-spaces* studied in Hyman (1968).

As far as the Examples are concerned, the following texts can be used for the relevant definitions:

Peano curve: Dugundji (1966).
Fréchet space: Engelking (1977).
Moore spaces: Moore (1955); Varadarajan (1966).

The technique of collaring is a basic tool for exhibiting topological properties of CW-complexes; it is intrinsically already contained in Whitehead (1949a), where the local contractibility of CW-complexes already appears. (The paracompactness of CW-complexes could be expected after this property was proved for simplicial complexes; it was shown for the first time in the case of CW-complexes in Miyazaki (1952), where also the simplicial sources are mentioned.) Local equiconnectivity of CW-complexes is due to Dyer & Eilenberg (1972).

Subcomplexes were extensively studied in Whitehead's original paper. The fact that a compact space dominated by a CW-complex is always dominated by a finite CW-complex (see Corollary 1.5.5) suggests the following question: under what conditions will a space, which is dominated by a finite CW-complex, have the type of a finite CW-complex? A subtle answer to this question relying on algebraic K-theory is given in Wall (1965). The proof of the fact that the closure of any cell of a regular CW-complex is a subcomplex (see Theorem 1.4.10) is inspired by the presentation in Lundell & Weingram (1969).

2

Categories of CW-complexes

The objective of this chapter is to study four categories whose objects are CW-complexes, namely:

(i) the category CW of CW-complexes and maps,
(ii) the category CW^c of CW-complexes and cellular maps,
(iii) the category CW^r of CW-complexes and regular maps and
(iv) the category RCW^c of relative CW-complexes and cellular maps.

2.1 Morphisms

Let Y, X be CW-complexes. A map $f: Y \to X$ is:

(i) *cellular*, if for all $n \in \mathbf{N}$, $f(Y^n) \subset X^n$;
(ii) *regular*, if f is cellular and takes every open cell of Y onto an open cell of X.[†]

Let Y, X be relative CW-complexes. A map $f : Y \to X$ is:

(iii) *cellular*, if for $n = -1$ and all $n \in \mathbf{N}, f(Y^n) \subset X^n$.

Note that, in all cases, the dimension of the cells is not necessarily maintained; indeed it may decrease.

Clearly, identity maps are cellular and regular with respect to the same CW-structure on domain and codomain, and composition of cellular (respectively, regular) maps yields a cellular (respectively, regular) map; indeed, cellular (respectively, regular) maps form subcategories of CW. Because regular maps are cellular,

$$CW^r \subset CW^c.$$

Proposition 2.1.1 *Let Y, X be CW-complexes and let $f: Y \to X$ be a regular map. Then,*

(i) *the image by f of a closed cell of Y is a closed cell of X;*
(ii) *$f(Y)$ is a subcomplex of X;*
(iii) *the induced map $Y \to f(Y)$ is an identification.*

Proof (i) Let e' be a cell of Y; then, by regularity, $e = f(e')$ is a cell of X. Since $\bar{e}'$ is compact, $f(\bar{e}')$ is a closed subset of X containing e, and thus $\bar{e} \subset f(\bar{e}')$.

[†] A better and more intuitive terminology would be to use the words 'skeletal' instead of 'cellular' and 'cellular' instead of 'regular'. However, we stick to the usual convention.

On the other hand, if $y \in \bar{e}'$ and $U \subset X$ is a neighbourhood of $f(y)$ then, by the continuity of f, the inverse image of U by f is a neighbourhood of y and meets e'; therefore U meets e. Since this holds for every neighbourhood of $f(y)$, it follows that $f(y) \in \bar{e}$.

(ii) By the definition of regularity, $f(Y)$ is a union of open cells of X. By (i), the closure of any cell in $f(Y)$ belongs also to $f(Y)$; thus $f(Y)$ is a subcomplex of X (see Proposition 1.4.4).

(iii) one has to show that a subset $C \subset f(Y)$ is closed if its inverse image by f is closed in Y. Now, C is closed in $f(Y)$ if $C \cap \bar{e}$ is closed for every cell e of $f(Y)$. But this is true because of the compactness of $f^{-1}(C) \cap \bar{e}'$, where e' is a cell of Y with $f(e') = e$ and the equation $f(f^{-1}(C) \cap \bar{e}') = C \cap \bar{e}$ which follows from (i). □

Corollary 2.1.2 *A surjective regular map between CW-complexes is an identification.* □

Example The covering projection

$$\mathbf{R} \to S^1, \qquad t \mapsto (\cos 2\pi t, \sin 2\pi t)$$

is a regular map with respect to the CW-structures of **R** and S^1 described in Example 10 in Section 1.2 and Example 2 in Section 1.1, respectively. □

2.2 Coproducts and products

The categories CW, CW^c, CW^r and RCW^c have arbitrary coproducts; more precisely:

Proposition 2.2.1 *Let*

$$\{\{X_\lambda; X_\lambda{}^n : n = -1, 0, 1, 2, \ldots\} : \lambda \in \Lambda\}$$

be a family of (relative) CW-complexes. Then,

$$\left\{\bigsqcup_{\lambda \in \Lambda} X_\lambda; \bigsqcup_{\lambda \in \Lambda} X_\lambda{}^n : n = -1, 0, 1, 2, \ldots\right\}$$

is a CW-complex, which, together with the canonical inclusion maps, satisfies the properties of a coproduct of the CW-complexes X_λ in the categories CW, CW^c, CW^r and RCW^c (in particular, this means that the inclusions $X_\lambda \to \bigsqcup_{\lambda \in \Lambda} X_\lambda$ are regular maps). Moreover, if all X_λ are regular CW-complexes, *$\bigsqcup_{\lambda \in \Lambda} X_\lambda$ is regular. Finally,*

$$\dim \bigsqcup_{\lambda \in \Lambda} X_\lambda = \max\{\dim X_\lambda : \lambda \in \Lambda\}.$$ □

The categories CW, CW^c and CW^r have also finite products, but these are harder to obtain.

Theorem 2.2.2 *Let* $\{X; X^n : n = 0, 1, 2, \ldots\}$ *and* $\{Y; Y^n : n = 0, 1, 2, \ldots\}$ *be CW-complexes. Then,*

(i) $\{X \times Y; \bigcup_{p+q=n} X^p \times Y^q : n = 0, 1, 2, \ldots\}$ *is a CW-complex, which, together with the canonical projection maps, satisfies the properties of a product of the CW-complexes X and Y in the categories CW, CW^c and CW^r;*

(ii) *the open (closed) cells of the CW-complex $X \times Y$ are the products of the open (closed) cells of X and Y respectively;*

(iii) *the projection maps $X \times Y \to X$ and $X \times Y \to Y$ are regular;*

(iv) *if X and Y are regular CW-complexes so is $X \times Y$;*

(v) *if* $\dim X = l \geqslant 0$, $\dim Y = m \geqslant 0$ *then,*

$$\dim (X \times Y) = l + m.$$

Proof Define

$$(X \times Y)^n = \bigcup_{p+q=n} X^p \times Y^p.$$

Clearly, $(X \times Y)^0$ is discrete.

Next, note that, for every pair (p, q) of natural numbers, $B^p \times B^q$ is a $(p+q)$-ball with boundary $B^p \times S^{q-1} \cup S^{p-1} \times B^q$ (see Proposition 1.0.2). Thus, the multiplication law (L5) for adjunction spaces shows that the pairs

$$(X^p \times Y^q, X^p \times Y^{q-1} \cup X^{p-1} \times Y^q)$$

are adjunctions of $(p+q)$-cells, whose attaching maps are denoted by $f_{p,q}$. Next, fix $n \in \mathbf{N}$. Then, for (p, q) with $p + q = n$, the union $X^p \times Y^{q-1} \cup X^{p-1} \times Y^q$ is a subspace of $(X \times Y)^{n-1}$ and the corresponding maps $f_{p,q}$ together determine a map f_n from the coproduct of their domains to $(X \times Y)^{n-1}$. Now, $(X \times Y)^n$ is obtained from $(X \times Y)^{n-1}$ by adjunction of n-cells via f_n. This construction of the map f_n shows implicitly statement (ii) of the theorem.

In order to prove that $X \times Y$ is determined by the family $\{(X \times Y)^n : n \in \mathbf{N}\}$, observe first that, for every $p \in \mathbf{N}$, $X^p \times Y$ is the union space of the expanding sequence $\{X^p \times Y^p : q = 0, 1, \ldots\}$ and that $X \times Y$ is the union space of the expanding sequence $\{X^p \times Y : p = 0, 1, \ldots\}$. Thus, $X \times Y$ is determined by the family $\{X^p \times Y^q : p, q \in \mathbf{N}\}$ and hence also by the family $\{(X \times Y)^n : n \in \mathbf{N}\}$. This completes the proof of statement (i).

Statements (iii) and (iv) are immediate consequences of (ii).

Finally, if $X = X^l$ and $Y = Y^m$, then $X \times Y = (X \times Y)^{l+m}$, which proves (v) (see Proposition 1.5.14). □

Examples 10 and 11 of Section 1.2 exhibit CW-structures for $\mathbf{R}^1$ and $\mathbf{R}^2$ respectively. Now one is able to extend this to higher-dimensional Euclidean spaces.

Example 1 The Euclidean space $\mathbf{R}^m$, $m \in \mathbf{N}$, is a locally finite CW-complex of dimension m. In fact, because $\mathbf{R}$ is locally compact, every Euclidean space $\mathbf{R}^m$ can be considered as an iterated product of $\mathbf{R}$ in the category of k-spaces; thus, by the previous theorem it inherits a CW-structure which is locally finite (see Proposition 1.5.10). Moreover, it follows from this consideration that $\mathbf{R}^m$ actually has covering dimension m. □

In the proof of Theorem 2.2.2, it is crucial to use the product in the category of k-spaces and not the usual Cartesian product, as one can see from the following example.

Example 2 Let S be the set of all sequences of non-zero natural numbers. Form a one-dimensional CW-complex X as follows. The 0-skeleton X^0 is equal to $S \cup \{0\}$. Next, for every $s \in S$ define the map $f_s : \delta B^1 = \{-1, 1\} \to X^0$ by taking $f_s(-1) = 0$ and $f_s(1) = s$. These maps f_s together define a map

$$f : \{-1, 1\} \times S \to X^0.$$

Then, a CW-complex X is obtained by the adjunction of 1-cells to X^0 via f; the characteristic map for the 1-cell corresponding to $s \in S$ will also be denoted by f_s.

A second one-dimensional CW-complex Y is defined as follows. Take Y^0 to be the set $\mathbf{N}$ of natural numbers. For every $i \in \mathbf{N} \backslash \{0\}$, define the map $g_i : \{-1, 1\} \to Y^0$ by $g_i(-1) = 0$ and $g_i(1) = i$. These define a map

$$g : \{-1, 1\} \times (\mathbf{N} \backslash \{0\}) \to Y^0;$$

then, Y is obtained by adjunction of 1-cells to Y^0 via g; the characteristic map for the 1-cell corresponding to $i \in \mathbf{N} \backslash \{0\}$ will also be denoted by g_i.

Now, if the Cartesian product of X and Y – which will be denoted by $X \times_c Y$ in the sequel – were a CW-complex in the sense of Theorem 2.2.2, it would have the topology determined by the products of the closed cells of X and Y i.e., the cells of $X \times Y$ (see Proposition 1.2.2). It will be shown that this is not the case. Take

$$K = \{(f_s(s_i^{-1}), g_i(s_i^{-1})) : (s, i) \in X^0 \times (Y^0 \backslash \{0\})\}.$$

Since K intersects any closed cell of $X \times Y$ in at most one element, K is a closed subset of $X \times Y$. However, K is not closed in $X \times_c Y$! This will be done by showing that the point $(0, 0)$, which does not belong to K, is a cluster point of K.

Any open neighbourhood of $(0,0)$ in $X \times_c Y$ contains an elementary neighbourhood $U \times V$ where:

$$U = \{f_s(t) : s \in \boldsymbol{S}, t < a_s\}$$
$$V = \{g_i(t) : i \in \mathbf{N} \setminus \{0\}, t < b_i\}$$

with $\{a_s : s \in \boldsymbol{S}\}$ a family of real numbers $0 < a_s \leqslant 1$ and $\{b_i : i \in \mathbf{N} \setminus \{0\}\}$ a sequence of real numbers $0 < b_i \leqslant 1$. Given an open neighbourhood of $(0,0)$ choose such an elementary neighbourhood and define a sequence $s \in \boldsymbol{S}$ by taking $s_i = 1 + [\max(i, b_i^{-1})]$ (here, for any $z \in \mathbf{R}$, $[z]$ means as usual the greatest integer contained in z); moreover, define the integer $i = 1 + [a_s^{-1}]$. Then

$$s_i^{-1} < \min(i^{-1}, b_i)$$

for all $i \in \mathbf{N} \setminus \{0\}$ and

$$s_i^{-1} < i^{-1} < a_s;$$

thus,

$$(f_s(s_i^{-1}), g_i(s_i^{-1})) \in (U \times V) \cap K.$$

This means that every neighbourhood of $(0,0)$ contains a point of K.

Since $X \times Y$ is the k-ification of $X \times_c Y$, one also obtains that $X \times_c Y$ fails to be a k-space; thus, there does not exist any CW-structure for $X \times_c Y$ (see Proposition 1.2.1)! □

Remark Whenever forming a product $X \times Y$ of CW-complexes, if at least one of X, Y is a locally finite CW-complex, then the k-topology on the product set coincides with the topology of the Cartesian product, because one factor is locally compact (see Proposition 1.5.10 and Section A.1). The same is also true under some other circumstances.

Proposition 2.2.3 *If X, Y are countable CW-complexes, then the Cartesian product of X and Y is homeomorphic to the product $X \times Y$ in the category $\mathcal{CW}$.*

Proof It is enough to show that the Cartesian product $X \times_c Y$ has the topology determined by the closed cells of the CW-complex $X \times Y$ (see Proposition 1.2.2). Take $U \subset X \times_c Y$ such that $U \cap (\bar{e} \times \bar{e}')$ is open, for all cells $e \subset X$, $e' \subset Y$. Consider a point $(x_0, y_0) \in U$ and enumerate the cells of X and Y so that e_0 is the carrier of x_0 and e'_0 is the carrier of y_0. Let X_i and Y_i denote the closures of the unions of the first i cells of X and Y respectively; the sets X_i and Y_i, being finite unions of closed cells, are compact and the intersections $U \cap (X_i \times Y_i)$ are open in $X_i \times Y_i$. Moreover, the spaces X, Y are determined by the families $\{X_i : i\in\mathbf{N}\}$, $\{Y_i : i\in\mathbf{N}\}$ (see Propositions 1.2.2 and A.2.1).

Because $U\cap(e_0\times e'_0)$ is open in $e_0\times e'_0$, one can find neighbourhoods V_0 and W_0 of x_0 and y_0 in X_0 and Y_0 respectively, such that the product of their compact closures is still contained in U. Suppose by induction that neighbourhoods V_i and W_i of x_0 and y_0 in X_i and Y_i respectively have been constructed so that $V_{i-1}\subset V_i$, $W_{i-1}\subset W_i$ and $\bar{V}_i\times\bar{W}_i\subset U$. To perform the induction, first construct an open set $V_{i+1}\subset X_{i+1}$ such that $\bar{V}_i\subset V_{i+1}$ and $\bar{V}_{i+1}\times\bar{W}_i\subset U$; the open set W_{i+1} can then can be constructed in a symmetric fashion. To begin with, for every point $(x,y)\in\bar{V}_i\times\bar{W}_i$, choose open neighbourhoods $V_{(x,y)}$, $W_{(x,y)}$ of x, y in X_{i+1}, Y_i respectively, such that their product is contained in U. For a fixed y, it is possible to find finitely many sets $V_{(x,y)}$ whose union V_y contains $\bar{V}_i$, in view of the compactness of the latter set. Let W_y denote the intersection of the corresponding finitely many $W_{(x,y)}$'s. Now finitely many sets W_y cover $\bar{W}_i$; the intersection V' of the corresponding V_y's is an open set containing $\bar{V}_i$ such that $V'\times\bar{W}_i\subset U$. Normality permits one to find an open set V_{i+1} whose compact closure is still contained in V'.

Finally, the sets $\bigcup_{i\in\mathbf{N}}V_i$, $\bigcup_{i\in\mathbf{N}}W_i$ are open neighbourhoods of x_0, y_0 in X, Y respectively, whose product is contained in U, and hence (x_0, y_0) is an interior point of U. □

A interesting application of products in the category CW is obtained by means of the telescope. To this end, consider an increasing sequence $\{X_n : n\in\mathbf{N}\}$ of subcomplexes of a CW-complex X whose union is X; note that any such sequence is an expanding sequence (see Corollary 1.4.7) and that X is actually its union space (see Propositions 1.2.2, A.2.1). Moreover, endow the half line $[0,\infty)$ with the CW-structure $\{\mathbf{N}, [0,\infty)\}$ (cf. Section 1.2, Example 10). The key to the desired result is given in the following statement:

Lemma 2.2.4 *Let X be a CW-complex and let $\{X_n : n\in\mathbf{N}\}$ be an increasing sequence of subcomplexes of X. Then, the telescope T of the expanding sequence $\{X_n\}$ is – up to homeomorphism – the subcomplex of the CW-complex $X\times[0,\infty)$ formed by the cells $e\times\{n\}$, $e\times(n,n+1)$ with e a cell of X_m, $m\leqslant n$; T is a regular CW-complex iff X is regular.* □

With this in mind one obtains:

Proposition 2.2.5 *Any countable CW-complex has the type of a locally finite and countable CW-complex.*

Proof Let $e_0, e_1, \ldots, e_j, \ldots$ be an enumeration of the cells of a countable CW-complex X. Then the subcomplexes $X_n = \bigcup_{j=0}^{n} X(e_j)$ are finite (see Corollary 1.5.3) and form an increasing sequence with union space X.[†] Observe that each of the products $X_n \times [n, n+1]$ is a finite subcomplex of the corresponding telescope T and each open cell of T meets only closed cells of at most two of these finite subcomplexes. Thus, T is locally finite. □

Remark The previous statement is not devoid of sense: Example 13 of Section 1.2 presents a countable, but not locally finite, CW-complex.

Exercise

Show that the cartesian product of two locally countable CW-complexes is homeomorphic to the product in the category CW. (Lundell & Weingram, 1969, Corollary 2.5.5)

2.3 Some special constructions in the category CW^c

Let Y, A be CW-complexes. A partial map $f : Y \dashrightarrow A$ is *cellular* if $D = \operatorname{dom} f$ is a subcomplex of Y and f is cellular as a map $D \to A$; in this situation, the restrictions of f to maps $D^n \to A^n$ are denoted by f^n. The basic result of this section is the following theorem.

Theorem 2.3.1 *Let Y, A be CW-complexes and let $f : Y \dashrightarrow A$ be a cellular partial map. Then, taking $X = A \sqcup_f Y$ and $X^n = A^n \sqcup_{f^n} Y^n$, the following are true:*

(i) *$\{X; X^n : n = 0, 1, \ldots\}$ is a CW-complex,*

(ii) *A is a subcomplex of X and*

(iii) *there exists a cellular characteristic map for the adjunction.*

Proof (i) Clearly, X^0 is a discrete space. To prove that the pair (X^n, X^{n-1}) is an adjunction of n-cells, for every $n > 0$, show the existence of a space X' containing X^{n-1} and contained in X^n such that the pairs (X', X^{n-1}) and (X^n, X') are adjunctions of n-cells, the latter with an attaching map factoring through X^{n-1}; then the addition law (L3) for adjunction spaces yields this claim.

[†] The reader should be aware of the difference in procedure between this proof and that of Proposition 1.5.11.

Set $X' = A^n \bigsqcup_{A^{n-1}} X^{n-1}$. Then,

$$X' = A^n \bigsqcup_{A^{n-1}} (A^{n-1} \bigsqcup_{f^{n-1}} Y^{n-1}) = A^n \bigsqcup_{f^n} (D^n \bigsqcup_{D^{n-1}} Y^{n-1})$$
$$= A^n \bigsqcup_{f^n} (D^n \cup Y^{n-1}),$$

using the law of horizontal composition (L1).

Since $D^n \cup Y^{n-1}$ is a subcomplex of the CW-complex Y^n, and D^n is a subcomplex of $D^n \cup Y^{n-1}$, the inclusions $D^n \to D^n \cup Y^{n-1}$, $D^n \cup Y^{n-1} \to Y^n$ are closed cofibrations (see Corollary 1.4.7); thus the law of vertical composition (L2) can be applied to give:

$$X^n = A^n \bigsqcup_{f^n} Y^n = (A^n \bigsqcup_{f^n} (D^n \cup Y^{n-1})) \bigsqcup_{\bar{f}} Y^n = X' \bigsqcup_{\bar{f}} Y^n,$$

where $\bar{f}: D^n \cup Y^{n-1} \to A^n \bigsqcup_{f^n} (D^n \cup Y^{n-1})$ denotes a suitable characteristic map.

Because the pairs (A^n, A^{n-1}), $(Y^n, D^n \cup Y^{n-1})$ are adjunctions of n-cells (the latter by Lemma 1.4.3 (ii)), so are the pairs $(X', X^{n-1}), (X^n, X')$. An attaching map for the latter pair has to factor through $D^n \cup Y^{n-1}$, thus through Y^{n-1}; but the induced map $Y^{n-1} \to X'$ factors through X^{n-1}, completing this part of the proof.

It remains to show that X has the final topology with respect to the canonical maps $j^n : X^n \to X$. A function g from X to a topological space Z, such that all compositions $g \circ j^n$ are continuous gives rise to compatible sequences of maps $A^n \to Z$, $Y^n \to Z$, thus to maps $A \to Z$, $Y \to Z$. These give a map which as a function coincides with g (apply the universal property of adjunction spaces).

(ii) follows immediately from the construction, which also implies that the characteristic maps for the adjunction spaces X^n together determine a characteristic map for the adjunction space X, thereby proving (iii). □

The theorem gives rise to some interesting constructions in the category CW^c, which will be described in the following examples.

Example 1 If A is a subcomplex of the CW-complex X, then X/A is a CW-complex. □

Example 2 If $f : Y \to X$ is a cellular map, then the mapping cylinder $M(f)$ is a CW-complex, containing both Y and X as subcomplexes. □

Example 3 If $f : Y \to X$ is a cellular map, then the mapping cone $C(f)$ is a CW-complex, containing X as a subcomplex. □

Example 4 If X is a CW-complex, then its cone CX and its suspension ΣX are CW-complexes. □

Examples 2, 3 and 4 have *reduced* analogues defined for based CW-complexes.

Example 5 The smash product of two based CW-complexes is a based CW-complex. □

Recall that the *wedge product* of a family of based spaces $\{(Y_\gamma, z_\gamma) : \gamma \in \Gamma\}$ is the based space $(\vee_\Gamma Y_\gamma, z)$ given by the set

$$\bigvee_\Gamma Y_\gamma = \left\{ (y_\gamma) \in \prod_{\gamma\in\Gamma} Y_\gamma : y_\gamma \neq z_\gamma, \text{ for at most one } \gamma \in \Gamma \right\},$$

endowed with the final topology with respect to the canonical map

$$p : \bigsqcup_{\gamma\in\Gamma} Y_\gamma \to \bigvee_{\gamma\in\Gamma} Y_\gamma,$$

and the point z taken to be the element (z_γ). If one takes $(Y_\gamma, z_\gamma) = (Y, z_0)$, for all γ in a given set Γ and a given based space (Y, z_0), then the wedge product $(\vee_\Gamma Y, z) = (\vee_\Gamma Y_\gamma, z)$ is called the Γ-fold wedge of (Y, z_0); in particular, the Γ-fold wedge of the based n-sphere (S^n, e_0), $n \in \mathbf{N}$, is referred to as a *bouquet of n-spheres* (see Section 1.1, Examples 9, 10).

Any wedge product of a family of based CW-complexes carries a CW-structure in the obvious manner. In particular, this appears in the following.

Example 6 For any CW-complex X and any $n \in \mathbf{N}$, the CW-complex X^{n+1}/X^n is a bouquet of $(n+1)$-spheres. □

If one wishes to construct a CW-complex out of a given CW-complex by the adjunction of further n-cells, these have to be attached via attaching maps into the $(n-1)$-skeleton; more precisely:

Proposition 2.3.2 *Let X be a CW-complex and let the pair (Y, X^{n-1}) be an adjunction of n-cells, n fixed. Then, the union $X^* = X \sqcup_{X^{n-1}} Y$ is a CW-complex such that*

(i) $X^{*n} = X^n \sqcup_{X^{n-1}} Y$

and

(ii) $X/X^n = X^*/X^{*n}$.

Proof From the hypotheses, it is clear that Y is a CW-complex containing X^{n-1} as a subcomplex, and it is trivial that the inclusion map $X^{n-1} \to X$ is cellular. Thus, X^* is a CW-complex and $X^n \sqcup_{X^{n-1}} Y$ is its n-skeleton (see Theorem 2.3.1).

To prove (ii), denote by j the inclusion $X^n \to X^{*n}$ and by c the constant map from X^{*n} to the singleton space $\{*\}$. Then:

$$X^*/X^{*n} = \{*\} \sqcup_c X^* = \{*\} \sqcup_c (X^* \sqcup_j X) = \{*\} \sqcup_{cj} X = X/X^n. \quad \square$$

The next proposition shows that a modification of the n-skeleton of a CW-complex does not alter its higher dimensional part.

Proposition 2.3.3 *Let A be a subcomplex of a CW-complex X, let A' be an n-dimensional CW-complex and let $f : A^n \to A'$ be a cellular map. Then, $X' = A' \sqcup_f X$ is a CW-complex such that*

$$X'/X'^n = X/X^n.$$

Proof The n-skeleton of X' is given by $X'^n = A' \sqcup_f X^n$ (see Theorem 2.3.1); let $\tilde{f}: X^n \to X'^n$ be a suitable characteristic map and let $c: X'^n \to \{*\}$ be the constant map. Then:

$$X'/X'^n = \{*\} \sqcup_c X' = \{*\} \sqcup_c (X'^n \sqcup_{\tilde{f}} X) = \{*\} \sqcup_{c\tilde{f}} X = X/X^n. \quad \square$$

The following is a technique to blow up a CW-complex within its type. Let X be a CW-complex and let $b : B^n \to X$ be a cellular map (with respect to the CW-structure of B^n described in Section 1.2, Example 1). Attach an n-cell to X, using $b|S^{n-1}$ as attaching map; the resulting space $\tilde{X}$ is a CW-complex consisting of the subcomplex X (see Theorem 2.3.1) and an extra n-cell. Let $\tilde{b}$ denote a characteristic map for the new n-cell, such that $\tilde{b}|S^{n-1} = b|S^{n-1}$. Define $c: S^n \to \tilde{X}$ by taking b on the lower hemisphere of S^n and $\tilde{b}$ on the upper one; this is again a cellular map. Attaching an $(n+1)$-cell to $\tilde{X}$ via c, one obtains the *elementary expansion of X along b*; this is again a CW-complex, denoted by X_b and containing both X and $\tilde{X}$ as subcomplexes.

Proposition 2.3.4 *Let X be a CW-complex and let $b : B^n \to X$ be a cellular map. Then, the elementary expansion X_b of X along b contains X as a strong deformation retract with a cellular retraction.*

Proof Let $\bar{c} : B^{n+1} \to X_b$ denote a characteristic map for the new $(n+1)$-cell of X_b, and define $H : B^{n+1} \times I \to X_b$ by taking

$$H(h^n(s,t), u) = \bar{c} \circ h^n(s, tu),$$

where h^n denotes the canonical homotopy deforming the lower hemisphere of S^n into the upper one (see page 4). Consider the composition $H' = (H|S^n \times I) \circ (i_+ \times 1_I) : B^n \times I \to X_b$; its restriction to $S^{n-1} \times I$ is nothing but the composition of the projection onto S^{n-1} with the map $b|S^{n-1}$ and the inclusion $X \to X_b$. Thus, H' gives rise to a homotopy rel. X, which retracts $\tilde{X}$ to X within X_b, and which in turn extends to a homotopy deforming the entire X_b into X, via the homotopy H. □

A CW-complex X' (respectively, its CW-structure) is a *subdivision* of a CW-complex X (respectively, its CW-structure) if X and X' coincide as spaces and every cell of X' is contained in a cell of X. This definition has the following immediate consequence.

Proposition 2.3.5 *Let the CW-complex X' be a subdivision of the CW-complex X; then:*

(i) *the identity map is cellular as a map $X \to X'$;*

(ii) *if X is finite, locally finite or countable the same holds true for X';*

(iii) $\dim X' = \dim X$. □

Remark If the CW-complex X' is a subdivision of the CW-complex X, then, as a consequence of the cellular approximation theorem (to be proved in the next section), the identity map regarded as a map $X' \to X$ is homotopic to a cellular map (but is itself cellular only if, also, the CW-structures of X and X' agree). □

Example 7 The CW-structure for the ball B^{m+1} described in Example 3 of Section 1.2 is a subdivision of the CW-structure given in Example 2 there, which in turn is a subdivision of the CW-structure in Example 1. □

Proposition 2.3.6 *Let the CW-complex X' be a subdivision of the CW-complex X and let A be a subcomplex of X. Then, there is a subcomplex A' of X' which is a subdivision of A.*

Proof Let Ω denote the set of all cells $e' \subset X'$ which meet A. Clearly, A is contained in the union of all the cells of Ω. If a cell $e' \in \Omega$ is contained in the cell e of X, then e is contained in A (see Corollary 1.4.2), and so $e' \subset A$. Thus, A is the union of all the cells of Ω, and because A is closed in X, these cells form the desired subcomplex A' of X' (see Proposition 1.4.4 (ii)). □

The requirement that the base point of a based CW-complex should be the only point of a 0-cell is not an essential restriction. This will be demonstrated by the following technical lemma which is a strengthening of the fact that, for any point x_o of a CW-complex X, the pair (X, x_o) is a well-pointed based space (see Corollary 1.3.7).

Lemma 2.3.7 *If x is a point of a CW-complex X, then there is a subdivision X' of X containing $\{x\}$ as a 0-cell.*

Proof Let X be a CW-complex, let x be a point of X and let the n-cell e be the carrier of x. The construction will be by induction on $n \geqslant 1$, leaving the skeleta X^m for $m \geqslant n$ unchanged. Let $c_e : S^{n-1} \to X^{n-1}$ denote an attaching map for e; by the induction hypothesis, one may assume it to be cellular with respect to the CW-structure

$$\{\{e_0\}, \ldots, \{e_0\}, S^{n-1}\}$$

for S^{n-1}. Now let $c: S^{n-1} \to X^n \backslash e$ be the composition of c_e with the inclusion $X^{n-1} \to X^n \backslash e$, which is again a cellular map. By Example 3, its mapping cone is a CW-complex; this CW-complex is homeomorphic to X^n where the homeomorphism can be chosen as the identity on $X^n \backslash e$ and transforming the peak of the mapping cone into x. Then, the image of the CW-structure of the mapping cone under this homeomorphism is the desired subdivision of X^n. □

Another special construction relates to covering projections. Their existence is no problem whenever one deals with CW-complexes.

Proposition 2.3.8 *Let X be a path-connected based CW-complex and let π be a subgroup of its fundamental group. Then, there is a based covering projection $p : \tilde{X} \to X$ such that π is the image of the fundamental group of $\tilde{X}$ under the homomorphism p_1.*

Proof See Proposition A.8.4 and Theorem 1.3.2. □

If the base of a covering projection is a CW-complex, its total space inherits a *canonical* CW-structure.

Proposition 2.3.9 *Let X be a CW-complex and let $p : \tilde{X} \to X$ be a covering projection. Then, the sequence $\{\tilde{X}^n = p^{-1}(X^n) : n \in \mathbf{N}\}$ is a CW-structure for the covering space $\tilde{X}$.*

Proof (0) Take $\tilde{x} \in \tilde{X}^0$. Then, $x = p(\tilde{x})$ belongs to X^0, which by assumption is discrete. Thus, there is an open neighbourhood U of x in X, which does not meet any other 0-cell of X and such that $p^{-1}(U) = \sqcup U_\lambda$ with each U_λ open in $\tilde{X}$ and containing exactly one inverse image of x. One of these U_λ contains $\tilde{x}$ and cannot contain any other point of $\tilde{X}^0$. Therefore, $\tilde{X}^0$ is a discrete space.

(1) The pairs $(\tilde{X}^n, \tilde{X}^{n-1})$ are adjunctions of n-cells, for every $n \in \mathbf{N} \setminus \{0\}$ (see Theorem 1.1.6(i)).

(2) It remains to prove that the covering space $\tilde{X}$ is the union space of the expanding sequence $\{\tilde{X}^n : n \in \mathbf{N}\}$. To this end, take $U \subset \tilde{X}$ such that $U \cap \tilde{X}^n$ is open in $\tilde{X}^n$, for every $n \in \mathbf{N}$. To show that U is open in $\tilde{X}$, it suffices to assume U sufficiently small, namely such that U is open in $\tilde{X}$ iff $p(U)$ is open in X (cf. last part of the proof of Theorem 1.1.6(i)). Because all the restrictions $p|\tilde{X}^n$ are open maps, $p(U) \cap X^n = p(U \cap \tilde{X}^n)$ is open in X^n, for all $n \in \mathbf{N}$; thus, $p(U)$ is open in X. □

In the context of covering projections, another fact deserves mentioning.

Corollary 2.3.10 *Let X be a CW-complex and let $p : \tilde{X} \to X$ be a covering projection. Then, p is a regular map in the strict sense that it maps open cells homeomorphically into open cells; the same holds true for covering transformations.* □

2.4 The cellular approximation theorem and some related topics

The categories CW and CW^c have a deeper relationship than just that given by the fact that CW^c is a subcategory of CW. This section is mostly devoted to showing that any map between two CW-complexes is homotopic to a cellular map.

Let X be a space, (Y, D) a pair of spaces and $f : Y \to X$ a map. An *approximation rel. D to f* is a map $g : Y \to X$, which is homotopic rel. D to f.

Lemma 2.4.1 *Let (Y, D) be an adjunction of n-cells, $n > 0$, and let (X, A) be a pair of spaces such that $\pi_n(X, A, x_o) = 0$, for every base point $x_o \in A$. Then any map $f : Y \to X$ with $f(D) \subset A$ has an approximation $g : Y \to X$ rel. D with $g(Y) \subset A$.*

Proof Suppose such a map f is given and let $\tilde{f} : D \to A$ denote the induced map. Take an n-cell $e \in \pi(Y \setminus D)$ and choose a characteristic map $\bar{c} : B_e = B^n \to Y$ for e with corresponding attaching map $c : S^{n-1} \to D$. The pair $(f \circ \bar{c}, \tilde{f} \circ c)$ represents an element of the relative homotopy group $\pi_n(X, A, x_0)$ with $x_0 = f \circ \bar{c}(e_0)$. By assumption, this group vanishes; thus, there exists a homotopy H_e rel. S^{n-1} between $f \circ \bar{c}$ and a map with image in A (see the description of the zero element of a relative homotopy group given in Section A.8). Now, because

$$X \times I = (A \times I) \sqcup \bigsqcup_{e \in \pi(Y \setminus D)} (B_e \times I),$$

the homotopies $H_e, e \in \pi(Y \setminus D)$ all together induce a homotopy $H : X \times I \to X'$ connecting f to the desired map g. □

From this, one obtains another characterization of n-connectivity for pairs of spaces.

Lemma 2.4.2 *A pair of spaces (X, A) is n-connected, $n \geqslant 0$, iff*

(i) *every path-component of X meets A and*

(ii) *for every map $f : Y \to X$, where Y is a relative CW-complex and $f(Y^{(-1)}) \subset A$, there is an approximation $g : Y \to X$ rel. $Y^{(-1)}$ to f with $g(Y^{(n)}) \subset A$.*

Proof '$\Leftarrow$': For $1 \leqslant k \leqslant n$, consider B^k provided with the usual CW-structure $\{\{e_0\}, \ldots, \{e_0\}, S^{k-1}, B^k\}$ and apply (ii) to a map $b : B^k \to X$ with $b(S^{k-1}) \subset A$.

'$\Rightarrow$': (i) is clear (ii) : let $f : Y \to X$ be a map with the required properties. One constructs inductively maps $g_k : Y \to X$, $-1 \leqslant k \leqslant n$, such that

(1) $g_{-1} = f$,

(2) $g_{k+1} \simeq g_k$ rel. $Y^{(k)}$, $-1 \leqslant k \leqslant n$,

(3) $g_k(Y^{(k)}) \subset A$,

as follows. Suppose g_k is given. Take an approximation $g' : Y^{(k+1)} \to X$ to $g_k | Y^{(k+1)}$ rel. $Y^{(k)}$ with $g'(Y^{(k+1)}) \subset A$ (see Lemma 2.4.1); let $H : Y^{(k+1)} \times I \to X$ denote the homotopy involved. Since the inclusion $Y^{(k+1)} \to Y$ is a closed cofibation (see Proposition 1.2.3 (iii)), there is an extension of H to a homotopy from g_k to a map g_{k+1} as desired.

Finally, take $g = g_n$. □

Lemma 2.4.3 *If the pair (X, A) is an adjunction of n-cells, $n > 0$, any map*

$$b : (B^k, S^{k-1}) \to (X, A)$$

with $k < n$ is homotopic rel. S^{k-1} to a map $b' : B^k \to A$.

In other words, *if the pair (X, A) is an adjunction of n-cells, $n > 0$, then it is $(n-1)$-connected.*

Before giving the proof of this lemma note the following:

Corollary 2.4.4 *For every natural number $k < n$,*

$$\pi_k(S^n, e_0) = 0.$$

Proof It is enough to take an element $\beta \in \pi_k(S^n, e_0) = \pi_k(S^n, \{e_0\}, e_0)$ represented by $b : (B^k, S^{k-1}) \to (S^n, \{e_0\})$. □

Proof of Lemma 2.4.3 The proof is done by induction on n. Since the corollary is valid whenever the lemma holds true, one can use Corollary 2.4.4 for the inductive step on the lemma itself.

If $n = 1$ and $k = 0$, both the lemma and the corollary are trivial; suppose that the lemma is true for n. Without loss of generality, one can assume $k > 0$. Let $\tilde{f}$ be a characteristic map for the adjunction of $(n+1)$-cells (X, A), and, for every $e \in \pi(X \setminus A)$, let the induced characteristic map be denoted by $\bar{c}_e$. Then take the open sets

$$U = C_{\tilde{f}}(A)$$

(the collar of A in X; see Section 1.1) and

$$U_e = \{\bar{c}_e(t) : 0 \leqslant |t| < \tfrac{3}{4}\},$$

for every $e \in \pi(X \setminus A)$. Clearly,

$$\Omega = \{U\} \cup \{U_e : e \in \pi(X \setminus A)\}$$

is an open covering of X.

Let B^* be the k-fold product of the ball B^1 with itself and let $b^* : B^* \to X$ be the composition of the map b with the canonical homeomorphism from B^* onto B^k (see Proposition 1.0.2). Let ε be the Lebesgue number of the covering $b^{*-1}(\Omega)$. Subdivide B^1 into finitely many closed intervals of finite length smaller than $2\varepsilon/\sqrt{k}$, giving rise to a new finite regular CW-structure on the product space B^*; in the sequel, it will be assumed that B^* is always endowed with this CW-structure. Its essential feature is that the covering of B^* by its closed cells refines the covering $b^{*-1}(\Omega)$.

Let V be the union of all open cells of B^* whose closures are contained in $b^{*-1}(U)$. The closure of each of these open cells is also contained in V, and thus, by Proposition 1.4.4, V is a subcomplex of B^*. Extend V dimensionwise to B^* be defining inductively $V_{-1} = V$ and $V_l = V_{l-1} \cup B^{*l}$, for $l = 0, \ldots, k$.

This is done in order to construct a map $\tilde{b} : B^* \to X$ which is homotopic to b^* rel. V and takes the sets $b^{*-1}(U_e)$ into $U \cap U_e$, for all $e \in \pi(X \setminus A)$. Assume $b_{l-1} = \tilde{b}|V_{l-1}$ defined, and look at an l-cell $e' \in \pi(V_l \setminus V_{l-1})$. Since its closure is not contained in $b^{*-1}(U)$, there is a (unique) n-cell $e \in \pi(X \setminus A)$ such that $\bar{e}' \subset b^{*-1}(U_e)$. Now $b_{l-1}(\bar{e}' \setminus e') \subset U \cap U_e$, thus $b_{l-1}|(\bar{e}' \setminus e')$ may be thought of as representing an element of $\pi_{l-1}(U \cap U_e, x_0)$, where x_0 is a suitable base point. But $U \cap U_e$ is homotopically equivalent to S^n, and, hence

$$\pi_{l-1}(U \cap U_e, x_0) \cong \pi_{l-1}(S^n, e_0).$$

The induction hypothesis on Corollary 2.4.4 implies that the group on the right-hand side of this isomorphism is trivial. So $b_{l-1}|(\bar{e}' \setminus e')$ extends to a map $\bar{e}' \to X$ whose image is contained in $U \cap U_e$.

This can be done for all the l-cells of B^* which do not belong to V, yielding the desired extension of b_{l-1} over V_l and ultimately the map $\tilde{b}$. It is still necessary to show that $\tilde{b}$ is homotopic rel. V to b^*. The desired homotopy $K^* : B^* \times I \to X$ is defined to be trivial over V and linearly on the cells. More precisely, take $K^*(v,t) = a^*(v) = \tilde{b}(v)$, for $v \in V$, and $K^*(v,t) = \bar{c}_e((1-t)u + tw)$, where $b^*(v) \in U_e$, $b^*(v) = \bar{c}_e(u)$ and $\tilde{b}(v) = \bar{c}_e(w)$. To show the continuity of this function $K^* : B^* \times I \to X$, it is enough to show that its restriction to each $\bar{e}' \times I$ is continuous, for each cell e' of B^*. If e' is a cell of V, there is nothing to prove; otherwise for $v \in \bar{e}'$ the assignments $v \mapsto u$, $v \mapsto w$ are continuous and so is their linear combination $(v,t) \mapsto (1-t)u + tw$.

Now switch back to B^k from B^*. Compose K^* with the canonical homeomorphism $B^k \times I \to B^* \times I$ to obtain a homotopy $K : B^k \times I \to X$ rel. a subset of B^k which contains S^{k-1}. Clearly, $K(-,0) = b$ and $K(B^k \times \{1\}) \subset U$. Finally, use the deformation from U to A to move $K(B^k \times \{1\})$ into A. □

Remark The lemma would be trivial if one could be sure, whenever, $k < n$, that every map $b : B^k \to X$ would miss at least one point in every n-cell of the adjunction space X, since then the image of b would be contained in a subspace of X having A as strong deformation retract. But the existence of 'Peano curves', i.e., of maps from B^1 *onto* B^n, shows the impossiblity of avoiding the deformation of the map b into a homotopic map which misses points in each of the $n+1$-cells. □

In case $k = n$, the proof just given leads also to an interesting result.

Lemma 2.4.5 *Let (X, A) be an adjunction of n-cells with simply connected A, $n \geqslant 2$; for every cell $e \in \pi(X \backslash A)$, let $\bar{c}_e$ denote a characteristic map. Then, the homomorphism*

$$\bigoplus_{e \in \pi(X\backslash A)} (\bar{c}_e)_n : \bigoplus_{e \in \pi(X\backslash A)} \pi_n(B^n, S^{n-1}, e_0) \to \pi_n(X, A, x_0)$$

is an epimorphism, for any choice of a base point $x_0 \in A$.

Proof Start the considerations in the proof of Lemma 2.4.3 with a map,

$$b : (B^n, S^{n-1}) \to (X, A).$$

The construction of the map $\tilde{b}$ can be carried out until $b_{n-1} : V_{n-1} \to U$ is reached. This map b_{n-1} is homotopic rel. V to $b^* | V_{n-1}$. By the homotopy extension property, b_{n-1} can be extended to a map $\tilde{b}$, taking values in X rather than in U. Since A is assumed to be simply connected, this map $\tilde{b}$

(up to composition with the canonical homeomorphism $B^n \to B^*$) represents the same element of $\pi_n(X, A, x_0)$ (see Proposition A.8.15). By the homotopy addition theorem (see Theorem A.8.16), this is the sum of the elements represented by the restrictions $\tilde{b}_{e'}$ of $\tilde{b}$ to the n-cells e' of B^*. Each e' either belongs to V, in which case $\tilde{b}_{e'}$ represents 0, or its image under $\tilde{b}$ is contained in exactly one n-cell $e \in \pi(X \backslash A)$; then, the corresponding element of $\pi_n(X, A, x_o)$ belongs to the image of $(\bar{c}_e)_n$. □

Lemma 2.4.3 has several consequences for CW-complexes.

Proposition 2.4.6 *If X is a relative CW-complex, then the pairs $(X, X^{(n)})$ are n-connected,* for every $n \in \mathbf{N}$.

Proof Note, first, that any compact subset $K \subset X$ is contained in some $X^{(m)}$. Indeed, the image of K under the projection $X \to X/X^{(-1)}$ is contained in some m-skeleton of the CW-complex $X/X^{(-1)}$ (see Proposition 1.5.2); but the inverse image of this skeleton is just $X^{(m)}$.

Let $b : B^k \to X$ be a map with $b(S^{k-1}) \subset X^{(n)}$, $0 \leqslant k \leqslant n$. Because $b(B^k)$ is compact, it is contained in a finite skeleton of X, say X^m. If $m > n$, the lemma allows to deform b homotopically rel. S^{k-1} into a map $B^k \to X^{(m-1)}$. The result is obtained by repeating this procedure finitely many times. □

Corollary 2.4.7 *If X is a relative CW-complex, then the pairs $(X^{(m)}, X^{(n)})$ are n-connected, for every $m \geqslant n \geqslant 0$.*

Proof Notice that $X^{(m)}$ is a relative CW-complex with $(X^{(m)})^{(n)} = X^{(n)}$. □

Proposition 2.4.6 and Corollary 2.4.7 can be reformulated in terms of homotopy groups.

Corollary 2.4.8 *If X is a relative CW-complex and $x_o \in X^{(-1)}$, then*

(i) *for $n > k \geqslant 0$*

$$\pi_k(X^{(n)}, x_o) \cong \pi_k(X, x_o);$$

(ii) *for $n \geqslant 0$, the canonical homomorphism*

$$\pi_n(X^{(n)}, x_o) \to \pi_n(X^{(n+1)}, x_o)$$

is an epimorphism;

(iii) *for $n \geqslant m \geqslant k > 0$,*

$$\pi_k(X^{(n)}, X^{(m)}, x_o) = \pi_k(X, X^{(m)}, x_o) = 0;$$

(iv) *for* $n > k > m \geqslant -1, k > 0,$

$$\pi_k(X^{(n)}, X^{(m)}, x_o) \cong \pi_k(X, X^{(m)}, x_o);$$

(v) *for* $n > m \geqslant -1$, $n > 0$, *the canonical homomorphism*

$$\pi_n(X^{(n)}, X^{(m)}, x_o) \to \pi_n(X, X^{(m)}, x_o)$$

is an epimorphism. □

In particular, choosing a base point and setting $n = 1$, $k = 0$ in (i), one obtains:

Corollary 2.4.9 *A CW-complex is path-connected iff its 1-skeleton is path-connected.* □

The previous two corollaries have an application to the study of covering projections onto CW-complexes.

Corollary 2.4.10 *Let* $p : \tilde{X} \to X$ *be a universal covering projection where* X *is a CW-complex and* $\tilde{X}$ *has the induced CW-structure* (*see Proposition* 2.3.9). *Then, for* $n \geqslant 2$, *the induced maps* $p^n : \tilde{X}^n \to X^n$ *are universal covering projections.*

Proof A map induced by a covering projection is always a covering projection (see Proposition A.4.17). Thus it suffices to show that the skeleta $\tilde{X}^n$ are simply connected. Since $\tilde{X}$ is path-connected, so are the $\tilde{X}^n$ (see Corollary 2.4.9). Moreover, for any choise of a base point $\tilde{x}_o \in \tilde{X}^n$ and $n \geqslant 2$, $\pi_1(\tilde{X}^n, \tilde{x}_o) \cong \pi_1(\tilde{X}, \tilde{x}_o) = 0$ (see Corollary 2.4.8(i)). □

The main theorem of this section is proved next.

Theorem 2.4.11 (*The cellular approximation theorem*) *Let* $f : Y \to X$ *be a map between CW-complexes* Y *and* X, *whose restriction to a subcomplex* D *of* Y *is cellular. Then there exists a cellular approximation* g *to* f *rel.* D. *Moreover, it is possible to choose a homotopy* $H : f \simeq g$, *so that, for every cell* $e \subset Y$, $H(Y(e) \times I) \subset X(f(Y(e)))$.

Proof Consider $Y \times I$ as the union space of the expanding sequence $\{Y_n = Y \times \{0\} \cup Y^{n-1} \times I \cup D \times I : n \in \mathbb{N}\}$. The proof is done by constructing a compatible sequence $\{H_n\}$ of maps $H_n : Y_n \to X$ satisfying the following properties (see Proposition A.5.7):

(1) $H_n(-, 0) = f$;

(2) $H_n(-, 1)$ is cellular;

(3) $H_n|D \times I$ is the composition of the projection onto D with $f|D$;
(4) for every cell $e' \subset Y^{n-1}$, $H_n(Y(e') \times I) \subset X(f(Y(e')))$.

For $n = 0$, take H_0 to be the map f on $Y \times \{0\}$ and the composition of the projection map $D \times I \to D$ with $f|D$ on $D \times I$. Assume that H_n has been suitably defined. Let e be an n-cell of $Y \backslash D$ and choose a characteristic map $\bar{c}_e : B^n \to Y$ for it. The composition

$$\bar{c}_e \times \mathbf{1}_I|(B^n \times \{0\} \cup S^{n-1} \times I) \circ r^n|B^n \times \{1\}$$

(where r^n denotes the map defined in Section 1.0) induces a map $b'_e : B^n \to Y_n$. Now observe that, according to property (4), the composition

$$b_e = H_n \circ b'_e : (B^n, S^{n-1}) \to (X, X^{n-1})$$

takes values only in the subcomplex $\hat{X} = X(f(Y(e)))$; since $Y(e)$ is finite (see Corollary 1.5.3), and therefore compact (see Proposition 1.5.8), $f(Y(e))$ is compact and thus $\hat{X}$ is finite (Corollary 1.5.4). Because the pair $(\hat{X}, \hat{X}^n)$ is n-connected (see Proposition 2.4.6), the map b_e is homotopic rel. S^{n-1} to a map $B^n \to \tilde{X}^n$. The homotopy involved factors through the restriction of the homotopy $R^n : B^n \times I \times I \to B^n \times I$ (see page 8) to $B^n \times \{1\} \times I = B^n \times I$, giving rise to an extension of

$$\bar{c}_e \times \mathbf{1}_I|B^n \times \{0\} \cup S^{n-1} \times I$$

over $B^n \times I$, which in turn factors through the characteristic map $\bar{c}_e \times \mathbf{1}_I$ for the $(n+1)$-cell $\bar{e} \times I$ of $Y \times I$, and therefore induces an extension $H_{n+1}|(Y_n \cup e)$ of H_n which takes values only on $\hat{X}$. This procedure, applied to all n-cells $e \subset Y \backslash D$, yields H_{n+1}. □

The statement of the cellular approximation theorem given here is sharper than the usual one; its advantage lies in the following fact:

Corollary 2.4.12 *Let $f : Y \to X$ be a map between CW-complexes Y and X. Then, there is a homotopy from f to a cellular map which deforms $f(D)$ only within A, for all given subcomplexes $D \subset Y$ and $A \subset X$ with $f(D) \subset A$.* □

Remark Thus the cellular approximation theorem implies that for any two homotopic cellular maps $f, g : Y \to X$ it is possible to choose a homotopy which deforms the image of Y^n entirely within X^{n+1}. This follows by considering $Y \times I$ as a CW-complex in the obvious manner and taking a cellular approximation to an arbitrarily given homotopy $Y \times I \to X$ from f to g rel. $Y \times \{0, 1\}$. □

This section is closed by two useful technical lemmas.

Lemma 2.4.13 *Let (X, A) be a pair of spaces with A a CW-complex. Then (X, A) is n-connected, $n \geqslant 0$, iff every path-component of X meets A, and, for $0 < k \leqslant n$, every map $b : B^k \to X$ with $b(S^{k-1}) \subset A^{k-1}$ is homotopic rel. S^{k-1} to some map $B^k \to A^k$, i.e., iff the canonical homomorphisms*

$$\pi_k(X, A^{k-1}, x_o) \to \pi_k(X, A^k, x_o)$$

are zero, for any 0-cell x_o of A.

Proof '⇒': If (X, A) is n-connected then a map b, as described in the statement, can be deformed rel. S^{k-1} into a map with target A. By Proposition 2.4.6, the pair (A, A^k) is k-connected, and thus the new map can further be deformed rel. S^{k-1} into a map of the desired kind.

'⇐': Since, by Corollary 2.4.8(iii), $\pi_{k-1}(A^k, A^{k-1}, x_o)$ is equal to zero, it follows from the hypothesis and the exact homotopy sequence of the triple (X, A^k, A^{k-1}, x_o) that the relative homotopy group $\pi_k(X, A^k, x_o)$ vanishes. Now use the exact homotopy sequence of the triple (X, A, A^k, x_o) to show that $\pi_k(X, A, x_o)$ is trivial. □

Lemma 2.4.14 *Let Y be a subcomplex of the finite CW-complex X such that the pair (X, Y) is $(n-1)$-connected, $n > 0$. Then, there exists a finite CW-complex Z containing Y as a subcomplex and satisfying the following conditions:*

(i) $Z^{n-1} = Y^{n-1}$;
(ii) *Z has the same number of cells as X in each dimension $> n + 1$:*
(iii) *Z is homotopy equivalent rel. Y to X.*

In particular, if X is a finite, n-dimensional CW-complex, Z is finite and, at most, $(n + 1)$-dimensional.

Proof Consider a cell e in $X \backslash Y$ of lowest dimension $r < n$. The following is a procedure to get rid of this cell at the expense of introducing one new $(r+2)$-cell. Doing this finitely many times one obtains the desired CW-complex Z.

Let $\bar{c}$ be a characteristic map for the cell e; by the cellular approximation theorem, one may assume $\bar{c}$ to be cellular with respect to the canonical CW-structure for B^r (see Theorem 2.4.11 and Section 1.1, Example 1). Since (X, Y) is $(n-1)$-connected, there is a cellular map $\hat{c} : B^r \to Y$ whose composition with the inclusion $Y \to X$ is homotopic rel. S^{r-1} to $\bar{c}$. Choose a corresponding homotopy and let $b : B^{r+1} \to X$ denote the map induced by factoring out h^r (see Section 1.0). Form the elementary expansions $Y_{\hat{c}}$ and X_b. Recall that $Y_{\hat{c}}$ is obtained from Y by attaching an r-cell and an $(r+1)$-cell; likewise, X_b contains one $(r+1)$-cell and one $(r+2)$-cell more than X. Now, $Y_{\hat{c}}$ can be considered as a subcomplex of X_b, by identifying the new r-cell of $Y_{\hat{c}}$ with the initial cell e, and the new $(r+1)$-cell of $Y_{\hat{c}}$ with the new $(r+1)$-cell of X_b. Let $d : Y_{\hat{c}} \to Y$ denote a cellular deformation retraction (see Proposition 2.3.4) and attach X_b to Y via d. Let $\bar{d} : X_b \to X^*$ denote a corresponding characteristic map; $\bar{d}$ is a homotopy equivalence (see Proposition A.4.11) and so is the composition $\tilde{d}$ of $\bar{d}$ with the inclusion $X \to X_b$. The cells of X^* are those of Y and also those of X_b which do not belong to $Y_{\hat{c}}$ (see Proposition 2.3.1); that shows the claim on the cells of the constructed space. Because $\tilde{d}|Y$ is just the inclusion of Y into X^*, $\tilde{d}$ is a homotopy equivalence, rel. Y (use Lemma A.5.10 with $\bar{f} = \tilde{d}$, $f = g = \mathbf{1}_Y$ and H the projection $Y \times I \to Y$). □

Exercises

1. Give an example of a universal covering projection onto a CW-complex such that the induced map between the 1-skeleta fails to be universal.
2. Prove the following version of Corollary 2.4.8: If X is a CW-complex, A is a subcomplex of X and x_o is a chosen base point belonging to A, then
   (i) for every $n > k > 0$,
   $$\pi_k(X^n, A^n, x_o) \cong \pi_k(X, A, x_o);$$
   (ii) for every $k > 0$, the canonical homomorphism
   $$\pi_k(X^k, A^k, x_o) \to \pi_k(X^{k+1}, A^{k+1}, x_o)$$
   is an epimorphism. (Note that the difficulty lies at $k = 1$!)

2.5 Whitehead's realizability theorem

Perhaps one of the best-known results of J.H.C. Whitehead is the following.

Theorem 2.5.1 *A weak homotopy equivalence between CW-complexes is a homotopy equivalence.*

The theorem states that, if X, Y are CW-complexes and the map $f : Y \to X$ induces isomorphisms between the corresponding homotopy groups at all levels, then there is a map $g : X \to Y$ – called a *realization* of the isomorphisms $\pi_n(X, f(y_o)) \to \pi_n(Y, y_o)$ – which is homotopy inverse to f.

Proof First of all, a weak homotopy equivalence induces a bijection between the sets of path-components. Therefore it suffices to consider path-connected CW-complexes. Let Y, X be CW-complexes and let $f : Y \to X$ be a weak homotopy equivalence; by the cellular approximation theorem 2.4.11, one can assume f to be cellular and thus the mapping cylinder $M(f)$ to be a CW-complex (see Section 2.3, Example 2). The pair $(M(f), Y)$ is n-connected for all $n \in \mathbf{N}$ (Proposition A.4.10(vi)); one must show that Y is a strong deformation retract of $M(f)$ (Proposition A.4.10(v)).

To this end, one has to deform the identity map of $M(f)$ into a deformation retraction $M(f) \to Y$. This is done in exactly the same way as the deformation of the map f into a cellular map in the proof of the cellular approximation theorem (see Theorem 2.4.11), just replacing the reference to the n-connectivity of the pair $(\hat{X}, \hat{X}^n)$ by the n-connectivity of the pair $(M(f), Y)$. □

Remark One might think that a sufficient condition for two CW-complexes to have the same homotopy type is that their homotopy groups are isomorphic. But it is essential that these isomorphisms are realized – at least in one direction – by a map. To clarify this point, the reader should look at the remark after Exercise 3, Section 4.4. □

There is a sharper version of Theorem 2.5.1 for finite-dimensional CW-complexes.

Theorem 2.5.2 *Let Y and X be finite-dimensional CW-complexes and let $f : Y \to X$ be a map inducing isomorphisms $f_k : \pi_k(Y, y_o) \to \pi_k(X, f(y_o))$ for every* 0-cell *y_o of Y and any $k = 0, 1, \dots, n$, where $n = \max\{\dim Y, \dim X\}$. Then f is a homotopy equivalence.*

Proof As in the proof of Theorem 2.5.1, assume X, Y to be path-connected, f to be cellular and the mapping cylinder $M(f)$ to be a CW-complex. The assumptions imply that the pair $(M(f), Y)$ is n-connected. Therefore, there is an approximation $g : M(f) \to M(f)$ to the identity map of $M(f)$ rel. Y with $g(M(f)^n) \subset Y$ (see Lemma 2.4.2). If $\dim M(f) = n$, i.e., if $\dim Y < n$, the induced map $M(f) \to Y$ is the desired homotopy inverse to the inclusion $i_Y : Y \to M(f)$.

Otherwise, $\dim M(f) = n + 1$. In this case, one first constructs a retraction $g' : M(f) \to Y$ as follows. On $M(f)^n$, one takes the map induced by g. Any $(n+1)$-cell e' of $M(f)$ is of the form $e' = e \times (0, 1)$, where e is an n-cell of Y. Let $\bar{c}_{e'}$ be a characteristic map for e'. Its composition with the map g represents an element of $\pi_{n+1}(M(f), Y, y_0)$, where y_o is a suitable base point. Since the induced homomorphism $\pi_n(Y, y_0) \to \pi_n(M(f), y_o)$ is an isomorphism, the boundary homomorphism $\pi_{n+1}(M(f), Y, y_o) \to \pi_n(Y, y_o)$ vanishes. Hence, $g \circ \bar{c}_{e'} | S^n$ is – as a map into Y – homotopically trivial, and thus it has an extension $g_{e'} : B^{n+1} \to Y$ which factors through $\bar{e}'$ via the map $\bar{c}_{e'}$. This defines the desired map $g'|\bar{e}'$. Note that $g'|Y = \mathbf{1}_Y$; thus g' is a retraction; but it still remains to show that $i_Y \circ g' \simeq \mathbf{1}_{M(f)}$, where i_Y denotes the inclusion of Y into $M(f)$. To this end, take an approximation $k : X \to M(f)$ to the inclusion of X into the mapping cylinder $M(f)$ rel. $\varnothing$ with $k(X) \subset Y$ (see again Lemma 2.4.2). Now recall that $f = r_f \circ i_Y$ (see Proposition A.4.10(iv)) and note that $k = i_Y \circ k'$ for a certain map $k' : X \to Y$. Since $k \circ r_f \simeq \mathbf{1}_{M(f)}$,

$$\begin{aligned} i_Y \circ g' &\simeq i_Y \circ g' \circ k \circ r_f \\ &\simeq i_Y \circ g' \circ i_Y \circ k' \circ r_f \\ &= i_Y \circ k' \circ r_f \\ &= k \circ r_f \\ &\simeq \mathbf{1}_{M(f)}. \end{aligned}$$ □

The following results are simple applications of the theorem.

Corollary 2.5.3 *Let Y and X be finite-dimensional CW-complexes and let $f : Y \to X$ be a map such that $f_0 : \pi_0(Y, y_o) \to \pi_0(X, f(y_o))$ is onto and $\pi_k(f, y_o) = 0$, for every 0-cell y_o of Y and any $k = 1, 2, \ldots, n+1$ where $n = \max\{\dim Y, \dim X\}$. Then f is a homotopy equivalence.*

Proof See Proposition A.8.9. □

Corollary 2.5.4 *A one-dimensional CW-complex is contractible iff it is simply connected.*

Proof The necessity of the condition is trivial; the sufficiency follows by application of the theorem to the constant map. □

2.6 Computation of the fundamental group

As a consequence of the cellular approximation theorem, the fundamental groups of CW-complexes depend only on their 2-skeleta (see Corollary 2.4.7). Since the fundamental groups of 0-dimensional CW-complexes are trivial, it is only necessary to inspect CW-complexes of dimensions 1 and 2. The basic result of this section concerns the sphere S^1 viewed as a CW-complex with one 0-cell and one 1-cell (see Section 1.1, Example 2).

Theorem 2.6.1 $\pi_1(S^1, e_0) \cong \mathbf{Z}$.

Proof The covering projection $\mathbf{R} \to S^1$ (see Section 2.1, Example) has a simply connected domain and the group of its covering transformations is **Z**. Thus, the fundamental group of its codomain is isomorphic to **Z** (see Theorem A.8.6). □

Corollary 2.6.2 *For any set Γ, the fundamental group of the Γ-fold wedge of S^1 is (up to isomorphism) the free group generated by Γ.*

Proof Let X denote the Γ-fold wedge of S^1. Take U as the collar of the base point of X, and, for every $\gamma \in \Gamma$, take U_γ as the union of U and the 1-sphere corresponding to γ. The family $\{U\} \cup \{U_\gamma : \gamma \in \Gamma\}$ is an open covering of X (see Lemma 1.1.7(iii)), closed under intersections. The space U is contractible (see Lemma 1.1.7(vi)), and so has trivial fundamental group; the spaces U_γ contain a 1-sphere as a strong deformation retract, so have fundamental group isomorphic to **Z**. Thus, the fundamental group of X is the free product of 'Γ' copies of **Z** (see Proposition A.8.20), i.e., a free group with one generator for each element of Γ. □

The two CW-complexes whose fundamental groups were just computed share the property that they have only one 0-cell. This fact does not represent a real restriction for computing the fundamental groups of CW-complexes, as will be seen in the sequel. To this end, a further notion is necessary. Given a CW-complex X, a *tree* of X is a non-empty, simply connected subcomplex of X with dimension at most 1. The set of trees in X is ordered by (set-theoretical) inclusion.

Lemma 2.6.3 *Each tree of a CW-complex is contained in a maximal tree.*

Proof Let $T_1 \subset T_2 \subset \cdots \subset T_k \subset \cdots$ be an increasing chain of trees of a CW-complex X containing a given tree T_0. Then, its union $T = \bigcup_{k=0}^{\infty} T_k$ is a CW-complex (see Corollary 1.4.5), which evidently is 1-dimensional. Since any pair of points of T is contained in some (path-connected) T_k, the space T is path-connected; thus, T is connected. A loop in T is a compact subset of T; therefore, it meets only finitely many cells of T, and, consequently, it is contained in a tree T_k, for some $k \in \mathbf{N}$. Since T_k is simply connected, the loop is trivial, and so T is simply connected; hence, T is itself a tree. Zorn's lemma now implies the existence of a maximal tree containing T_0. □

There is a useful characterization of maximality for trees.

Lemma 2.6.4 *A tree T in a connected CW-complex X is maximal iff it contains all 0-cells of X, i.e., iff $T^0 = X^0$.*

Proof Since a connected CW-complex is path-connected (see Corollary 1.4.12), and since a CW-complex is path-connected iff its 1-skeleton is path-connected (Corollary 2.4.9), one may assume $\dim X = 1$.

'$\Rightarrow$': Let T be any tree in X. If there are 0-cells in X not belonging to T, there is one of them, say e_0, that is in the boundary of a 1-cell e_1 meeting T. Then, $T' = T \cup e_1 \cup e_0$ is a connected, 1-dimensional subcomplex of X, which contracts to T, and therefore to a point. Thus, T' is a tree in X larger than T, and so T cannot be maximal.

'$\Leftarrow$': Assume the tree T with $T^0 = X^0$ is contained in a tree T'. The quotient T'/T is a bouquet of 1-spheres whose fundamental group is a free group (see Corollary 2.6.2). On the other hand, this quotient can be thought out as obtained by attaching T' to a point $*$ via the constant map $T \to *$. Since T is contractible (see Corollary 2.5.4), this attaching map is a homotopy equivalence, and then so is the projection $T' \to T'/T$, which can be viewed as a characteristic map for the adjunction (see Proposition A.4.11). But T' is simply connected; thus, the free group above is trivial, implying that the quotient T'/T can consist only of the base point, i.e., $T = T'$. □

Now, if X is a connected CW-complex and if T is a maximal tree of X then, X/T is a CW-complex with exactly one 0-cell and the (homotopy) type of X (for a more precise exposition see Corollary 2.6.10). This leads to the following 'omnibus' theorem.

Theorem 2.6.5 *Let X be a connected,* 1*-dimensional based CW-complex. Then*:

(i) *The fundamental group $\pi_1(X, x_0)$ is a free group generated by a set of cardinality smaller than or equal to the cardinality of the set of* 1*-cells of X*;

(ii) *if X has only finitely many* 1*-cells then $\pi_1(X, x_0)$ is finitely generated*;

(iii) *if $\pi_1(X, x_0)$ is finitely generated then X is homotopically equivalent to a CW-complex with finitely many* 1*-cells.* □

The fundamental group of a connected CW-complex of arbitrary dimension > 1 is an epimorphic image of the fundamental group of its 1-skeleton (see Corollary 2.4.7); this implies the following.

Corollary 2.6.6 *Let X be a connected, based CW-complex with only finitely many* 1*-cells. Then, $\pi_1(X, x_0)$ is finitely generated.* □

Now turn to 2-dimensional CW-complexes. To this end, recall that a pair of sets (Γ, R) is a *presentation of the group* G if Γ is a subset of G, R is a subset of the free group $F(\Gamma)$ generated by Γ, and $G \cong F(\Gamma)/N$, where N is the intersection of all normal subgroups of $F(\Gamma)$ that contain the set R; the elements of Γ are said to be *generators* of G and the elements of R are the *relations*. For later use, recall also that a group G is *finitely presented* if it has a presentation (Γ, R), where both sets Γ and R are finite.

Lemma 2.6.7 *Let (X, A) be an adjunction of exactly one 2-cell with A a path-connected space. Then, for any $x_0 \in A$, the group $\pi_1(X, x_0)$ is a factor group of $\pi_1(A, x_0)$ produced by just one relation.*

Let $c : B^2 -/\rightarrow A$ be a partial map generating the adjunction (X, A). This gives rise to the pushout

$$\begin{array}{ccc} \pi_1(B^2, e_0) = \{1\} & \rightarrow & \pi_1(X, c(e_0)) \\ \uparrow & & \uparrow \\ \pi_1(S^1, e_0) = \mathbf{Z} & \rightarrow & \pi_1(A, c(e_0)) \end{array}$$

(see Theorem A.8.19). So the claim is proved for $x_0 = c(e_0)$. The general case is proved using the isomorphisms induced from paths connecting this specific base point to arbitrary ones (see Section A.8, page 287).

Theorem 2.6.8 *The fundamental group of a connected CW-complex X has a presentation (X_1, R), where X_1 is a set of 1-cells outside of a maximal tree and R is in a bijective correspondence with the set of 2-cells of X.*

Proof As mentioned in the beginning of this section, assume $\dim X = 2$. Let $\bar{f}$ be a characteristic map for the 2-cell adjunction (X, X^1) and let X_2 denote the set of 2-cells of X. Take U to be the $\bar{f}$-collar of X^1 that is open in X, and, for each $e \in X_2$, take the open set $U_e = e \cup U$. The family $\{U\} \cup \{U_e : e \in X_2\}$ is an open covering of X, closed under intersections. U contains X^1 as a strong deformation retract; thus, its fundamental group is isomorphic to the free group $F(X_1)$. Each pair (U_e, U) is an adjunction of just one 2-cell and gives rise to a relation in $F(X_1)$ (by the previous lemma). The fundamental group of X is now obtained by taking all these relations together (see Proposition A.8.20). □

The following is the CW-version of the Seifert–van Kampen Theorem.

Theorem 2.6.9 *Let (X, x_0) be a connected, based CW-complex and let $\{(X_\lambda, x_0) : \lambda \in \Lambda\}$ be a family of connected based subcomplexes which covers (X, x_0) and is closed under intersections. Then, the group $\pi_1(X, x_0)$ is isomorphic to the colimit group of the diagram whose objects are all groups $\pi_1(X_\lambda, x_0)$ and whose morphisms are the homomorphisms induced by all possible inclusions $X_\lambda \subset X_\mu$.*

Proof Take a sequence $\{\bar{f}^n : n \in \mathbf{N}\}$ of characteristic maps for the adjunctions (X^n, X^{n-1}). For every $n \in \mathbf{N}$ and every $\lambda \in \Lambda$, let $\bar{f}^n_\lambda$ denote the induced characteristic map for the n-cell adjunction $(X_\lambda \cup X^n, X_\lambda \cup X^{n-1})$. Then, define inductively open sets $U_{\lambda,n} \subset X_\lambda \cup X^n$ by taking $U_{\lambda,0} = X_\lambda \cup X^0$ and $U_{\lambda,n+1}$ as the $\bar{f}^{n+1}_\lambda$-collar of $U_{\lambda,n}$. Then, for every $\lambda \in \Lambda$, the union $\bigcup_{n=0}^{\infty} U_{\lambda,n}$ is an open set in X containing X_λ as a strong deformation retract (cf. the proof of Proposition 1.3.1). Moreover, the family $\{U_\lambda : \lambda \in \Lambda\}$ is a covering of X, satisfying the property $U_\lambda \cap U_\mu = U_\gamma$, whenever $X_\lambda \cap X_\mu = X_\gamma$. Thus, the diagram described in the statement of the theorem does not change (up to isomorphism) if, for its construction, one uses the open sets U_λ instead of the subcomplexes X_λ; this implies the result (see Proposition A.8.20). □

Lemma 2.6.4 also has a nice consequence, which is independent of any considerations on fundamental groups.

Corollary 2.6.10 *Any non-empty, connected CW-complex X is homotopy equivalent to a CW-complex Z with exactly one 0-cell and whose higher-dimensional cells have based characteristic maps i.e. all pairs (Z^n, Z^{n-1}) are based adjunctions of n-cells.*

Proof A non-empty, connected CW-complex X contains a 0-cell (see Proposition 1.4.15) which may be considered as a trivial tree. This is contained in a maximal tree T (see Lemma 2.6.3) which covers the 0-skeleton of X. The quotient X/T is a CW-complex (see Section 2.3, Example 1) which contains only one 0-cell. As in the proof of the sufficiency part of the Lemma 2.6.4, it can be viewed as obtained by attaching X to a singleton space $*$ via the constant map $T \to *$ and the same argument shows that the projection $X \to X/T$ is a homotopy equivalence.

Now, assume X to be a CW-complex with exactly one 0-cell. Construct inductively the n-skeleta Z^n of the desired CW-complex Z (allowing based characteristic maps) and homotopy equivalences $h^{(n)} : X^n \to Z^n$ which fit into a commutative ladder in order to give a homotopy equivalence $h : X \to Z$ (see Proposition A.5.11). The induction starts with $Z^1 = X^1$ and

$h^{(1)} = 1$. Assume Z^{n-1} and $h^{(n-1)}$ have been constructed and satisfy the required conditions. Let f denote an attaching map for the adjunction of n-cells (X^n, X^{n-1}); its domain is a coproduct of $(n-1)$-spheres and can be viewed as a CW-complex possessing only 0-cells and $(n-1)$-cells. Then, form

$$Z' = Z^{n-1} \bigsqcup\nolimits_{h^{(n-1)}} X^n;$$

the characteristic map $h' : X^n \to Z'$ is a homotopy equivalence (see Proposition A.4.11) and the pair (Z', Z^{n-1}) is an adjunction of n-cells, for which the composition $h^{n-1} \circ f$ may be chosen as attaching map (see the horizontal composition law (L1) in Section A.4). Now, approximate $h^{(n-1)} \circ f$ by a cellular map f' (see Theorem 2.4.11) and attach n-cells to Z^{n-1} via f' to obtain the CW-complex Z^n and a homotopy equivalence $h'' : Z' \to Z^n$ rel. Z^{n-1} (see Proposition A.4.15). Finally, take $h^{(n)} = h'' \circ h'$. □

2.7 Increasing the connectivity of maps

In this section a technique is developed to transform a map between CW-complexes into a homotopy equivalence by attaching cells to its domain. To begin with, one has to inspect the lower dimensions.

Lemma 2.7.1 *Let $f : Y \to X$ be any map. Then there is a* 0*-connected map $f' : Y' \to X$ such that:*
(i) *Y' is obtained from Y by an adjunction of* 0*-cells;*
(ii) *$f' | Y = f$; and*
(iii) *$Y' \backslash Y$ is finite if $\pi_0(X, f(y_o)) \backslash f_0(\pi_0(Y, y_o))$ is finite, for any choice of a base point $y_o \in Y$.*

Proof Choose one point in each path component of X that does not meet $f(Y)$ and add it to Y as a 0-cell. This gives Y' and induces the desired map f'. □

Once a map is 0-connected, one can consider it as a sum (= coproduct) of based maps with path-connected codomain. Therefore one may restrict the attention to such maps in the sequel.

Lemma 2.7.2 *Let X be a path-connected space and let $f : Y \to X$ be a* 0*-connected map. Then there is a* 1*-connected map $f' : Y' \to X$ such that:*
(i) *Y' is obtained from Y by an adjunction of* 1*-cells;*
(ii) *$f' | Y = f$; and*
(iii) *$Y' \backslash Y$ has finitely many path components (i.e., $Y' \backslash Y$ consists of finitely*

many open 1-*cells only) if* $\pi_1(f, y_o)$ *is finite, for any choice of a base point* $y_o \in Y$.

Remark The assumption on the path-connectivity of X implies that the requirement on f to be 0-connected excludes the case $Y = \varnothing$.

Proof of Lemma 2.7.2 There are two types of elements in $\pi_1(f, y_o)$: those that are in the image of the function $\pi_1(X, f(y_o)) \to \pi_1(f, y_o)$ and those that are not.

The latter correspond to the path-components of Y, and, if the condition of (iii) holds true, Y consists of finitely many path-components only. Choose a base point y_o whose path-component will be denoted by λ_o and a family $\{y_\lambda : \lambda \in \pi_0(Y, y_o) \backslash \{\lambda_o\}\}$ of points such that $y_\lambda \in \lambda$ for every $\lambda \in \pi_0(Y, y_o) \backslash \{\lambda_o\}$. For each such λ, attach a 1-cell with boundary $\{y_o, y_\lambda\}$ to the space Y and extend the map f over these 1-cells by choosing a path in X from $f(y_o)$ to $f(y_\lambda)$.

Thus, one may assume Y to be path-connected. The function $\pi_1(X, f(y_o)) \to \pi_1(f, y_o)$ is surjective, and so the elements of $\pi_1(f, y_o)$ can be represented by loops in X. For each such element γ, choose a representative $f_\gamma : S^1_\gamma = S^1 \to X$. Then, take Y' as the wedge product of (Y, y_o) and the bouquet of the circles (S^1_γ, e_0), and define f' as the wedge of all the maps f, f_γ. □

Lemma 2.7.3 *Let* X *be a path-connected space and let* $f : Y \to X$ *be a* 1-*connected map. Then there is a map* $f' : Y' \to X$ *inducing an isomorphism of fundamental groups such that*:

(i) Y' *is obtained from* Y *by an adjunction of* 2-*cells*;

(ii) $f' | Y = f$; *and*

(iii) $Y' \backslash Y$ *has finitely many path components* (*i.e.*, $Y' \backslash Y$ *consists of finitely many open* 2-*cells only) if* $\ker(f_1)$ *is finitely generated.*

Remark f_1 denotes the induced homomorphism $\pi_1(Y, y_o) \to \pi_1(X, f(y_o))$ obtained after the choice of a base point $y_o \in Y$. Moreover, '$\ker(f_1)$ finitely generated' means that there is a finite set $\Gamma \subset \pi_1(Y, y_o)$ such that $\ker(f_1)$ is the smallest normal subgroup of $\pi_1(Y, y_o)$ containing Γ; it does *not* mean that $\ker(f_1)$ is finitely generated as a group.

Proof of Lemma 2.7.3 Choose a base point $y_o \in Y$. Since the map f is assumed to be 1-connected, the homomorphism f_1 is an epimorphism. Let $\{a_\lambda\}$ be a family of based maps $a_\lambda : S^1_\lambda = S^1 \to Y$ whose homotopy classes generate the kernel of f_1. Because a composition $f \circ a_\lambda$ represents the neutral element of $\pi_1(X, f(y_o))$ one can extend it to a map $b_\lambda : B^2 \to X$. Now use the maps a_λ to attach 2-cells to Y, thus obtaining a space Y', and the maps b_λ to extend the map f over the new 2-cells, which yields a map $f' : Y' \to X$.

Let $\bar{\imath} : Y \to Y'$ denote the inclusion. Turn to the induced homomorphisms between the corresponding fundamental groups. Firstly, f'_1 is an epimorphism, since its composition with $\bar{\imath}_1$ yields the epimorphism f_1. Secondly, the homomorphism $\bar{\imath}_1$ itself is an epimorphism, since the pair (Y', Y) is 1-connected (Lemma 2.4.3); this implies $\ker(\bar{\imath}_1) \subset \ker(f_1)$. On the other hand, the characteristic maps of the attached cells allow to deform the compositions $\bar{\imath} \circ a_\lambda$ into constant maps (within Y'); therefore, $\ker(\bar{\imath}_1)$ contains a system of generators of $\ker(f_1)$. Thus, $\ker(\bar{\imath}_1) = \ker(f_1)$, and consequently f'_1 is an isomorphism. □

Lemma 2.7.4 *Let Y, X be path-connected spaces and let $f : Y \to X$ be a map that induces an isomorphism of fundamental groups. Then, there is a 2-connected map $f' : Y' \to X$ such that:*

(i) *Y' is obtained from Y by an adjunction of 2-cells;*

(ii) *$f' | Y = f$; and*

(iii) *$Y' \setminus Y$ has finitely many path components* (i.e., *$Y' \setminus Y$ consists of finitely many open 2-cells only*) if *$\pi_2(f, y_o)$ is a finitely generated Λ-module, for any choice of a base point y_o.*

($\Lambda = \mathbf{Z}\pi_1(X, f(y_o))$ denotes the integral group ring of the fundamental group of the based space $(X, f(y_o))$.)

Proof Let $\{(b_\lambda, a_\lambda)\}$ be a family of representatives for a system of Λ-generators for $\pi_2(f, y_o)$. As in the preceding proof, use the maps a_λ as attaching maps to get the space Y' and the maps b_λ to extend the given map f to a map $f' : Y' \to X$. Again, let $\bar{\imath} : Y \to Y'$ denote the inclusion. The same argument as before shows that the induced homomorphism f'_1 is an epimorphism; thus, the map f' is at least 1-connected.

Now look at the exact homotopy sequence of the pair $(f', \bar{\imath})$. Its essential part is

$$\pi_2(\bar{\imath}, y_o) \to \pi_2(f, y_o) \to \pi_2(f', y_o) \to \pi_1(\bar{\imath}, y_o).$$

The construction of the space Y' and the map f' shows that the image of the homomorphism on the left contains a set of generators of its codomain; thus, it is an epimorphism. On the other hand, the right end of the display is trivial since it comes from an attaching of 2-cells (once more Lemma 2.4.3). Now the exactness forces $\pi_2(f', y_o)$ to be also trivial. Thus, f' is 2-connected. □

Lemma 2.7.5 *Let Y, X be path-connected spaces and let $f : Y \to X$ be an $(n-1)$-connected map, $n \geqslant 3$. Then there is an n-connected map $f' : Y' \to X$ such that:*

(i) *Y' is obtained from Y by an adjunction of n-cells;*

(ii) *$f' | Y = f$; and*

(iii) *$Y' \setminus Y$ has finitely many path components (i.e., $Y' \setminus Y$ consists of finitely many open n-cells only) if $\pi_n(f, y_0)$ is a finitely generated Λ-module, for any choice of a base point y_0.*

(The assumption here automatically implies that the map f induces an isomorphism of fundamental groups.)

Proof Construct the space Y' and the map f' as in the preceding proofs. Once more, let $\bar{\imath} : Y \to Y'$ denote the inclusion. Then, for $k < n-1$, the induced homomorphisms $\bar{\imath}_K$ are isomorphisms, since the pair (Y', Y) is $(n-1)$-connected (Lemma 2.4.3). By assumption, the same holds for the homomorphisms f_k. Thus, the equation $f = f' \circ \bar{\imath}$ yields that also the homomorphisms f'_k are isomorphisms. Again by assumption, the homomorphism f_{n-1} is an epimorphism, and so is the homomorphism f'_{n-1}. Therefore, the map f' is at least $(n-1)$-connected. The same considerations on the exact homotopy sequence of the pair $(f', \bar{\imath})$ as before show that the map f' is also n-connected, as desired. □

Addendum 2.7.6 *If the space Y considered in the statements of the Lemmas 2.7.1 to 2.7.5 is provided with a CW-structure, then the spaces Y' may be constructed in such a way to get CW-complexes containing the given ones as subcomplexes.*

Proof The maps used to attach the necessary cells can be taken as cellular (use the cellular approximation theorem 2.4.11 and the homotopy extension property of the pairs (B^n, S^{n-1})), thus giving CW-structures to the spaces Y' (see Theorem 2.3.1). □

These considerations are summed up in the following statement:

Theorem 2.7.7 *Let Y be a CW-complex, X be a space and $f : Y \to X$ be a map. Then, there are a CW-complex Y' containing Y as a subcomplex and a map $f' : Y' \to X$ which extends f and is a weak homotopy equivalence.*

Proof As described in the previous results, construct CW-complexes $Y^{(n)}$ containing $Y^{(n-1)}$ as subcomplexes and n-connected extensions $f^{(n)} : Y^{(n)} \to X$ of f (starting with $Y^{(-1)} = Y$). Define Y' to be the union space of the expanding sequence $\{Y^{(n)}\}$ and take $f' : Y' \to X$ as the unique map induced by the sequence $\{f^{(n)}\}$. □

Taking $Y = \varnothing$ in the preceding theorem, one obtains:

Corollary 2.7.8 *Any space is the codomain of a weak homotopy equivalence whose domain is a CW-complex.* □

The assumption that Y should be a CW-complex in Theorem 2.7.7 is needed only in order to obtain Y' as a CW-complex again. Dropping this assumption and taking X as a singleton space yields:

Corollary 2.7.9 *Any space can be enlarged by means of cell adjunctions to a weakly contractible space, i.e. space with only trivial homotopy groups.* □

Exercises

1. Let π be a group. Construct a based CW-complex with fundamental group π and vanishing higher homotopy groups (*Eilenberg–MacLane space of type* $(\pi, 1)$).
2. Let π be an abelian group and $n > 1$. Construct a based CW-complex X with $\pi_n(X) = \pi$ and $\pi_k(X) = 0$, for all $k \neq n$ (*Eilenberg–MacLane space of type* (π, n)).
3. Let $\{\pi_n : n \in \mathbf{N}\}$ be a sequence of groups such that, for every $n \geqslant 2$, π_n is abelian and provided with an action of π_1. Construct a based CW-complex whose homotopy groups are the given ones with the prescribed action of the fundamental group. Observe that this CW-complex may be taken as locally finite if all the groups are countable. (Whitehead, 1949b)

Notes to Chapter 2

The example of two CW-complexes whose cartesian product fails to be a CW-complex (see Section 2.2, Example 2) is due to Dowker (1952). The product of two CW-complexes, one of them locally finite (see Remark before Proposition 2.2.3), was already handled by J. H. C. Whitehead; the case of two countable factors can be found in Milnor (1956).

The technique of elementary expansions (see Propositions 2.3.5, 2.3.6 and Lemma 2.4.14) is another of J. H. C. Whitehead's masterful contributions to combinatorial topology; it was actually started long before the CW-theory itself (Whitehead, 1939, 1950). The cellular approximation theorem (see Theorem 2.4.11) is already contained in Whitehead's basic paper (Whitehead, 1949a); the proof given here relies on the expositions in Schubert (1964, 1968) – exploiting the idea of '*Freimachen eines Punktes*' (see Remark after the proof of Lemma 2.4.3) – and Brown (1988), where it is credited to ideas of J. F. Adams. The classical approach to the fundamental group of simplicial complexes by means of the 'edge path group', due to Poincaré (1895), was, with time, carried over to CW-complexes; for special CW-complexes, one could consult Schubert (1964, 1968), and, for the general case, Massey (1984). The ideas about increasing the connectivity of maps were first published in Wall (1965), where credit is given to unpublished work of J. Milnor.

3

Combinatorial complexes

3.1 Geometric simplices and cubes

While balls and spheres may be thought of as made out of rubber, and therefore as easily deformable, simplices form the solid bricks of the spaces under consideration. Recall, first, that a finite family $\{s_0, \ldots, s_k\}$ of 'points' in $\mathbf{R}^{n+1}$ is *affinely independent* iff the family $\{s_1 - s_0, \ldots, s_k - s_0\}$ of 'vectors' is linearly independent. A (*geometric*) *simplex* in $\mathbf{R}^{n+1}$ is simply the convex hull

$$\Delta = H(\{s_0, \ldots, s_k\})$$

of an affinely independent family $\{s_0, \ldots, s_k\}$, whose members are called *vertices* of the simplex. A simplex Δ with $k+1$ vertices will be given the specific name of *k-simplex*.

Example 1 Define a binary, reflexive and antisymmetric relation R on $\mathbf{Z}^{n+1} \subset \mathbf{R}^{n+1}$ by

$$sRs' \Leftrightarrow s_i \leqslant s'_i \leqslant s_i + 1, \qquad i = 0, 1, \ldots, n.$$

Every subset of $\mathbf{Z}^{n+1}$ that is totally ordered with respect to the relation R is an affinely independent family of $\mathbf{R}^{n+1}$ whose convex hull is a geometric simplex in $\mathbf{R}^{n+1}$. □

A 0-simplex is a *point*, a 1-simplex is an *interval* or an *edge*, a 2-simplex is a *triangle* and a 3-simplex is a *tetrahedron*; sometimes it is also convenient to consider the empty set as a (-1)-simplex. Given a k-simplex Δ, any set L of vertices of Δ forms a *face* of the simplex by taking its convex hull $\Delta_+ = H(L)$, which is again a simplex; if L does not contain all vertices, one speaks of a *proper* face. Given a face Δ_+ of a simplex Δ, the vertices of Δ outside Δ_+ form the *complementary* face Δ_- to Δ_+; if Δ is a k-simplex and Δ_+ is an l-simplex, then Δ_- is a $(k-l-1)$-simplex.

The union of all proper faces of a simplex Δ is its boundary $\delta\Delta$; note that this combinatorial boundary is the topological boundary of Δ in its affine hull and it is the topological boundary in $\mathbf{R}^{n+1}$ only if $k = n+1$. The difference $\mathring{\Delta} = \Delta \backslash \delta\Delta$ is the *interior of* Δ, the *open simplex*; a point of Δ is an *interior point* if it belongs to $\mathring{\Delta}$.

If $s_0, \dots, s_k$ are the vertices of a simplex, its points can be uniquely described in the form

$$s = \sum_{i=0}^{k} t_i s_i,$$

with $t_i \in \mathbf{I}$ for all i and

$$\sum_{i=0}^{k} t_i = 1;$$

the real numbers t_i are called *barycentric coordinates* of the point s.

A point of a simplex Δ is an interior point if all its barycentric coordinates are different from zero. Every point $s \in \Delta$ determines a unique face of Δ, for which s is an interior point; hence, $\Delta = \bigsqcup (\Delta_+)^\circ$ where Δ_+ runs through all faces of Δ. Given a point $s \in \Delta$, the unique face containing s as an interior point is called the *carrier* of s and is denoted by Δ_s.

Let Δ_1 be a geometric simplex and let Δ_0 be a proper face of Δ_1. A *simplicial retraction from Δ_1 to Δ_0* is a retraction $\Delta_1 \to \Delta_0$ that is induced by the composition of a linear map and a translation mapping vertices onto vertices. Simplicial retractions performed within a larger simplex do not really alter the simplex.

Lemma 3.1.1 *Let Δ be a simplex, let Δ_1 be a proper face of Δ, let Δ_0 be a proper face of Δ_1 and let $\phi : \Delta_1 \to \Delta_0$ be a simplicial retraction. Let the space X be obtained from Δ_0 by attaching Δ via ϕ. Then there is a homeomorphism $h : X \to \Delta$ extending the inclusion $\Delta_0 \subset \Delta$.*

Proof Take $n + 1 = \dim \Delta$ and $m + 1 = \dim \Delta_1$; without loss of generality, assume $\dim \Delta_0 = m$. All of the following considerations are staged in the one-point compactification $\mathbf{R}^{n+1}_\infty$ of

$$\mathbf{R}^{n+1}_+ = \{(t_0, \dots, t_n) : t_n \geqslant 0\}.$$

Choose an n-dimensional face Δ_2 of Δ containing Δ_1 and make the identifications:

$$\begin{aligned}
&\Delta_0 = H(\{\mathbf{0}, e_0, e_1, \dots, e_{m-1}\}),\\
&\Delta_1 = H(\{\mathbf{0}, e_0, e_1, \dots, e_m\}),\\
&\Delta_2 = H(\{\mathbf{0}, e_0, e_1, \dots, e_{n-1}\}),\\
&\Delta = \mathbf{R}^{n+1}_\infty,\\
&\phi(t_0, \dots, t_m) = (t_0, \dots, t_{m-1}),\\
&\text{for all } (t_0, \dots, t_m) \in \Delta_1.
\end{aligned}$$

Define a map $\psi : \mathbf{R}^{n+1}_+ \to \Delta_0$ by assigning to each point $t \in \mathbf{R}^{n+1}_+$ the nearest point of Δ_0. The existence of such a point follows from the compactness of Δ_0, its uniqueness from convexity and Pythagoras' theorem; the continuity of ψ is due to the fact that it is a contraction; i.e.,

$$|\psi(t)-\psi(t')| \leqslant |t-t'|,$$

for all pairs $t, t' \in \mathbf{R}_+^{n+1}$. Note that $\psi|\Delta_1 = \phi$. Next define maps $f : \mathbf{R}_+^{n+1} \to \mathbf{R}_+$ by assigning to each point $t \in \mathbf{R}_+^{n+1}$ its distance from Δ_1 and $g : \mathbf{R}_+^{n+1} \backslash \Delta_0 \to I$ by taking

$$g(t) = \frac{f(t)}{|t-\psi(t)|}.$$

Next, take $h : \mathbf{R}_\infty^{n+1} \to \mathbf{R}_\infty^{n+1}$ given by

$$h(t) = \begin{cases} \psi(t) + g(t)\cdot(t-\psi(t)), & \text{if } t \in \mathbf{R}_+^{n+1} \backslash \Delta_0 \\ t, & \text{otherwise.} \end{cases}$$

The continuity of h has to be checked only near Δ_0 and at infinity. In the first case, one has again a contraction; in the second, note that $|t| \to \infty$ implies $g(t) \to 1$, and consequently $h(t) \to t$. Furthermore, one still retains the equality $h|\Delta_1 = \phi$. It remains to show that h maps $\mathbf{R}_+^{n+1} \backslash \Delta_1$ homeomorphically onto $\mathbf{R}_+^{n+1} \backslash \Delta_0$. This is a consequence of the following observations. Given a point $t \in \mathbf{R}_+^{n+1} \backslash \Delta_1$, the half line L starting at $\psi(t)$ in the direction of t is mapped surjectively onto itself by h. For the injectivity, assume t' to be another point on L outside Δ_1, and suppose $h(t') = h(t)$; then $t' - \psi(t')$ is a real multiple of $t - \psi(t)$ and, consequently, $f(t') = f(t)$, implying $t' = t$. □

An *affine embedding* is a map $\mathbf{R}^{n+1} \to \mathbf{R}^{m+1}$, which is the composition of an injective linear map and a translation. Such maps preserve affine independence. Therefore the image of a simplex under an affine embedding is again a simplex.

Recall that for any subset S of a metric space the *diameter* of S – notation: diam(S) – is defined to be the supremum of the set of the distances between any two points of S. The diameter of a geometric simplex can be easily computed.

Lemma 3.1.2 *The diameter of a geometric simplex is equal to the maximum of the lengths of all its edges.*

Proof The assertion ensues from a double application of the following fact. Let $\hat{s}$ be a point of a simplex Δ. Then, for any point $s \in \Delta, s \neq \hat{s}$, there exists a vertex of Δ which is not nearer to $\hat{s}$ then to s. To see this, let $s_0, \ldots, s_k$ denote the vertices of Δ and let $t_0, \ldots, t_k$ be the respective barycentric coordinates of s. Without loss of generality, we may assume $s \neq s_0, \ldots, s_k$. Then it is enough to show that there is a vertex $s_j \neq \hat{s}$ such that the (possibly degenerate) triangle $\hat{s}ss_j$ has a non-acute angle at the vertex s, i.e., such that the scalar product $(\hat{s}-s)\cdot(s_j-s)$ is not positive. But this follows from the equation

$$\sum_{i=0}^{k} t_i(\hat{s}-s)\cdot(s_i-s)=(\hat{s}-s)\cdot(s-s)=0$$

since all t_i's are non-negative and some of them are positive. □

Every k-simplex has a distinguished interior point, its *barycentre*; that is, the point with all barycentric coordinates equal to $1/(k+1)$. In the next section, the notion of 'barycentric subdivision' will be discussed; one of its crucial properties comes out of the following fact.

Lemma 3.1.3 *Let Δ_0, Δ_1 be non-empty faces of a k-simplex Δ with $\Delta_0 \subset \Delta_1$ and let $\boldsymbol{b}_0, \boldsymbol{b}_1$ denote their respective barycentres. Then, the distance of $\boldsymbol{b}_0$ and $\boldsymbol{b}_1$ is smaller than or equal to $k/(k+1)\cdot \operatorname{diam}(\Delta)$.*

Proof Let $\boldsymbol{b}_-$ denote the barycentre of the complementary face Δ_- to Δ_0 in Δ_1. Then, $\boldsymbol{b}_1$ can be represented in the form

$$\boldsymbol{b}_1=\frac{1+\dim\Delta_-}{1+\dim\Delta_1}\boldsymbol{b}_- + \frac{1+\dim\Delta_0}{1+\dim\Delta_1}\boldsymbol{b}_0,$$

which leads to

$$\boldsymbol{b}_1-\boldsymbol{b}_0=\frac{1+\dim\Delta_-}{1+\dim\Delta_1}(\boldsymbol{b}_- - \boldsymbol{b}_0).$$

The result now follows from

$$\frac{1+\dim\Delta_-}{1+\dim\Delta_1}\leqslant\frac{\dim\Delta_1}{1+\dim\Delta_1}\leqslant\frac{k}{k+1}$$

(since $\Delta_0 \neq \varnothing$ implies $\dim\Delta_- < \dim\Delta_1$) and

$$|\boldsymbol{b}_- - \boldsymbol{b}_0|\leqslant \dim\Delta.$$ □

In Euclidean spaces, there is a more specific concept of cone than that introduced in Section A.4, after Corollary A.4.16. A (*Euclidean*) *cone* in $\mathbf{R}^{n+1}$ is a triple $(C, B, \boldsymbol{p})$, consisting of subsets $C, B \subset \mathbf{R}^{n+1}$ and a point $\boldsymbol{p} \in \mathbf{R}^{n+1}$ such that the map $B\times I\to\mathbf{R}^{n+1}, (s,t)\mapsto(1-t)s+t\boldsymbol{p}$, induces a homeomorphism $B\times I/B\times\{0\}\to C$. If the triple $(C, B, \boldsymbol{p})$ is a cone then the set C is its *global set*, the set B is its *base* and the point $\boldsymbol{p}$ is its *peak*. By abuse of language, a subset C of $\mathbf{R}^{n+1}$ is also called a cone if it is the global set of a cone. Conversely, if a set B and a point $\boldsymbol{p}$ are such that they may form the base and the peak of a cone respectively, then the corresponding global set is uniquely determined. In this case, the point $\boldsymbol{p}$ and the set B are said *to form a cone* whose global set is denoted by $\boldsymbol{p}B$. Note that the base of a cone might be empty, i.e., one-point sets are also considered as cones: $\boldsymbol{p}=\boldsymbol{p}\varnothing$ for all points $\boldsymbol{p}\in\mathbf{R}^{n+1}$.

Example 2 Let Δ be a geometric simplex, $\boldsymbol{s}_0$ a vertex of Δ and Δ_0 the complementary face of Δ. Then the triple $(\Delta, \Delta_0, \boldsymbol{s}_0)$ is a cone: $\Delta=\boldsymbol{s}_0\Delta_0$. □

Example 3 Let Δ be a geometric simplex and $\boldsymbol{s}$ an interior point of Δ. Then the triple $(\Delta, \delta\Delta, \boldsymbol{s})$ is a cone: $\Delta=\boldsymbol{s}(\delta\Delta)$. □

This is a special case of a more general situation.

Example 4 Let C be any compact convex set in $\mathbf{R}^{n+1}$, let B denote its boundary (in its affine hull) and let $\boldsymbol{p}$ be an interior point of C. Then, the triple $(C, B, \boldsymbol{p})$ is a cone. □

Example 5 Let Δ be a k-simplex in $\mathbf{R}^{n+1}$ and let the point $\boldsymbol{p}\in\mathbf{R}^{n+1}$ be affinely independent of the vertices of Δ. Then, $\boldsymbol{p}$ and Δ form the cone $\boldsymbol{p}\Delta$ which is a $(k+1)$-simplex. □

The *standard-n-simplex* in $\mathbf{R}^{n+1}$ is the n-simplex

$$\Delta^n=\left\{(t_0,\ldots,t_n) : t_i\in I \text{ for all } i, \sum_{i=0}^{n} t_i=1\right\},$$

whose vertices are the vectors $\boldsymbol{e}_i, i=0,1,\ldots,n$, which form the canonical basis of $\mathbf{R}^{n+1}$. Every k-simplex Δ is homeomorphic to the standard-k-simplex Δ^k; if $\{\boldsymbol{s}_0,\ldots,\boldsymbol{s}_k\}$ is the set of vertices of Δ, the linear map

$$\mathbf{R}^{k+1}\to\mathbf{R}^{n+1}, \quad \boldsymbol{e}_i\mapsto\boldsymbol{s}_i$$

induces a homeomorphism $\Delta^k\to\Delta$.

Simplices are balls; suitable homeomorphisms can be easily constructed (Exercise). For the needs of homology a family of maps $\psi^n : \Delta^n\to B^n$ satisfying a certain coherence condition is presented here. The construction of this maps is done inductively. The spaces Δ^0 and B^0 are singleton spaces; so there exists one and only one map $\Delta^0\to B^0$ which is taken as ψ^0. Now assume ψ^n to be given and define a map $\psi' : I\times\Delta^n\to\mathbf{R}^{n+1}$ by taking

$$\psi(t,s)=t\boldsymbol{e}_0+(1-t)b^n(\psi^n(s)).$$

This map factors through the identification

$$I\times\Delta^n\to\Delta^{n+1}, \quad (t,s)\mapsto(t,(1-t)s)$$

and its image is contained in the ball B^{n+1}. Thus, ψ' induces a map $\Delta^{n+1}\to B^{n+1}$ which is taken as ψ^{n+1}. The properties of these map are listed in the following statement.

Lemma 3.1.4 *For all $n \in \mathbf{N}$*

(i) ψ^n maps the interior of Δ^n homeomorphically onto the interior of B^n;

(ii) ψ^{n+1} maps the boundary of Δ^{n+1} onto S^n;

(iii) ψ^{n+1} maps the interior of the face opposite to the vertex $\mathbf{e}_0$ in Δ^{n+1} homeomorphically onto $S^n \backslash \{\mathbf{e}_0\}$;

(iv) ψ^{n+1} maps the faces of Δ^{n+1} containing the vertex $\mathbf{e}_0$ constantly onto the base point $\mathbf{e}_0$;

(v) the diagram

$$\begin{array}{ccccc} B^n & \xrightarrow{b^n} & S^n & \hookrightarrow & B^{n+1} \\ {\scriptstyle \psi^n}\uparrow & & & & \uparrow{\scriptstyle \psi^{n+1}} \\ \Delta^n & & \hookrightarrow & & \Delta^{n+1} \\ & s & \mapsto & (0, s) & \end{array}$$

commutes.

Proof by induction and direct computation. □

The following remark will be useful to the readers familiar with the basic facts of singular homology. These are briefly reviewed in Section A.7.

Remark It follows from (ii) that the 'singular simplex' ψ^{n+1} represents a cycle and thus a homology class (indeed a generator) of the relative homology group $H_{n+1}(B^{n+1}, S^n)$ while the composition $b^n \circ \psi^n$ represents a generator of $H_n(S^n)$. The diagram shows that these generators are related in a neat way, that is, they are 'coherent'. □

The considerations on single simplices are closed with a theorem which, although not related to other subjects in this book, has a certain importance and deserves to be mentioned.

Theorem 3.1.5 *Let $\Delta_0 \supset \Delta_1 \supset \cdots \supset \Delta_j \supset \cdots$ be a decreasing sequence of simplices in $\mathbf{R}^{n+1}$. Then,*

$$\Delta = \bigcap_{j=0}^{\infty} \Delta_j$$

is a simplex in $\mathbf{R}^{n+1}$.

Proof The sequence $\{\dim \Delta_j\}$ of natural numbers is decreasing; thus it becomes stationary. Therefore, one may assume $\dim \Delta_j = n+1$ for all $j \in \mathbf{N}$. Let $\mathbf{s}_{j,0}, \ldots, \mathbf{s}_{j,n+1}$ denote the vertices of Δ_j. The sequence

$$\{\{\mathbf{s}_{j,0}, \ldots, \mathbf{s}_{j,n+1}\}\}_{j \in \mathbf{N}}$$

contains a convergent subsequence; thus, assume that this sequence is itself convergent. If one takes, for $0 \leqslant i \leqslant n+1$,

$$s_i = \lim_{j\to\infty} s_{j,i},$$

then Δ is the convex hull of the points s_i. Clearly, all s_i are contained in the intersection Δ; the converse requires a preliminary consideration. Let $s \in \Delta$ be given; then there is, for every $j \in \mathbf{N}$, a representation

$$s = \sum_{i=0}^{n+1} t_{j,i} s_{j,i},$$

with $t_{j,i} \in \boldsymbol{I}$ for all i and

$$\sum_{i=0}^{n+1} t_{j,i} = 1.$$

Let $(t_0, \ldots, t_{n+1})$ be a cluster point of the sequence $\{(t_{j,0}, \ldots, t_{j,n+1})\}$; then, $\sum_{i=0}^{n+1} t_i s_i$ is a convex combination of the s_i representing s.

Now assume the elements of the family $\{s_0, \ldots, s_{n+1}\}$ to be labelled in such a way that, for some $k \in \mathbf{N}$ with $0 \leqslant k \leqslant n+1$, the family $\{s_0, \ldots, s_k\}$ is a maximal *convexly* independent subfamily. It is necessary to show that the family $\{s_0, \ldots, s_k\}$ is *affinely* independent.

To this end, define maps $\phi_j : \Delta_j \to \Delta^{k+1}$ by taking the restrictions of the affine maps which send $s_{j,i}$ to e_h, if $s_j = s_h$ with $0 \leqslant h \leqslant k$, and to e_{k+1} otherwise. The main step of the argument consists in proving that

$$\lim_{j\to\infty} \phi_j(s_h) = e_h$$

for $0 \leqslant h \leqslant k$. Fix h with $0 \leqslant h \leqslant k$, and denote by Δ_{j+}, Δ_{j-} the face of Δ_j spanned by vertices $s_{j,i}$ with $s_j = s_h$ and its complementary face, respectively. Then, for every $j \in \mathbf{N}$, there is a unique representation

$$s_h = t'_j s'_j + t''_j s''_j$$

with $(t'_j, t''_j) \in \Delta^1, s'_j \in \Delta_{j+}$ and $s''_j \in \Delta_{j-}$. If $j \to \infty$, the simplices Δ_{j+} contract to the point s_h, which implies

$$\lim |s_h - s'_j| = 0.$$

On the other hand, the simplices Δ_{j-} converge to the convex hull of all $s_i \neq s_h$; since s_h is no convex combination of these s_i's, one finds

$$\liminf |s''_j - s'_j| > 0.$$

This implies

$$0 = \lim_{j\to\infty} \frac{|s_h - s'_j|}{|s''_j - s'_j|} = \lim_{j\to\infty} t''_j = \lim_{j\to\infty} (1 - t'_j),$$

and so

$$1 = \lim_{j\to\infty} t'_j,$$

which gives finally

$$\lim_{j\to\infty} \phi_j(s_h) = \lim_{j\to\infty} (t'_j\phi_j(s'_j) + t''_j\phi_j(s''_j)) = \left(\lim_{j\to\infty} t'_j\right) \cdot e_h = e_h.$$

Now assume that one of the $s_h, 0 \leqslant h \leqslant k$, say s_k, is affinely dependent of the remaining, i.e.,

$$s_k = \sum_{i=0}^{k-1} t_i s_i,$$

with

$$\sum_{i=0}^{k-1} t_i = 1.$$

Then, it follows that

$$\phi_j(s_k) = \sum_{i=0}^{k-1} t_i \phi_j(s_i)$$

and thus,

$$e_k = \lim_{j\to\infty} \phi_j(s_k) = \sum_{i=0}^{k-1} t_i \lim_{j\to\infty} \phi_j(s_i) = \sum_{i=0}^{k-1} t_i e_i,$$

contradicting the affine independence of the family $\{e_0, \ldots, e_k\}$. □

There is another type of solid brick in these constructions which is needed only in a very standardized form. Given a point $s = (s_0, \ldots, s_n) \in \mathbf{R}^{n+1}$ and a real number $\varepsilon > 0$, the set

$$W(s; \varepsilon) = \{x \in \mathbf{R}^{n+1} : |x_i - s_i| \leqslant \varepsilon \quad \text{for all } i\}$$

is called the $(n+1)$-*cube with centre s and edge length* 2ε. A subset W of $\mathbf{R}^{n+1}$ is a *cube* if it is of the form $W = W(s; \varepsilon)$ for a point $s \in \mathbf{R}^{n+1}$ and a real number $\varepsilon > 0$; in this case, one also says that the cube W is *centred at s*. Given a cube $W = W(s; \varepsilon)$, a subset of the form

$$W_{i+} = \{x \in W : t_i = s_i + \varepsilon\}$$

or

$$W_{i-} = \{x \in W : t_i = s_i - \varepsilon\}$$

(i fixed) is called an (*n-dimensional*) *face* of W; note that it can be viewed as the image of a cube in $\mathbf{R}^n$ with respect to an affine embedding $\mathbf{R}^n \to \mathbf{R}^{n+1}$. The union of all n-dimensional faces of a cube is its boundary (in the topological sense, as a subspace of $\mathbf{R}^{n+1}$). A cube can be considered as a cone with its boundary as base and its centre as peak. Cubes are balls: the composition of the translation sending s to the origin and the multiplication by the scalar $1/\varepsilon$ induces a homeomorphism $W(s; \varepsilon) \to (B^1)^{n+1}$; but $(B^1)^{n+1}$ is an $(n+1)$-ball (see Proposition 1.0.2). Consequently, the boundary of a cube in $\mathbf{R}^{n+1}$ is an n-sphere. Finally, observe that cubes may be *arbitrarily small*, i.e., given a point s in an open set $U \subset \mathbf{R}^{n+1}$ there exists a cube centred at s and completely contained in U.

Exercise

Describe homeomorphisms $\Psi^n : \Delta^n \to B^n$ by means of coordinate functions such that there is a based homotopy of pairs $(\Delta^n, \delta\Delta^n) \to (B^n, S^{n-1})$ connecting these homeomorphisms and the standard maps ψ^n of Lemma 3.1.4.

3.2 Euclidean complexes

The previous section was devoted to an analysis of single simplices. Conglomerates of these bricks are now going to be studied. A *Euclidean complex* is a set K of simplices in a fixed $\mathbf{R}^{n+1}$, such that

(1) K contains all faces of all members of K;
(2) the intersection of two members of K is a face of both; and
(3) every member of K has a neighbourhood which intersects only finitely many members of K.

A set of simplices satisfying only the second and the third condition *generates* a Euclidean complex; it is obtained by adding to the given set all faces of its elements.

The vertices of the simplices of a Euclidean complex K will be called *vertices* of K. The third condition in the definition of Euclidean complexes implies that the set $\Lambda = \Lambda_K$ of vertices of K forms a discrete subspace of $\mathbf{R}^{n+1}$. Since a simplex is uniquely determined by the set of its vertices, there is an injective function from K to the set of finite subsets of Λ associating to each simplex the set of its vertices. The image of K with respect to this function is called the *vertex scheme* of K.

The union of all the simplices of a Euclidean complex K – taken in $\mathbf{R}^{n+1}$ – is the *underlying polyhedron*; it will be denoted by $|K|$. More generally, a (*Euclidean*) *polyhedron* is a subspace of some $\mathbf{R}^{n+1}$, which is the underlying polyhedron of some Euclidean complex. Given a polyhedron P, a Euclidean complex K with $|K| = P$ is called a *simplicial decomposition* of P.

Example 1 The set of *all* faces of a simplex $\Delta \subset \mathbf{R}^{n+1}$ forms a Euclidean complex whose underlying polyhedron is the simplex itself and which – by abuse of notation – will also be denoted by Δ. Again, the set of all *proper* faces of Δ is a Euclidean complex whose underlying polyhedron is the boundary $\delta\Delta$ of Δ and which is also denoted by $\delta\Delta$. □

Example 2 In $\mathbf{R}^{n+1}$, take all the simplices described in Example 1 of Section 3.1. They form a Euclidean complex whose underlying polyhedron is the space $\mathbf{R}^{n+1}$ itself. □

Clearly, a Euclidean complex is *finite* if it consists of finitely many simplices only; in this case, the third condition in the definition of Euclidean complexes is meaningless. Examples are the Euclidean complexes Δ and $\delta\Delta$, derived as in Example 1 from a geometric simplex Δ in $\mathbf{R}^{n+1}$. The underlying polyhedron of a finite Euclidean complex is a finite union of compact sets and therefore compact.

Lemma 3.2.1 *Let K be a Euclidean complex. Then every point $s \in |K|$ is*
(i) *contained in only finitely many simplices of K; and*
(ii) *an interior point of exactly one simplex of K.*

Proof Let a point $s \in |K|$ be given. Among the simplices of K containing s, let Δ be one with minimal dimension. To prove (i), take a neighbourhood of Δ that meets only finitely many simplices of K. Clearly, any simplex of K containing s meets this neighbourhood; so there can be only finitely many of those.

If s would not be an interior point of Δ it would be contained in a proper face of Δ, i.e., in a lower-dimensional member of K, in contradiction to the choice of Δ. To obtain uniqueness, it suffices to observe that two simplices of a Euclidean complex having an interior point in common must be equal. □

Given a Euclidean complex K and a point $s \in |K|$, the unique simplex of K that contains s as an interior point is called the *carrier* of s in K and is denoted by Δ_s.

A Euclidean complex L is a *subcomplex* of the Euclidean complex K if $L \subset K$, i.e., if every simplex of L belongs to K. The *k-skeleton* K^k of K is the subcomplex consisting of all simplices $\Delta \in K$ with $\dim \Delta \leqslant k, 0 \leqslant k$; if $n+1$ is the dimension of the ambient Euclidean space, then one has $K^k = K$ for every $k \geqslant n+1$. The boundary of every $(k+1)$-simplex of K can be considered as a subcomplex of K^k. In contrast to the situation for CW-complexes (see Corollary 1.4.5), the following result is evident.

Proposition 3.2.2 *An arbitrary union or intersection of subcomplexes is a subcomplex.* □

Given a Euclidean complex K and a subset $L \subset K$, the intersection $K(L)$ of all subcomplexes of K containing L is the *subcomplex generated by* L; $K(L)$ consists of all faces of all the members of L.

Since any simplex has only finitely many faces, this implies:

Lemma 3.2.3 *A finite subset of a Euclidean complex generates a finite subcomplex.* □

The image of a polyhedron under an affine embedding is again a polyhedron; more precisely:

Lemma 3.2.4 *Let K be a Euclidean complex in $\mathbf{R}^{n+1}$ and let $f : \mathbf{R}^{n+1} \to \mathbf{R}^{m+1}$ be an affine embedding. Then, $\{f(\Delta) : \Delta \in K\}$ is a Euclidean complex in $\mathbf{R}^{m+1}$.*

Proof The set $\{f(\Delta) : \Delta \in K\}$ is a collection of simplices satisfying the axioms of a Euclidean complex. □

Corollary 3.2.5 *Given any simplex $\Delta \subset \mathbf{R}^{n+1}$, there exists a Euclidean complex K with $\Delta \in K$ and underlying polyhedron $\mathbf{R}^{n+1}$.*

Proof Let L denote the simplicial decomposition of $\mathbf{R}^{n+1}$ described in Example 2 and choose an affine homeomorphism $\mathbf{R}^{n+1} \to \mathbf{R}^{n+1}$, taking one simplex of L onto Δ. □

Corollary 3.2.6 *Cubes are polyhedra.*

Proof The described simplicial decomposition of $\mathbf{R}^{n+1}$ contains a subcomplex whose underlying polyhedron is the cube $(B^1)^{n+1}$. Any other cube can be obtained from this as the image under an affine homeomorphism $\mathbf{R}^{n+1} \to \mathbf{R}^{n+1}$. □

In order to make a connection between Euclidean complexes and CW-complexes the following fact is needed.

Lemma 3.2.7 *The underlying polyhedron of a Euclidean complex K is determined by the family of all the simplices of K.*

Proof Firstly, assume K to be finite and let V be a subset of $|K|$, which intersects every simplex Δ of K in a closed set. Since a simplex is closed in $\mathbf{R}^{n+1}$ and only finitely many simplices are under consideration, V is a closed subset of $\mathbf{R}^{n+1}$ and thus of $|K|$.

Secondly, let K be infinite. Let U be a subset of $|K|$ which intersects every simplex Δ of K in an open set. Choose for every Δ an open set U_Δ in $|K|$, which meets only finitely many simplices of K. Let L_Δ denote the finite subcomplex of K which is generated by these simplices. Then, in view of the part of the assertion already proved, $U \cap |L_\Delta|$ is open in $|L_\Delta|$. Now, $U \cap U_\Delta = U \cap |L_\Delta| \cap U_\Delta$ is open in U_Δ and thus open in $|K|$. Therefore, U being the union of the open sets $U \cap U_\Delta$, it is open in $|K|$. □

Theorem 3.2.8 *A simplicial decomposition provides a Euclidean polyhedron with the structure of a finite-dimensional, countable, locally finite and regular CW-complex.*

Proof Let K be a Euclidean complex in $\mathbf{R}^{n+1}$. Its underlying polyhedron $|K|$ is Hausdorff as a subspace of $\mathbf{R}^{n+1}$. The interiors of the simplices of K provide $|K|$ with a cell decomposition (see Lemma 3.2.1 (ii)) whose k-skeleton is just $|K^k|$. If Δ is any k-simplex in K, any homeomorphism $B^k \to \Delta$ (for the existence of such homeomorphisms use an appropriate modification of Lemma 3.1.4) maps the boundary sphere S^{k-1} of B^k into the boundary $\delta\Delta$ of Δ, contained in the $(k-1)$-skeleton of this cell decomposition. Now, the choice of such a homeomorphism for each $\Delta \in K$ completes the structure of a cell complex for $|K|$. This structure is closure finite because a simplex has only finitely many faces and the space $|K|$ has the right topology (see Lemma 3.2.7). Thus, $|K|$ is a Whitehead complex, and, consequently, a CW-complex (see Theorem 1.6.3).

The CW-complex obtained, being embedded in the Euclidean space $\mathbf{R}^{n+1}$, is locally finite, countable and of dimension $\leqslant n+1$ (see Theorem 1.5.18 (ii)). Moreover, its closed cells are simplices, i.e., balls, and thus the CW-structure is regular. □

This is the theorem that brings into light the interplay between CW-complexes and Euclidean complexes.

Corollary 3.2.9 *Let K be a Euclidean complex. Then:*

(i) *a (CW-)subcomplex of $|K|$ is the underlying polyhedron of a (Euclidean) subcomplex of K;*

(ii) *K is finite iff $|K|$ is compact; and*

(iii) $\dim|K| = \min\{k \in \mathbf{N} : K^k = K\} = \max\{\dim\Delta : \Delta \in K\}$. □

Statement (ii) appears here as a consequence of the corresponding fact for CW-complexes (see Proposition 1.5.8); note that an easy direct proof also is possible.

In view of (iii), the dimension of a Euclidean complex K is defined as the dimension of its underlying polyhedron:

$$\dim K = \dim|K|.$$

Corollary 3.2.10 *All simplicial decompositions of a polyhedron have the same dimension.* □

Cubes and simplices are *ANRs* (see Proposition A.6.3). More generally:

Proposition 3.2.11 *A compact polyhedron is an ANR.*

Proof Given a compact polyhedron, choose for it a simplicial decomposition K which is a finite Euclidean complex (Corollary 3.2.9 (ii)). The proposition is proved by induction on the number k of simplices in K.

If $k=0$, then $|K|=\varnothing$ and the statement holds true. Assume the proposition to be correct for $k<m$. From a Euclidean complex with m simplices, take a simplex $\Delta \in K$ of maximal dimension. Then $\tilde{K}=K\backslash\{\Delta\}$ and $\delta\Delta$ have less than m simplices, and therefore, by the induction hypothesis, $|\tilde{K}|$ and $\delta\Delta$ are ANRs. Since also Δ is an ANR, one concludes finally that the union $|K|=|\tilde{K}|\bigsqcup_{\delta\Delta}\Delta$ is an ANR (see Proposition A.6.6).

□

Remark A certain converse statement is true: any compact ANR has the homotopy type of a compact polyhedron. This fact is beyond the scope of this book. □

Let $\boldsymbol{p}$ be a point in the underlying polyhedron $|K|$ of a Euclidean complex K. The finitely many simplices of K that contain the point $\boldsymbol{p}$ generate a finite subcomplex of K, the star of $\boldsymbol{p}$, denoted by $\mathrm{st}_K\boldsymbol{p}$; the simplices of $\mathrm{st}_K\boldsymbol{p}$ that do not contain $\boldsymbol{p}$ form a subcomplex of $\mathrm{st}_K\boldsymbol{p}$, called the *link* of $\boldsymbol{p}$ in K. The notions 'star' here and in Section 1.4 are related by

$$|\mathrm{st}_K\boldsymbol{p}|=\mathrm{St}(\{\boldsymbol{p}\}),$$

where that star St is taken in the CW-complex $|K|$. It follows that $|\mathrm{st}_K\boldsymbol{p}|$ is a compact neighbourhood of $\boldsymbol{p}$ in $|K|$. Unlike the stars of points in CW-complexes, the stars in Euclidean complexes are cones.

Proposition 3.2.12 *Let $\boldsymbol{p}$ be a point in the underlying polyhedron $|K|$ of a Euclidean complex K, and let L denote the link of $\boldsymbol{p}$ in K. Then the triple $(|\mathrm{st}_K\boldsymbol{p}|,|L|,\boldsymbol{p})$ is a cone.*

Proof One has to show that the map

$$f:|L|\times I\to\mathbf{R}^{n+1},(\boldsymbol{s},t)\mapsto t\boldsymbol{s}+(1-t)\boldsymbol{p}$$

has $|\mathrm{st}_K\boldsymbol{p}|$ as image and is injective on $|L|\times(I\backslash\{0\})$.

'image $f\subset|\mathrm{st}_K\boldsymbol{p}|$': take $(\boldsymbol{s},t)\in|L|\times I$. The carrier Δ_s of $\boldsymbol{s}$ in L is a face of a simplex $\Delta\in K$, which contains $\boldsymbol{p}$ (Lemma 3.2.1). Since Δ is convex, it contains the whole interval $\{t\boldsymbol{s}+(1-t)\boldsymbol{p}:t\in I\}$, and, in particular, the point $f(\boldsymbol{s},t)$.

'$|\mathrm{st}_K\boldsymbol{p}|\subset$ image f': clearly $\boldsymbol{p}\in$image f; take $\boldsymbol{s}\in|\mathrm{st}_K\boldsymbol{p}|\backslash\{\boldsymbol{p}\}$. If its carrier Δ_s belongs to L, then one has $\boldsymbol{s}'=f(\boldsymbol{s}',1)$. Otherwise, the array $\{\boldsymbol{p}+t'(\boldsymbol{s}'-\boldsymbol{p}):t'>0\}$ meets the boundary of $\Delta_{s'}$ in a point $\boldsymbol{s}=\boldsymbol{p}+t'(\boldsymbol{s}'-\boldsymbol{p})$ with $t'>1$, whose carrier belongs to L; then one has $\boldsymbol{s}'=f(\boldsymbol{s},1/t')$.

Injectivity: assume that $f(\boldsymbol{s}_0,t_0)=f(\boldsymbol{s}_1,t_1)$, and, without loss of generality, that $0 < t_0 \leqslant t_1$. Then compute $s_1 = f(s_0, t_0/t_1)$. The carrier of s_0 is a face of a simplex Δ in K which contains p (Lemma 3.2.1). Choose Δ so that it has minimal dimension. Then the points p and s_0 belong to Δ but no proper face of Δ contains both of them. Therefore the open interval $\{ts + (1-t)p : 0 < t < 1\}$ is completely contained in the interior of Δ and does not meet $|L|$. In view of $s_1 \in |L|$, this implies $t_0/t_1 = 1$, i.e., $t_0 = t_1$, and, consequently, $s_0 = s_1$. □

From the general theory of CW-complexes, one derives that a Euclidean complex K' is a *subdivision* of a Euclidean complex K if K' is a refinement of K; i.e., if every simplex of K' is completely contained in a simplex of K (see Section 2.3). If K' is a subdivision of K, and L is a subcomplex of K, then the simplices of K' that are contained in $|L|$ form a subcomplex of K' that is the induced subdivision of L.

For the explicit construction of subdivisions, the following combinatorial analogue of the CW cone construction (see Section 2.3, Example 4) is very helpful.

Proposition 3.2.13 *Let (C, B, p) be a cone where B is the underlying polyhedron of a finite Euclidean complex L. Then, the global set C is the underlying polyhedron of the finite Euclidean complex*

$$pL = L \cup \{p\Delta : \Delta \in L\}.$$

In other words, cones with a compact polyhedral base are polyhedra. □

Example 3 Let K be a Euclidean complex in $\mathbf{R}^{n+1}$. Define inductively the *barycentric subdivisions* $\mathrm{Sd}\, K^k$ of the skeleta K^k by taking $\mathrm{Sd}\, K^0 = K^0$ and $\mathrm{Sd}\, K^{k+1} = \mathrm{Sd}\, K^k \cup \bigcup \boldsymbol{b}_\Delta(\delta\Delta)'$, where the union '$\cup$' runs through all the $(k+1)$-simplices $\Delta \in K$, $\boldsymbol{b}_\Delta$ is barycentre of Δ and $(\delta\Delta)'$ denotes the induced subdivision of $\delta\Delta$ with respect to $\mathrm{Sd}\, K^k$. The Euclidean complex $\mathrm{Sd}\, K = \mathrm{Sd}\, K^n$ obtained in this way is called the *barycentric subdivision* of K. Its vertices are the barycentres of the simplices in K; its k-simplices correspond to the strongly increasing sequences $\Delta_0 \subset \Delta_1 \subset \cdots \subset \Delta_k$ of simplices of K: the corresponding barycentres span a simplex. Since the distances between two barycentres forming an edge of $\mathrm{Sd}\, K$ are smaller than the diameter of their carrier in K (see Lemma 3.1.3), and the diameters of the simplices are determined by their edges (see Lemma 3.1.2), it follows that the simplices of $\mathrm{Sd}\, K$ are 'smaller' than the simplices of K; more precisely,

$$\sup\{\mathrm{diam}\, \Delta : \Delta \in \mathrm{Sd}\, K\} \leqslant \left(\frac{n+1}{n+2}\right) \cdot \sup\{\mathrm{diam}\, \Delta : \Delta \in K\}.$$ □

This implies, for finite Euclidean complexes, that by repeated barycentric subdivision one can obtain a Euclidean complex with sufficiently small simplices, a fact that will be used mainly in the following form:

Proposition 3.2.14 *Let K be a finite Euclidean complex and let* $\{U_\lambda : \lambda \in \Lambda\}$ *be an open covering of* $|K|$. *Then, there is a subdivision* K' *of* K *such that the covering* $\{|\mathrm{st}_{K'}\kappa| : \kappa \in K'_0\}$ *refines* $\{U_\lambda : \lambda \in \Lambda\}$.

Proof Take $d = \max\{\mathrm{diam}\,\Delta : \Delta \in K\}$. Then, choose $r \in \mathbf{N}$ such that $2\cdot((n+1)/(n+2))^r \cdot d$ becomes smaller than the Lebesgue number of the covering. The r-fold subdivision $\mathrm{Sd}^r K$ has the desired properties, since the diameter of each $|\mathrm{st}_{(\mathrm{Sd}^r K)}\kappa|$ is less than or equal to $2\cdot\max\{\mathrm{diam}\,\Delta : \Delta \in \mathrm{Sd}^r K\}$. □

There is a slightly different type of subdivision process based on an analogue of Lemma 2.3.8:

Lemma 3.2.15 *Given a Euclidean complex K and a point* $\boldsymbol{p} \in |K|$, *there is a subdivision of K for which* $\boldsymbol{p}$ *is a vertex.*

Proof Let L denote the link of $\boldsymbol{p}$ in K and take the subdivision of K given by $(K \backslash \mathrm{st}_K \boldsymbol{p}) \cup \boldsymbol{p}L$. □

The construction of the subdivision in this proof will be referred to as the *starring of K at* $\boldsymbol{p}$. Iterated used of it will be made in announced subdivision process.

Example 4 Let K be a Euclidean complex and let Q be a finite subset of $|K|$. Enumerate the elements of Q say, $\boldsymbol{p}_1, \ldots, \boldsymbol{p}_k$, so that the corresponding sequence of dimensions of the carriers is weakly decreasing. Then perform the starring operation, first at $\boldsymbol{p}_1$, next at $\boldsymbol{p}_2$, and so on. This process is the *starring at a finite set of points in order of decreasing dimension.* Note that the barycentric subdivision of a finite Euclidean complex can be viewed as a starring at the set of the barycentres of all simplices in order of decreasing dimension. □

Proposition 3.2.2 stated that arbitrary unions and intersections of subcomplexes of a Euclidean complex are Euclidean complexes; the situation is far different if one deals with arbitrary Euclidean complexes not contained in a 'supercomplex', even if these are supposed to live in the same Euclidean space.

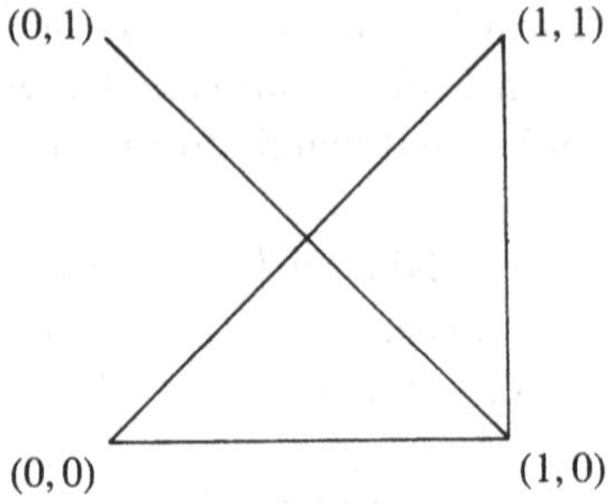

Figure 8

Example 5 In $\mathbf{R}^2$, let K_0 denote the Euclidean complex formed by the standard-simplex Δ^1 and its vertices, and let K_1 denote the Euclidean complex formed by the triangle $H(\{(0,0),(1,0),(1,1)\})$, and its faces (see Figure 8). Then, $K_0 \cap K_1$ consists of the 0-simplex $(1,0)$ and is a Euclidean complex; on the other hand, since Δ^1 and $H(\{(0,0),(1,1)\})$ are 1-simplices in $K_0 \cup K_1$, whose intersection is not a common face of both, the union $K_0 \cup K_1$ fails to be a Euclidean complex. □

This example reflects the general situation. Given two Euclidean complexes K_0, K_1 in the same space R^{n+1}, their intersection $K_0 \cap K_1$ is again a Euclidean complex with $|K_0 \cap K_1| \subset |K_0| \cap |K_1|$; in general, one has a proper inclusion and no equality. By chance, the union $K_0 \cup K_1$ could be a Euclidean complex. A necessary condition for this to happen is the following:

> the intersection of a simplex belonging to K_0 and a simplex belonging to K_1 is a face of both.

If the union is a Euclidean complex, then one has $|K_0 \cup K_1| = |K_0| \cup |K_1|$, as well as $|K_0 \cap K_1| = |K_0| \cap |K_1|$.

Figure 8 suggests that although the union of the Euclidean complexes in question is not a Euclidean complex, the union of their underlying polyhedra is still a polyhedron. However, this is also not true in general, as is seen in the next example.

Example 6 The subsets $\{0\}$ and $\{1/n : n \in \mathbf{N} \setminus \{0\}\}$ of $\mathbf{R}$ are 0-dimensional polyhedra, but their union fails to be a polyhedron. □

Certain classes of polyhedra allow unions. To see this, one needs a somewhat technical fact.

Proposition 3.2.16 *Let K_0, K_1 be finite Euclidean complexes in $\mathbf{R}^{n+1}$ and let L be a subcomplex of K_0 such that $|L| \cap |K_1| = \varnothing$. Then there are* *subdivisions K'_0, K'_1 of K_0, K_1 respectively, such that K'_0 contains L as a subcomplex and $K'_0 \cup K'_1$ is again a Euclidean complex whose underlying polyhedron is $P = |K_0| \cup |K_1|$.*

Proof Let $\Delta_1, \Delta_2, \ldots, \Delta_r$ be any enumeration of the simplices in $K_0 \cup K_1$. Choose, for every index j, a Euclidean complex K_j with $\Delta_j \in K_j$ and $P \subset |K_j|$ (Corollary 3.2.5). Now, P is covered by the intersections of the form $\Delta'_1 \cap \cdots \cap \Delta'_j \cap \cdots \cap \Delta'_r$, with $\Delta'_j \in K_j$ for all j, and $\Delta'_j = \Delta_j$ for at least one j. These sets are convex, and any intersection of them is again of this form. The boundary of any such set (in its affine hull) is a union of such sets of lower dimension. Now assume, by an induction on dimension, that a simplicial decomposition of such a boundary is given; choose an interior point and get a simplicial decomposition of the whole set by means of the cone construction (see Section 3.1, Example 4, and Proposition 3.2.13). Collecting all the simplices obtained in this way, one obtains a simplicial decomposition K of P containing subdivisions K''_0, K''_1 of K_0, K_1, respectively, as subcomplexes. At this point, it is already clear that the union of two compact polyhedra is again a polyhedron.

Let $f : |K_0| \to I$ denote the map which sends every vertex of L to 0, every other vertex to 1, and maps the higher-dimensional simplices linearly. Choose a real number $\varepsilon > 0$ such that $f^{-1}[0, \varepsilon]$ does not contain any vertex of K not belonging to the induced subdivision of L. Thus, every simplex of K meeting $f^{-1}[0, \varepsilon]$ meets also $|L|$; if such a simplex does not belong to $|L|$ then it has an interior point in common with the set $f^{-1}\{\varepsilon\}$. Choose such a point p_Δ in every corresponding simplex Δ, and apply the operation of starring at these points in order of decreasing dimension to K; let K' denote the resulting Euclidean complex. K' contains a simplicial decomposition $\tilde{K}$ of $f^{-1}[\varepsilon, 1] \cup |K_1|$ as a subcomplex, which in turn contains the subdivision K''_1 of K_1 (see Figure 9). Clearly, $L \cup \tilde{K}$ is a

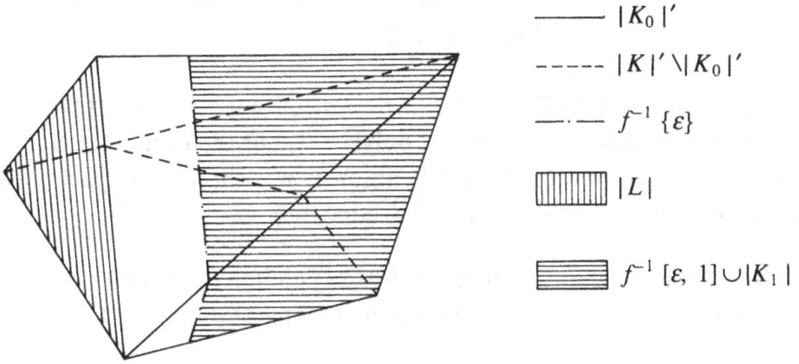

Figure 9

Euclidean complex; it remains to describe a suitable simplicial decomposition of the remaining part of $|K_0| \cup |K_1|$. This is the union of the intersections $\Delta_f = \Delta \cup f^{-1}[0, \varepsilon]$, where Δ runs through the simplices of $K_0 \backslash L$ that have at least one vertex in L. Each such set is convex; by induction on the dimension, one may assume a simplicial decomposition $\dot{\Delta}_f$ of its boundary which does not subdivide the touched simplices of $L, \tilde{K}$ respectively. Taking an interior point p, one gets the simplicial decomposition $p\dot{\Delta}_f$ of Δ_f (see Proposition 3.2.13). Because of the inductive procedure, these simplicial decompositions of all the sets Δ_f are compatible with each other and yield, all together, a Euclidean complex of the desired kind. □

The first part of this proof and the ideas expressed therein allow two further conclusions.

Corollary 3.2.17 *The union of finitely many compact polyhedra in the same Euclidean space is a compact polyhedron.* □

Corollary 3.2.18 *Two Euclidean complexes with the same underlying polyhedron have a common subdivision.*

Proof Let K_0 and K_1 be Euclidean complexes, with $|K_0| = |K_1| = P$. The intersections $\Delta_0 \cap \Delta_1$ with $\Delta_0 \in K_0$ and $\Delta_1 \in K_1$ form a covering of P by compact convex sets, which can inductively be refined to a simplicial decomposition as in the previous proof. □

Remark This result is to be seen in contrast to the so-called 'Hauptvermutung', which will be explained in the Remark following Example 2 of the next section. □

Compact polyhedra are not closed under infinite unions, not even under infinite unions of expanding sequences.

Example 7 The subsets $P_k = \{0\} \cup \{1/n : n \in \mathbf{N}, 0 < n \leqslant k\}, k \in \mathbf{N}$, of $\mathbf{R}^1$ form an expanding sequence of compact polyhedra, but their union in $\mathbf{R}^1$, fails to be a polyhedron. Note that this union is not the union space of the expanding sequence in the sense of Section A.5. □

However, under an additional hypothesis, expanding sequences of compact polyhedra converge to a (not necessarily compact) polyhedron.

Proposition 3.2.19 *Let $P_0 \subset P_1 \subset \cdots \subset P_j \subset \cdots$ be an expanding sequence of compact polyhedra in $\mathbf{R}^n$, such that, for every $j \in \mathbf{N}$, P_j is contained in the* *interior* $(P_{j+1})^\circ$ *of* P_{j+1} *with respect to the union* $P=\bigcup_{j=0}^{\infty}P_j$. *Then, P is again a Euclidean polyhedron.*

Proof Let K_j be a simplicial decomposition of P_j, $j\in\mathbf{N}$. The assumption $P_j\subset(P_{j+1})^\circ$ implies that the pair $\{P_{j+2}\backslash P_j,(P_{j+1})^\circ\}$ is an open covering of P_{j+2}; therefore, one may assume the appearing simplices to be so small that the carrier of any point in $P_{j+2}\backslash P_{j+1}$ with respect to K_{j+2} does not meet P_j (see Proposition 3.2.14). Then, define subcomplexes L_j of K_j by taking $L_0=K_0$, $L_1=K_1$, $L_j=\{\Delta\in K_j:\Delta\cap P_{j-2}=\varnothing\}$ for $j>1$. This implies immediately that $P_{j+1}=P_j\cup|L_{j+1}|$, for all j, and $P_{j-1}\cap|L_{j+1}|=\varnothing$, for $j>0$.

Take a subdivision K_1'' of $L_1=K_1$ containing a subdivision K_0' of K_0 as a subcomplex (using Proposition 3.2.16 with $L=\varnothing$). This starts an inductive construction of subdivisions K_j'',K_{j-1}' of the Euclidean complexes K_j,K_{j-1} respectively, such that K_{j-1}' is a subcomplex of K_j'' and contains (for $j>1$) K_{j-2}' as a subcomplex as follows. Assume the construction is done up to the natural number $j>0$. Then, apply Proposition 3.2.16 to $K_0=K_j''$, $K_1=L_{j+1}$ and the subcomplex $L=K_{j-1}'$ of K_j'' and take $K_{j+1}''=K_0'\cup K_1'$, $K_j'=K_0'$. Now, $K_0'\subset K_1'\subset\cdots\subset K_j'\subset\cdots$ is an expanding sequence of finite Euclidean complexes. For every simplex in K_j' the finitely many simplices of K_{j+1}' form a neighbourhood; thus, the union of the sequence is again a Euclidean complex: it is a simplicial decomposition of P. □

Remark The previous proposition covers all polyhedra, i.e., every polyhedron may be viewed as the union space of an expanding sequence of compact polyhedra, each of them contained in the interior of the subsequent one. This follows from the fact that polyhedra are locally finite and countable CW-complexes (see Theorem 3.2.8), that such CW-complexes are unions of compact subcomplexes in the described manner (see Proposition 1.5.13), and that (CW-)subcomplexes of the underlying polyhedron of a Euclidean complex are polyhedra (see Corollary 3.2.9 (i)). □

It is now possible to derive a local characterization of polyhedra in $\mathbf{R}^{n+1}$.

Theorem 3.2.20 *A subspace P of* $\mathbf{R}^{n+1}$ *is a polyhedron iff every point* $\boldsymbol{p}\in P$ *is the peak of a cone with compact base which is a neighbourhood of* $\boldsymbol{p}$ *in P.*

Proof '⇒': If K is a Euclidean complex with $|K|=P$, then a point $\boldsymbol{p}\in P$ is the peak of the cone $|\mathrm{st}_K\boldsymbol{p}|$ (see Proposition 3.2.12) whose base is compact, as the underlying polyhedron of the link of $\boldsymbol{p}$ in K, which is a finite Euclidean complex.

'$\Leftarrow$': Some preliminary considerations are needed. Let P be a subspace of $\mathbf{R}^{n+1}$ and let $(C, B, \boldsymbol{p})$ be a cone with its peak $\boldsymbol{p}$ in P, its base B compact and its global set C a neighbourhood of $\boldsymbol{p}$ in P. Because of compactness, the set B and the point $\boldsymbol{p}$ have a positive distance; thus there are small $(n+1)$-cubes centred at $\boldsymbol{p}$, which do not meet B. If W is such a cube, and δW denotes its boundary, then $(W \cap C, \delta W \cap B, \boldsymbol{p})$ is a cone with the same properties as the original one. Thus, in this context, one can always deal with cones whose bases are contained in the boundaries of small cubes centred at their peaks. (Given two cones with the same peak and the bases in the boundary of the same cube centred at the common peak, their intersection is again a cone with these properties. This shows that the intersection of two subspaces satisfying the hypothesis has the same property.)

What follows is the main part of the proof. First, assume P to be compact. Perform an induction on the dimension of the affine hull $A(P)$ of P. In dimension 0, there is just a point and nothing to prove. Assume that $\dim A(P) = k+1$. For every point $\boldsymbol{p} \in P$, choose a cone as described in the beginning and with the particular property that its base is contained in the boundary of a small cube centred at $\boldsymbol{p}$. By compactness, finitely many of these cones cover P; thus, it suffices to show that each of them is a polyhedron (Corollary 3.2.17). The intersection of P with any proper face of these cubes has dimension at most k; thus, one can use the induction hypothesis and obtain that this intersection is a polyhedron. The boundary of a cube is the union of its finitely many proper faces. Thus, the intersections of P with the boundaries of the cubes in question are polyhedra. Since cones with compact polyhedral bases are polyhedra (see Proposition 3.2.13), the desired result follows in this case. Note that this already shows that an intersection of two compact polyhedra is a polyhedron.

Now assume P to be non-compact. Since the Euclidean space $\mathbf{R}^{n+1}$ satisfies the second axiom of countability, there is a sequence $\{W_i\}$ of $(n+1)$-cubes such that every cube $W \subset \mathbf{R}^{n+1}$ contains at least one cube W_i, which in turn contains the centre of W as an interior point. For a given point $\boldsymbol{p} \in P$, take a cube W centred at $\boldsymbol{p}$ such that $W \cap P$ is a compact neighbourhood of $\boldsymbol{p}$ in P; by the previous argument, $W \cap P$ is a polyhedron. But then, there is a cube W_i whose intersection $W_i \cap P$ with P is also a compact neighbourhood of $\boldsymbol{p}$, and, moreover, a polyhedron. Let $\{W'_i\}$ denote the subsequence of the sequence $\{W_i\}$ consisting of the cubes appearing in this form. Then, clearly, all $P'_i = W'_i \cap P$ are compact polyhedra whose interiors (with respect to P) cover P. But – as explained before – the fact that P is the union of the P'_i's is not enough! In view of the preceding proposition, one has to construct an expanding sequence

$P_0 \subset P_1 \subset \cdots \subset P_j \subset \cdots$ of compact polyhedra contained in P such that, for every $j \in \mathbf{N}$, P_j is contained in the interior of P_{j+1} with respect to P and such that $P = \bigcup_{j=0}^{\infty} P_j$. This will be done inductively, starting with $P_0 = P'_0$. Assume P_j already constructed. The boundary of P_j is compact; thus it is covered by the interiors of finitely many polyhedra P'_i. Adding these and the P'_i having the smallest index which is not yet contained in P_j, one obtains P_{j+1}. □

This local description, together with the technique of using cones whose base is contained in a cube, yields also the general intersection property.

Corollary 3.2.21 *The intersection of finitely many polyhedra in the same ambient space is again a polyhedron.* □

Moreover, as a consequence of the theorem there is a whole mass of further examples for polyhedra.

Corollary 3.2.22 *Open subspaces of $\mathbf{R}^{n+1}$ or a Euclidean polyhedron are polyhedra.*

Exercises

1. Prove that the carrier of a point s in the underlying polyhedron of a Euclidean complex K is the intersection of all simplices of K which contain the point.
2. Show that cones with a non-compact base can never be polyhedra.
3. Develop a technique for performing a starring at infinitely many points.

3.3 Simplicial complexes

The underlying polyhedra are, up to homeomorphism, determined by the vertex schemes of the corresponding Euclidean complexes; more precisely:

Proposition 3.3.1 *Let K and L be Euclidean complexes and let $f : \Lambda_K \to \Lambda_L$ be a bijection between the corresponding vertex sets such that a set $x \subset \Lambda_K$ spans a simplex of K iff its image $f(x) \subset \Lambda_L$ spans a simplex of L. Then f extends to a homeomorphism $|K| \approx |L|$.*

Proof Take a point $s \in |K|$; let $s_0, \ldots, s_k$ denote the vertices of its carrier Δ_s and let $t_0, \ldots, t_k$ denote its barycentric coordinates with respect to Δ_s. Then define

$$f(s) = \sum_{i=0}^{k} t_i f(s_i).$$ □

This fact leads to a more abstract notion. A *simplicial complex* is a set K of finite sets closed under the formation of subsets, i.e., any subset of a member of K is also a member of K; more formally:

$$x \in K \wedge y \subset x \Rightarrow y \in K.$$

The vertex schemes of Euclidean complexes are examples of simplicial complexes; one should keep them in the back of one's mind for the following considerations.

The members of a simplicial complex K are again called *simplices* of K; more precisely, one has a *simplex of dimension k*, or a *k-simplex* for short, if the simplex has exactly $k+1$ elements. In this abstract setting, denote simplices as above just by small italic letters as $x, y, \ldots$, and write

$$\dim x = k$$

if x is a k-simplex. If x, y are simplices of K with $y \subset x$, then – in accordance with intuition – y is called a *face* of x. The simplices of K are sets, and so one may form their union; this is the set $\Lambda = \Lambda_K$ of *vertices* of K or the *vertex set* of K for short. Considering the vertices as singletons, i.e., as one-element sets, identify the set Λ with the set K_0 of 0-simplices of K. More generally, denote by K_k the set of all k-simplices of K:

$$K_k = \{x \in K : \dim x = k\}$$

and by K^k the *k-skeleton* of K:

$$K^k = \{x \in K : \dim x \leqslant k\};$$

the skeleta are *subcomplexes* of K, i.e., subsets of K which are simplicial complexes themselves.

Other classes of interesting examples for simplicial complexes are the nerves of coverings and ordered simplicial complexes defined presently.

Example 1 Let $\{U_\lambda : \lambda \in \Lambda\}$ be a family of arbitrary sets; then the set $K(\Lambda)$ of all finite subsets of Λ such that

$$\bigcap_{\lambda \in x} U_\lambda \neq \emptyset$$

is a simplicial complex. Note that in general the vertex set of this simplicial complex $K(\Lambda)$ is not the index set Λ itself, but only its subset consisting of the indices λ with $U_\lambda \neq \emptyset$. Now, if Z is a space and $\{U_\lambda : \lambda \in \Lambda\}$ is a covering of Z (see Section A.3), then the simplicial complex $K(\Lambda)$ obtained in this way is called the *nerve* of the covering $\{U_\lambda : \lambda \in \Lambda\}$. □

Example 2 This generalizes Example 1 of Section 3.1. Let Γ be a set and let R be a binary, reflexive and antisymmetric (in general non-transitive) relation on Γ. Then the set $K = K(\Gamma, R)$ of all finite subsets $x \subset \Gamma$ such that $R \cap x \times x$ is a total order on x forms a simplicial complex. The pair (K, R) is called an *ordered simplicial complex.* If (K, R) is an ordered simplicial complex, then the relation R is called a *local vertex ordering* on K. Note that any simplicial complex K can be turned into an ordered simplicial complex (in many ways) by means of the following procedure: choose a total order $\tilde{R}$ on the vertex set of K and take

$$R = \bigcup_{x \in K} \tilde{R} \cap x \times x. \qquad \square$$

Given two simplicial complexes K and L, a *simplicial map* $f : K \to L$ is defined as a function $f : K_0 \to L_0$ which maps every simplex of K onto a simplex of L, i.e. satisfying the condition

$$x \in K \Rightarrow f(x) \in L.$$

Simplicial complexes and simplicial maps form a category that will be denoted by *SiCo* in the sequel. If local vertex orderings R and $\tilde{R}$ are given for K and L respectively, a simplicial map $f : K \to L$ such that

$$(\lambda, \gamma) \in R \Rightarrow (f(\lambda), f(\gamma) \in \tilde{R}$$

is called *order preserving.* Again, ordered simplicial complexes and order-preserving simplicial maps form a category, denoted by *OSiCo.* Although the procedure of ordering an arbitrary simplicial complex described in Example 2 is not at all functorial, there is an interesting functor Sd : *SiCo* → *OSiCo*, called *barycentric subdivision.* It associates to a simplicial complex K the ordered simplicial complex (K', R) having the vertex set $\Gamma' = K$ and local vertex ordering

$$R = \{(x, y) \in K \times K : x \subset y\};$$

it associates to a simplicial map $f : K \to L$ the order preserving simplicial map $f' : K' \to L'$ given by

$$f'(x) = f(x)$$

where, on the left-hand side, x is an element of the domain of the function f', which, on the right-hand side, is interpreted as a subset of the domain on the function f. The notation Sd for this functor reflects its geometrical meaning as a kind of barycentric subdivision, which will be exhibited later in this section. (Note that instead of the relation R one could also have taken the opposite relation

$$R^{\text{op}} = \{(x, y) \in K \times K : y \subset x\}.)$$

Remark Clearly, two simplicial complexes K and L are *isomorphic* if there is a bijection $f : K_0 \to L_0$ such that

$$x \in K \Leftrightarrow f(x) \in L$$

for every subset $x \subset K_0$, i.e., such that f and f^{-1} can be considered as simplicial maps. In this terminology, Proposition 3.3.1 says that the underlying polyhedra of two Euclidean complexes having isomorphic vertex schemes are homeomorphic. Evidently, the converse statement fails to be true: the underlying polyhedra of two Euclidean complexes may be homeomorphic without the corresponding vertex schemes being isomorphic. In view of Corollary 3.2.18, one might expect two Euclidean complexes with homeomorphic underlying polyhedra to have subdivisions with isomorphic vertex schemes. That is the famous 'Hauptvermutung' (main conjecture) of algebraic topology. But this also turned out to be wrong (cf. the notes at the end of this chapter)! □

Now let us turn from combinatorics to geometry. Let K be a simplicial complex. To each (abstract) simplex $x \in K$ associate the (concrete, geometric) simplex formed by the set

$$\Delta_x = \left\{ s = (s_\lambda : \lambda \in x) \in \boldsymbol{I}^x : \sum_{\lambda \in x} s_\lambda = 1 \right\}$$

and the subspace topology with respect to $\boldsymbol{I}^x$; thus Δ_x is homeomorphic to the standard-dim x-simplex. View Δ_x as a subset of $\boldsymbol{I}^\Lambda$. Define the *geometric realization* $|K|$ of K to be the union $\cup \Delta_x$ in $\boldsymbol{I}^\Lambda$, with the topology determined by the family $\{\Delta_x : x \in K\}$ of all simplices. By abuse of language, one often refers to the geometric realization of a simplicial complex simply as a *simplicial complex* and omits the bars in the notation. In this sense, simplicial complexes are CW-complexes – like Euclidean polyhedra.

Theorem 3.3.2 *If K is any simplicial complex, the sequence $\{|K^n| : n \in \mathbf{N}\}$ provides $|K|$ with the structure of a regular CW-complex.*

Proof Clearly, the chosen topology on $|K|$ is finer than the subspace topology with respect to the inclusion of $|K|$ into the product space $\boldsymbol{I}^\Lambda$, and therefore it is Hausdorff. The remainder of the proof is the same as in the corresponding part of Theorem 3.2.8. □

In the sequel, the geometric realization $|K|$ of a simplicial complex K will be tacitly assumed to be provided with the CW-structure described in the preceding proposition. Its closed cells are just the sets Δ_x; thus, they correspond bijectively to the simplices of K. So, given a point $s \in |K|$, the simplex $x = x_s$ is said to be the *carrier* of s if the cell Δ_x is the carrier of s in the sense of Section 1.2.

In the proof of Proposition 3.3.2, already another topology has been considered on the underlying set of $|K|$, namely the *trace of the product topology*. One may also think about a third topology, namely the *metric topology*; i.e., the topology induced by the metric $d : |K| \times |K| \to \mathbf{R}$,

$$d(s, \tilde{s}) = \sqrt{\sum_{\lambda \in \Lambda} (s_\lambda - \tilde{s}_\lambda)^2}.$$

But this is nothing new:

Proposition 3.3.3 *For a simplicial complex K, the metric topology and the trace of the product topology on the underlying set of its geometric realization $|K|$ coincide.*

Proof First observe that the metric topology is finer than the trace of the product topology, since, for every $\lambda \in \Lambda$, the restriction of the corresponding projection $p_\lambda : \boldsymbol{I}^\Lambda \to \boldsymbol{I}$ to $|K|$ provided with the metric topology is continuous.

Conversely, it will be shown that, for any positive real number ε, an ε-neighbourhood U of a point $s \in |K|$ in the metric topology is also a neighbourhood of s in the trace of the product topology. For s fixed, $k = \dim x_s$, define the positive real

$$r = \frac{\varepsilon}{\sqrt{(k+1)(k+2)}}$$

and the set

$$\tilde{U} = \{\tilde{s} \in |K| : \lambda \in x_s \Rightarrow |s_\lambda - \tilde{s}_\lambda| < r\}.$$

Then $\tilde{U}$ is evidently open in the trace of the product topology. Next, consider a single point $\tilde{s} \in \tilde{U}$. From $\sum_{\lambda \in \Lambda} s_\lambda = \sum_{\lambda \in \Lambda} \tilde{s}_\lambda = 1$, it follows that

$$\sum_{\lambda \notin x_s} \tilde{s}_\lambda = \sum_{\lambda \in x_s} (s_\lambda - \tilde{s}_\lambda) \leqslant \sum_{\lambda \in x_s} |s_\lambda - \tilde{s}_\lambda| \leqslant (k+1)r$$

and this allows us to estimate

$$d(s, \tilde{s})^2 = \sum_{\lambda \in x_s} (s_\lambda - \tilde{s}_\lambda)^2 + \sum_{\lambda \notin x_s} \tilde{s}_\lambda^2 \leqslant (k+1)r^2 + (k+1)^2 r^2 = \varepsilon^2,$$

which implies $\tilde{s} \in U$. Since this holds for all $\tilde{s} \in \tilde{U}$, one concludes that $\tilde{U} \subset U$. □

The equivalent metric or product topologies on $|K|$ are often referred to as the *strong topology* of the simplicial complex K. The underlying set of $|K|$ endowed with the strong topology will be denoted by $|K|_{\mathrm{m}}$. The following is an example in which the strong topology is really different from the topology determined by all simplices.

Example 3 Let $\Delta\mathbf{N}$ be the simplicial complex consisting of all finite subsets of the set $\mathbf{N}$ of natural numbers; its geometric realization $\Delta_{\mathbf{N}} = |\Delta\mathbf{N}|$ is often referred to as the *infinite simplex.* A basis for the set of open sets of $\Delta_{\mathbf{N}}$ is given by all sets

$$U = \Delta_{\mathbf{N}} \cap \mathop{\times}_{\lambda=0}^{\infty} U_\lambda,$$

where every set U_λ is of one of the following types

$$\begin{aligned} &(a_\lambda, b_\lambda), && 0 \leqslant a_\lambda < b_\lambda \leqslant 1,\\ &[0, b_\lambda), && 0 < b_\lambda \leqslant 1,\\ &(a_\lambda, 1], && 0 \leqslant a_\lambda < 1,\\ &[0,1]. && \end{aligned}$$

On the other hand, a basis for the open sets in the strong topology is given by those sets U which satisfy the added condition that $U_\lambda = [0,1]$ for almost all λ. Thus,

$$\Delta_{\mathbf{N}} \cap \mathop{\times}_{\lambda=0}^{\infty} [0, \tfrac{1}{2})$$

is open in $\Delta_{\mathbf{N}}$, but not in $|\Delta\mathbf{N}|_{\mathrm{m}}$. □

As will be proved in the sequel, the difference between these two topologies is hardly important, as in fact they agree up to homotopy. What follows, while included here as preparation for this, is also of more general interest. Given a space X and a partition of unity $\{\mu_\lambda : \lambda \in \Lambda\}$ on X, there is an interesting function $\psi_\Lambda : X \to |K(\Lambda)|$, where $K(\Lambda)$ denotes the nerve of the induced open covering of X. It is defined by

$$\psi_\Lambda(x) = (\mu_\lambda(x))$$

for all $x \in X$. Moreover, if this partition of unity is subordinated to a covering $\{U_\lambda\}$ with nerve $\tilde{K}(\Lambda)$, then $K(\Lambda)$ is a subcomplex of $\tilde{K}(\Lambda)$ and ψ_Λ can also be considered as a function with values in $|\tilde{K}(\Lambda)|$.

Lemma 3.3.4 *Let X be a space and let the family $\{u_\lambda : \lambda \in \Lambda\}$ be a partition of unity on X subordinated to the covering $\{U_\lambda\}$. Then,*

(i) $\psi_\Lambda : X \to |K(\Lambda)|_{\mathrm{m}} \subset |\tilde{K}(\Lambda)|_{\mathrm{m}}$ *is continuous; and*

(ii) $\psi_\Lambda : X \to |K(\Lambda)| \subset |\tilde{K}(\Lambda)|$ *is continuous provided $\{\mu_\lambda\}$ is locally finite.*

Proof (i) is trivial. As for (ii), take a point $x \in X$ and a neighbourhood U of x in X that meets only finitely many U_λ. Then $\tilde{x} = \{\lambda \in \Lambda : \mu_\lambda | U \neq 0\}$ is a simplex of $K(\Lambda)$ and the function $\psi_\Lambda | U$ factors through the geometric simplex $\Delta_{\tilde{x}}$. The induced function $U \to \Delta_{\tilde{x}} \subset I^{\tilde{x}}$ has the continuous components $\mu_\lambda | U$, for every $\lambda \in \tilde{x}$; thus, it is continuous. Clearly, the inclusion of $\Delta_{\tilde{x}}$ into $|K(\Lambda)|$ is continuous and thus, the same holds for the function $\psi_\Lambda | U$.

This shows that the function ψ_Λ is continuous at x, and, since this is true for every point $x \in X$, it is globally continuous. □

If K is a simplicial complex, the projections $p_\lambda : |K|_{\mathrm{m}} \to I$, $\lambda \in \Lambda_K$ form a point-finite, but in general not locally finite, partition of unity for $|K|_{\mathrm{m}}$ as well as for $|K|$ (see Section A.3). For every $\lambda \in \Lambda_K$, the interior of the star of the 0-cell λ, namely

$$\mathrm{St}(\lambda)^\circ = \{s \in |K|_{\mathrm{m}} : p_\lambda(s) \neq 0\}$$

is the *open star* of λ; the family $\{\mathrm{St}(\lambda)^\circ : \lambda \in \Lambda_K\}$ is an open covering of both $|K|_{\mathrm{m}}$ and $|K|$, called the *star covering.*

Proposition 3.3.5 *The nerve of the star covering of a simplicial complex K is the simplicial complex K itself. Moreover, the canonical function induced by the partition of unity on* $|K|_{\mathrm{m}}$ *given by the projections* $p_\lambda : |K|_{\mathrm{m}} \to I$, $\lambda \in \Lambda_K$, *is nothing but the identity function* $|K|_{\mathrm{m}} \to |K|$.

Proof For the first statement, one has to show that a finite, but non-empty, subset $x \subset \Lambda_K$ is a simplex of K iff the open stars of its elements have a non-empty intersection. Now, if x is a simplex then the intersection of the stars of its vertices contains the non-empty interior of Δ_x. On the other hand, take a point $s \in |K|_{\mathrm{m}}$ with $p_\lambda(s) \neq 0$ for all $\lambda \in x$; then, x is a subset of the carrier of s, and thus a simplex itself.

The second statement of the proposition is trivial. □

Corollary 3.3.6 *For a simplicial complex K, the identity function* $|K|_{\mathrm{m}} \to |K|$ *is continuous if its star covering is a locally finite covering of* $|K|_{\mathrm{m}}$. □

Later on (see Proposition 3.3.14), it will be seen that the given condition is not only sufficient, but also necessary for the continuity of this identity map.

One is now ready for the actual comparison between the topologies of $|K|$ and $|K|_{\mathrm{m}}$.

Proposition 3.3.7 *The geometric realization* $|K|$ *of a simplicial complex K is homotopy equivalent to* $|K|_{\mathrm{m}}$.

Proof There is a trick leading towards a locally finite partition of unity.

To this end, define functions $p : |K|_{\mathrm{m}} \to \boldsymbol{I}$ and $\tilde{p}_\lambda : |K|_{\mathrm{m}} \to \boldsymbol{I}$, for all $\lambda \in \Lambda = K_0$, by taking

$$p(\boldsymbol{s}) = \max\{s_\lambda : \lambda \in \Lambda\},$$
$$\tilde{p}_\lambda(\boldsymbol{s}) = \max\{0, 2s_\lambda - p(\boldsymbol{s})\}.$$

The function $\tilde{p}_\lambda$ will be continuous if the function p can be shown to be continuous. This latter claim is proved by showing that p is continuous at any fixed point $\boldsymbol{s} \in |K|_{\mathrm{m}}$. Take the set

$$\tilde{U} = \left\{ \tilde{\boldsymbol{s}} \in |K|_{\mathrm{m}} : \sum_{\lambda \in X_s} \tilde{s}_\lambda > 1 - \tfrac{1}{4} \cdot p(\boldsymbol{s}) \right\},$$

and, for every $\lambda \in \Lambda$, the set

$$U_\lambda = \{\tilde{\boldsymbol{s}} \in |K|_{\mathrm{m}} : \tilde{s}_\lambda > \tfrac{1}{2} \cdot p(\boldsymbol{s})\}.$$

Now $\tilde{U}$ and all U_λ are open in $|K|_{\mathrm{m}}$, so is

$$U = \tilde{U} \cap \bigcup_{\lambda \in \Lambda} U_\lambda.$$

Clearly, $\boldsymbol{s} \in U$, and, for all $\tilde{\boldsymbol{s}} \in U$,

$$p(\tilde{\boldsymbol{s}}) = \max\{\tilde{s}_\lambda \in I : \lambda \in x_s\}.$$

Thus the restriction $p|U$ can be viewed as the maximum of finitely many continuous functions and therefore is also continuous.

Next, take $\boldsymbol{s}$ and U as just defined, and note that, for all $\lambda \in \Lambda \backslash x_s$, $\tilde{p}_\lambda | U = 0$. Thus, U is a neighbourhood of $\boldsymbol{s}$ with $\tilde{p}_\lambda | U \neq 0$ for only finitely many $\lambda \in \Lambda$. Then, the maps $\bar{p}_\lambda : |K|_{\mathrm{m}} \to I$ given by

$$\bar{p}_\lambda(\boldsymbol{s}) = \tilde{p}_\lambda(\boldsymbol{s}) \cdot [\textstyle\sum \tilde{p}_\lambda(\boldsymbol{s})]^{-1},$$

where the summation runs over all $\lambda \in \Lambda$, form a locally finite partition of unity on $|K|_{\mathrm{m}}$.

Now observe that the nerve of the covering that is induced by this partition of unity is again the given simplicial complex K itself. Indeed,

the vertex set is Λ;

if a subset $x \subset \Lambda$ is a simplex of K, take the barycentre $\boldsymbol{b}$ of Δ_x and find $\bar{p}_\lambda(\boldsymbol{b}) \neq 0$, for all $\lambda \in x$, showing that x belongs to the nerve;

if a subset $x \subset \Lambda$ belongs to the nerve, then there is a point $\boldsymbol{s} \in |K|_{\mathrm{m}}$ such that $\bar{p}_\lambda(\boldsymbol{s}) \neq 0$, for all $\lambda \in x$, which yields $\tilde{p}_\lambda(\boldsymbol{s}) \neq 0$, for all $\lambda \in x$, by the construction of the maps $\bar{p}_\lambda$, and thus $x \in K$.

The canonical function $\bar{p} : |K|_{\mathrm{m}} \to |K|$ given by

$$\bar{p}(\boldsymbol{s}) = \{\bar{p}_\lambda(\boldsymbol{s}) : \lambda \in \Lambda\}$$

is not only continuous (see Lemma 3.3.4 (ii)), but also a homotopy inverse for the identity map $id : |K| \to |K|_{\mathrm{m}}$. The maps $H : |K| \times I \to |K|$ and

$H_m : |K|_m \times I \to |K|_m$, given by

$$H(s,t) = t \cdot (\bar{p} \circ id(s)) + (1-t) \cdot s$$

and

$$H_m(s,t) = t \cdot (id \circ \bar{p}(s)) + (1-t) \cdot s$$

respectively, show that *id* and $\bar{p}$ are homotopy inverse to each other. □

CW-complexes, in particular simplicial complexes, are LEC spaces (see Theorem 1.3.6). Although the main thrust of this book is directed towards cellular structures with the topology determined by the 'closed cells', sometimes the strong topology has to be taken into account. This is why the next result is included here.

Proposition 3.3.8 *A simplicial complex K with the strong topology is an LEC space.*

Proof Let X denote $|K|_m$. The proof consists in exhibiting a neighbourhood U of $\Delta X \subset X \times X$, which is deformable to ΔX in $X \times X$ rel. ΔX. Since $X \times X$ is a metric space, and therefore perfectly normal, there is also a map $\alpha : X \times X \to I$ such that $\alpha^{-1}(0) = \Delta X$ and $\alpha|(X \times X \backslash U) = 1$, and therefore the diagonal map $\Delta : X \to X \times X$ is a closed cofibration (see Proposition A.4.1 (iv)).

Take U to be the union of the sets $U_\lambda = \mathrm{St}(\lambda)^\circ \times \mathrm{St}(\lambda)^\circ$, for all $\lambda \in \Lambda$. Note that the nerve of the covering $\{U_\lambda\}$ of U is just the simplicial complex K itself. Indeed, if a finite intersection of U_λ's is non-empty, then the intersection of the corresponding open stars is non-empty, and therefore the vertices involved form a simplex of K; on the other hand, given a simplex x of K with barycentre $\boldsymbol{b}$ (in X), the point $(\boldsymbol{b}, \boldsymbol{b})$ belongs to all the U_λ with $\lambda \in X$, and so the intersection of these U_λ's is non-empty.

Now construct a partition of unity $\{\mu_\lambda\}$ on U, subordinated to the covering $\{U_\lambda\}$, as follows. Take the function $\mu : U \to \mathbf{R}$ given by

$$\mu(s, s') = \sum \min\{s_\lambda, s'_\lambda\},$$

where the sum runs over all the vertices λ of K. This function is nowhere zero and continuous; the former statement follows because $(s, s') \in U_\lambda$ implies $\min\{s_\lambda, s'_\lambda\} > 0$. To prove continuity at the point (s_0, s'_0), decompose μ in the form $\mu = \mu' + \mu''$, where μ' takes care of the finitely many summands corresponding to the vertices of the carriers of $\boldsymbol{s}, \boldsymbol{s}'$, and μ'' collects the other ones. Clearly, μ' is continuous, as a sum of finitely many continuous functions. The function μ'' takes the value 0 at the point $(\boldsymbol{s}_0, \boldsymbol{s}'_0)$; to show its continuity at this point, let $\varepsilon > 0$ be a given real number and observe that μ'' takes only values $\leqslant \varepsilon$ on the open set

$\{(s, s') : \sum s_\lambda > 1 - \varepsilon, \sum s'_\lambda > 1 - \varepsilon, \lambda$ running through the finitely many vertices of the carriers of $s, s'\}$. Thus μ is a map with strictly positive real values and allows to define maps $\mu_\lambda : U \to I$, for each λ, by taking

$$\mu_\lambda(s, s') = \min\{s_\lambda, s'_\lambda\}/\mu(s, s')$$

which form the announced partition of unity on U.

This partition of unity induces a map $\psi : U \to X$ (see Lemma 3.3.4 (i)), which in turn leads to the homotopy $H : U \times I \to X \times X$ given by

$$H(s, s', t) = \begin{cases} ((1-2t)s + 2t\psi(s, s'), s'), & 0 \leqslant t \leqslant \frac{1}{2}, \\ (2(1-t)\psi(s, s') + (2t-1)s', s'), & \frac{1}{2} \leqslant t \leqslant 1. \end{cases}$$

This homotopy yields the desired deformation of U into ΔX. □

In the preceding proof, another structure on a simplicial complex with the strong topology became transparent; this is based on the convexity of balls and simplices, used often and fruitfully. The essential property of convex sets in linear spaces is that they are not only path-connected but also allow a canonical choice for paths $\sigma_{x,y}$ connecting two points x, y, which depends continuously on these points and becomes constant if the points coincide; this is done by taking $\sigma_{x,y}(t) = (1-t)x + ty$, for all $t \in I$. The applicability of this idea leads to the consideration of metric spaces in which it can be imitated at least locally. More precisely, an *equilocally convex structure* – or *ELCX structure* for short – on a metric space X consists of an open covering $\{V_\gamma : \gamma \in \Gamma\}$ of X and a homotopy $E : U \times I \to X$ such that

(1) $U = \cup V_\gamma \times V_\gamma \subset X \times X$;
(2) E is a homotopy from the restriction of the first projection to U to the restriction of the second projection to U rel. to the diagonal $\Delta X \subset U$, i.e., $E(x, y, 0) = x, E(x, y, 1) = y$, for all $(x, y) \in U$, $E(x, x, t) = x$, for all $x \in X$ and all $t \in I$; and
(3) $E(V_\gamma \times V_\gamma \times I) \subset V_\gamma$, for all $\gamma \in \Gamma$.

An *ELCX-space* is defined to be a metric space provided with an ELCX-structure. If one wishes to be perfectly clear, one should use the notation

$$\{X; \{V_\gamma\}, E\}$$

to describe the ELCX-space consisting of the metric space X, the *convex* covering $\{V_\gamma\}$ and the *equiconnecting* homotopy E; otherwise, if the ELCX-structure is clearly understood, just write X instead of the previous lengthy expression. By abuse of language, a space X is said to be an ELCX-space if it is metric and an ELCX-structure for it is implicitly understood.

A subspace A of an ELCX-space X is an *ELCX-subspace of* X if

(1) A is a closed subset of X; and
(2) $E((A \times A)\cap U \times I)\subset A$;

in this case, the family $\{A\cap V_\gamma\}$ and the induced homotopy $(A \times A)\cap U \times I \to A$ form an ELCX-structure on A.

In light of these definitions the proof of Proposition 3.3.8 shows that:

Corollary 3.3.9 *Any simplicial complex with the strong topology has an ELCX-structure for which every subcomplex is an ELCX-subspace.*

Proof Let K be a simplicial complex. Take the covering by the open stars of the vertices of K as in the proof of Proposition 3.3.8 and define

$$E(s, s', t) = \begin{cases} (1-2t)s + 2t\psi(s, s'), & 0 \leqslant t \leqslant \frac{1}{2}, \\ 2(1-t)\psi(s, s') + (2t-1)s', & \frac{1}{2} \leqslant t \leqslant 1. \end{cases} \qquad \square$$

Simplicial complexes with the strong topology possess another intersting property.

Theorem 3.3.10 *A simplicial complex with the metric topology is an absolute neighbourhood retract.*

Proof Let K be a simplicial complex, let Λ denote its vertex set, let Λ' be the union of Λ and one extra element ω and let $\Delta\Lambda'$ denote the simplicial complex formed by all finite subsets of Λ'. Furthermore, let $\mathbf{R}(\Lambda')$ denote the vector space consisting of all functions $s : \Lambda' \to \mathbf{R}$ which vanish almost everywhere, endowed with the Euclidean norm

$$|s| = \sqrt{\sum_{\lambda\in\Lambda'} s(\lambda)^2}.$$

Then, $|\Delta\Lambda'|_{\mathrm{m}}$ is a convex subspace of the normed linear space $\mathbf{R}(\Lambda')$, and, hence, an ANR (see Proposition A.5.3). Now consider the subspace $\mathbf{C}K \subset |\Delta\Lambda'|_{\mathrm{m}}$ consisting of all points s such that $\{\lambda\in\Lambda : s(\lambda) \neq 0\}$ is a simplex of K; geometrically, one can view $\mathbf{C}K$ as a cone with base $|K|_{\mathrm{m}}$ and peak ω. Define a retraction $r : |\Delta\Lambda'|_{\mathrm{m}} \to \mathbf{C}K$ as follows. Clearly, one must set $r(\omega) = \omega$. Assume that $s \in |\Delta\Lambda'|_m$ with $s \neq \omega$ is given. Take the carrier x_s of s and choose an ordering $\lambda_0 < \cdots < \lambda_n$ of $x_s \backslash \{\omega\}$ such that $s(\lambda_0) \geqslant s(\lambda_1) \geqslant \cdots \geqslant s(\lambda_n)$. Take the maximal index k such that $r(s) = \tilde{s}$ by the formulae

$$\begin{aligned}
\tilde{s}(\lambda) &= s(\lambda_i), && \text{if } \lambda = \lambda_i, 0 \leqslant i \leqslant k = n;\\
\tilde{s}(\lambda) &= s(\lambda_i) - s(\lambda_{k+1}), && \text{if } \lambda = \lambda_i, 0 \leqslant i \leqslant k < n;\\
\tilde{s}(\lambda) &= 0, && \text{if } \lambda \in \Lambda \setminus \{\lambda_i : 0 \leqslant i \leqslant k\};\\
\tilde{s}(\omega) &= 1 - \sum \tilde{s}(\lambda), && \text{the sum taken over all } \lambda \in \Lambda.
\end{aligned}$$

The definition is independent of the choice of the ordering for $x_s \setminus \{\omega\}$; however, a little effort is needed to show that all the coordinate functions $r_\lambda : |\Delta\Lambda'|_m \to \mathbf{R}$, $s \mapsto \tilde{s}(\lambda)$ are continuous.

This will be proved by showing the continuity of each r_λ at any fixed point $s_0 \in |\Delta\Lambda'|_m$. Assume $s_0(\lambda_0) \geqslant s_0(\lambda_1) \geqslant \cdots \geqslant s_0(\lambda_n) > 0$ and $s_0(\lambda) = 0$, for $\lambda \in \Lambda \setminus \{\lambda_0, \ldots, \lambda_n\}$. To begin with, consider the open set

$$\begin{aligned}
\tilde{U} = \{s \in |\Delta\Lambda'| : |s(\lambda_i) - s_0(\lambda_i)| < s_0(\lambda_n)/[4(n+1)],\\
\text{for } 0 \leqslant i \leqslant n, \quad \text{and} \quad |s(\omega) - s_0(\omega)| < s_0(\lambda_n)\}.
\end{aligned}$$

Let $i(1) < \cdots < i(m) < n$ denote all the indices with $s_0(\lambda_{i(p)}) > s_0(\lambda_{i(p)+1})$, for $p = 1, 2, \ldots, m$ and define t_p as the arithmetic mean of $s_0(\lambda_{i(p)})$ and $s_0(\lambda_{i(p)+1})$. Take

$$U_0 = \{s \in U : s(\lambda_i) > t_1 \quad \text{for } i \leqslant i(0)\};$$

then, for every $p = 1, 2, \ldots, m-1$, take

$$U_p = \{s \in U : t_p > s(\lambda_i) > t_{p+1}, \quad \text{for } i(p) < i \leqslant i(p+1)\},$$

and, finally, take

$$U_m = \{s \in U : t_m > s(\lambda_i), \quad \text{for } i(m) < i \leqslant n\}.$$

For each $p = 0, 1, \ldots, m$, the sets U_p are open and so is their intersection

$$U = \bigcap_{p=0}^{m} U_p.$$

If all $s_0(\lambda_i)$ are equal, i.e., if there are no indices $i(p)$, then take $U = \tilde{U}$.

Now the restrictions $r_\lambda | U$ are continuous at the point s_0, thus completing the proof of the continuity.

This establishes $\mathbf{C}K$ as a retract of $|\Delta\Lambda'|_m$, and, consequently, as an ANR (see Proposition A.6.4). Next, $\mathbf{C}K \setminus \{\omega\}$ is an open subspace of $\mathbf{C}K$, and so is ANR (see again Proposition A.6.4). Finally, the retraction $\mathbf{C}K \setminus \{\omega\} \to |K|_m$, $s \mapsto [1/(1 - s_\omega)](s - s_\omega \omega)$ establishes $|K|_m$ as an ANR (see once more Proposition A.6.4). □

The idea of geometric realization extends to a functor. Let K and L be simplicial complexes with vertex sets Λ and Γ respectively, and let $f : K \to L$ be a simplicial map. One considers $|K|$ and $|L|$ as subsets of the vector spaces $\mathbf{R}^\Lambda$ and $\mathbf{R}^\Gamma$ respectively, and forms the linear function $\hat{f} : \mathbf{R}^\Lambda \to \mathbf{R}^\Gamma$, taking the basis vector e_λ, $\lambda \in \Lambda$, to the basis vector $e_{f(\lambda)} \in \mathbf{R}^\Gamma$.

Then $\hat{f}$ maps every simplex Δ_x continuously into the simplex $\Delta_{f(x)}$; thus, it induces a map $|f| : |K| \to |L|$, called the *geometric realization* of f; with respect to the CW-structures given in Proposition 3.3.2 the map $|f|$ is regular in the sense of Section 2.1. This altogether establishes geometric realization as a functor from the category *SiCo* to the full subcategory of the category CW^r generated by the regular CW-complexes. As for simplicial complexes, one often refers to the geometric realization of a simplicial map simply as a *simplicial map*, and omits the bars in the notation.

The finiteness notions for CW-complexes (Section 1.5) have translations in the context of simplicial complexes. A simplicial complex K is said to be

finite, if it contains only finitely many simplices;
locally finite, if every simplex of K is a face of only finitely many simplices of K, which is the same as requiring that every vertex belongs to only finitely many simplices;
countable, if it contains only countably many simplices;
finite-dimensional, if $K = K^k$ for some natural number k (in this case the natural number

$$\dim K = \dim |K|$$

is called the *dimension* of K).

The following fact is evident:

Proposition 3.3.11 *The functor 'geometric realization'* $|{-}| : SiCo \to CW^r$ *preserves and reflects finiteness, countability, local finiteness and finite-dimensionality.* □

There is also a slightly more delicate statement:

Proposition 3.3.12 *The nerve of a covering of a space is locally finite iff the covering itself is star-finite.* □

Local finiteness delivers a criterion for the coincidence of the two topologies on a simplicial complex. First, note

Lemma 3.3.13 *If the simplicial complex K is locally finite, the star covering of K is a locally finite covering of* $|K|_m$.

(As in Corollary 3.3.6, the sufficient condition given here turns out to be necessary also; this is a consequence of the next proposition.)

Proof Take a point $s \in |K|$ and consider the open neighbourhood $U = \cap\{\mathrm{St}(\lambda_s)^\circ : \lambda_s \in x_s\}$, where x_s is the carrier of s. By the local finiteness of K – being the nerve of its star covering (see Proposition 3.3.5) – each of the finitely many open stars $\mathrm{St}(\lambda_s)^\circ$, $\lambda_s \in x_s$ meets only finitely many stars $\mathrm{St}(\lambda)^\circ$, $\lambda \in \Lambda \backslash x_s$. Thus, U is a neighbourhood of s, meeting only finitely many members of the star covering of $|K|_m$. □

Proposition 3.3.14 *The topology determined by all simplices agrees with the strong topology on a simplicial complex iff the simplicial complex is locally finite.*

Proof '⇒': The assumption implies that the corresponding CW-complex is metrizable. Thus the result follows from Proposition 1.5.17.

'⇐': This follows immediately from Lemma 3.3.13 and Corollary 3.3.6. □

Moreover, the finiteness notions permit the comparison of simplicial complexes to Euclidean complexes. A Euclidean complex $\tilde{K}$ is called a *Euclidean realization* of the simplicial complex K, if the vertex scheme of $\tilde{K}$ is isomorphic to K.

Theorem 3.3.15 *A simplicial complex has a Euclidean realization iff it is finite-dimensional, countable and locally finite; if the dimension of such a simplicial complex is n, then its Euclidean realization can be taken in* $\mathbf{R}^{2n+1}$.

Proof It is a consequence of Theorem 3.2.9 that the given conditions on K are necessary. Conversely, let K be a countable and locally finite simplicial complex of dimension n. In order to construct a Euclidean realization $\tilde{K}$ of K, choose first a sequence $\{\boldsymbol{v}_j : j \in \mathbf{N}\}$ of points in $\mathbf{R}^{2n+1}$ such that every $2n+2$ members of the sequence are affinely independent and such that the sequence $\{(\boldsymbol{v}_j)_0, j \in \mathbf{N}\}$ of the 0th coordinates is monotonically increasing with $(\boldsymbol{v}_{j+1})_0 \geqslant (\boldsymbol{v}_j)_0 + 1$, for all $j \in \mathbf{N}$. This can be done by means of the following inductive process. Start with $\boldsymbol{v}_0 = 0$, and take for $j = 1, 2, \ldots, 2n$,

$$\boldsymbol{v}_j = \boldsymbol{e}_j + j\boldsymbol{e}_0$$

where $\boldsymbol{e}_0, \boldsymbol{e}_1, \ldots, \boldsymbol{e}_{2n}$ denotes the canonical basis of the vector space $\mathbf{R}^{2n+1}$. Now assume that $\boldsymbol{v}_k$ is chosen up to $k \geqslant 2n$. Every $2n+1$ of the points $\boldsymbol{v}_0, \ldots, \boldsymbol{v}_k$ span an (affine) hyperplane of $\mathbf{R}^{2n+1}$. But there are only finitely many of those hyperplanes in $\mathbf{R}^{2n+1}$, thus their union does not cover the total space $\mathbf{R}^{2n+1}$, and one may choose $\boldsymbol{v}_{k+1}$ outside of this union, so that the extra condition on the 0th coordinates is satisfied.

Since K is assumed to be countable, its vertex set K_0 is countable. Therefore, there is an injective function $f : K_0 \to \mathbf{R}^{2n+1}$ taking values only in the set $\{v_j\}$. Then, for every simplex $x \in K$, the convex hull $\tilde{x}$ of the set $f(x)$ is a geometric simplex in $\mathbf{R}^{2n+1}$, with $\dim \tilde{x} = \dim x$. The following claim is now made: the set $\tilde{K} = \{\tilde{x} \subset \mathbf{R}^{2n+1} : x \in K\}$ of geometric simplices is an Euclidean complex. The first condition is clear: a face of a geometric simplex $\tilde{x}$ is spanned by the image of a subset of x under the function f; this subset is a simplex of K, because K is a simplicial complex. Looking at an intersection $\tilde{x} \cap \tilde{y}$, one notes that the total number of vertices involved, i.e., the cardinality of $x \cup y$, is not greater than $2n+2$, since, under the assumption $\dim K = n$, every simplex of K has at most $n+1$ vertices. Thus, $f(x \cup y)$ is an affinely independent set in $\mathbf{R}^{2n+1}$ and spans a simplex, of which both $\tilde{x}$ and $\tilde{y}$ are faces. But the intersection of two faces of a geometric simplex is a common face of both.

It remains to verify the third condition of the definition of Euclidean complexes; this says that every element of $\tilde{K}$ has a neighbourhood meeting only finitely many elements of $\tilde{K}$. Since geometric simplices are compact, this is equivalent to the requirement that every point of $s \in \bigcup \tilde{K}$ has a neighbourhood meeting only finitely many elements of $\tilde{K}$. Take such a point $s = (s_0, \ldots, s_{2n})$ and consider the cube $W(s; 1)$. It is a neighbourhood of s and contains only points of those members of $\tilde{K}$ which have at least one vertex with the 0th coordinate less than $s_0 + 1$. But there are only finitely many vertices of this kind in this game, and by the local finiteness of K each of them belongs to only finitely many simplices of $\tilde{K}$.

Thus, $\tilde{K}$ is a Euclidean complex, and, by construction, its vertex scheme is isomorphic to K, thus proving the theorem. □

In general, a simplicial complex L is called a *subdivision* of the simplicial complex K if there is a *piecewise linear* homeomorhism $h : |L| \to |K|$, i.e., a homeomorphism mapping each simplex (= closed cell) of $|L|$ by a restriction of an affine embedding into a simplex of $|K|$. If K is the vertex scheme of a Euclidean complex $\tilde{K}$, and $\tilde{L}$ is a subdivision of $\tilde{K}$ (see Section 3.2), then the vertex scheme L of $\tilde{L}$ is a subdivision of K; the homeomorphism required by the definition can be taken as induced from the identity on the underlying polyhedron $|\tilde{K}| = |\tilde{L}|$. In view of this definition, the next statement is not a tautology.

Proposition 3.3.16 *For any simplicial complex K, the barycentric subdivision K' is a subdivision of K.*

Proof Define a function $h : \Lambda' = K \to |K|$ by associating to each vertex

x of K', which is a simplex of K the barycentre of the simplex Δ_x, which is a closed cell in $|K|$. Let $\boldsymbol{x}=\{x_0, x_1, \ldots, x_k\}$ be an arbitrary k-simplex of K'; assume, without loss of generality

$$x_0 \subset x_1 \subset \cdots \subset x_k,$$

and let $\Delta_0, \Delta_1, \ldots, \Delta_k$ denote the simplices (= cells) of $|K|$ corresponding to $x_0, x_1, \ldots, x_k$ respectively. Then, all Δ_i, $0 \leqslant i \leqslant k$, are faces of Δ_k. Now interpret $h|\boldsymbol{x}$ as a function defined on the vertices of $\Delta_{\boldsymbol{x}} \subset |K'|$ (a basis of the vector space $\mathbf{R}^{\boldsymbol{x}}$) with values in Δ_k (in the vector space $\mathbf{R}^{x_k}$ and extend it to a map $h_{\boldsymbol{x}} : \Delta_{\boldsymbol{x}} \to \Delta_k \subset |K|$, which is the restriction of an affine embedding. These maps $h_{\boldsymbol{x}}$, taken for all simplices $\boldsymbol{x} \in K'$, fit together to define a map $h : |K'| \to |K|$, which is continuous because its restriction to each closed cell of $|K'|$ is continuous.

In order to recognize h as a homeomorphism, one exhibits its inverse map. Take a point $\boldsymbol{s} \in |K|$. Let $\lambda_0, \lambda_1, \ldots, \lambda_k$ denote the vertices of $x_{\boldsymbol{s}}$, the carrier of $\boldsymbol{s}$, numbered in such a way that $s_0 \geqslant s_1 \geqslant \cdots \geqslant s_k$, for the corresponding barycentric coordinates. Take $x_j = \{\lambda_0, \lambda_1, \ldots, \lambda_j\}$ and let $\boldsymbol{b}_j$ denote the barycentre of the cell of $|K|$ corresponding to the simplex x_j, for $0 \leqslant j \leqslant k$. Then $\boldsymbol{s}$ has a unique barycentric representation

$$\boldsymbol{s} = \sum_{j=0}^{k} s'_j \cdot \boldsymbol{b}_j.$$

Now $\boldsymbol{x} = (x_0, x_1, \ldots, x_k)$ is a simplex of K' and gives rise to the closed cell $\Delta_{\boldsymbol{x}}$ in $|K'|$. By assigning to $\boldsymbol{s}$ the point of $\Delta_{\boldsymbol{x}}$ whose coordinate at the place x_j is just s'_j, one obtains a well-defined and continuous function $|K| \to |K'|$ which is an inverse map to h. □

Remark The homeomorphism h is canonical but not natural. Moreover, the following example shows that there cannot be any natural equivalence between the functors 'geometric realization' and 'geometric realization composed with barycentric subdivision'!

Example 4 Take K to be the power set of $\{0, 1, 2\}$ and L the power set of $\{0, 1\}$. Clearly, $|K|, |K'|$ can be identified with the standard-2-simplex Δ^2 and $|L|$, $|L'|$ with the standard-1-simplex Δ^1. Let $f, g : K \to L$ denote the simplicial maps given by

$$f(0) = g(0) = 0, \quad f(1) = 0, \quad g(1) = 1, \quad f(2) = g(2) = 1.$$

A natural equivalence between the two functors described above would require homeomorphisms (see Figure 10) $h_2 : |K'| \to |K|$ and $h_1 : |L'| \to |L|$, such that

$$|f| \circ h_2 = h_1 \circ |f'|, \qquad |g| \circ h_2 = h_1 \circ |g'|.$$

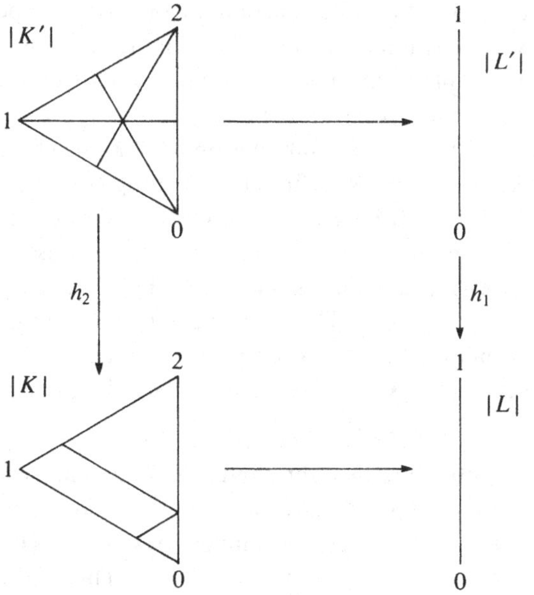

Figure 10

Now consider the vertex v of K' corresponding to the simplex $\{0, 1, 2\}$ of $|K|$. In the geometric realization, it yields an interior point $\tilde{v}$ of the geometric simplex Δ^2. Thus $h_2(\tilde{v})$ also has to be an interior point of Δ^2. But note that $f'(v) = g'(v)$! This implies $h_1 \circ |f'|(\tilde{v}) = h_1 \circ |g'|(\tilde{v}) = w$, forcing

$$h_2(\tilde{v}) \in |f|^{-1}(w) \cap |g|^{-1}(w).$$

But this intersection contains only a single point, which belongs to the boundary and is not at all an interior point. □

As an application of barycentric subdivision, one can prove the following classical theorem.

Theorem 3.3.17 (*Simplical approximation theorem*) *Let K, L be simplicial complexes with K finite and let $g : |K| \to |L|$ be a map. Then there is a subdivision K' of K together with a piecewise linear homeomorphism $h : |K'| \to |K|$ and a simplicial map $f : K' \to L$, such that*

(i) *$g \circ h \simeq |f|$; and*

(ii) *for every $x \in |K'|$, $|f|(x)$ belongs to the carrier of $g \circ h(x)$ in L (the homotopy between these two points will be given by the line segment connecting them in this carrier).*

Proof Since $|K|$ is compact, its image by g in $|L|$ is contained in the geometric realization of a finite subcomplex of L (see Proposition 1.5.2); thus, one may also assume L as finite, and, moreover, one may take K, L as Euclidean complexes in some $\mathbf{R}^n$ (see Theorem 3.3.15). For any vertex $\lambda \in L_0$, let V_λ denote its open star; the family $\{g^{-1}(V_\lambda) : \lambda \in L_0\}$ is an open covering of $|K|$ and thus there is subdivision K' of K such that the covering $\{|\mathrm{st}_{K'}\boldsymbol{\kappa}| : \boldsymbol{\kappa} \in K'_0\}$ of $|K'| = |K|$ refines the covering $\{g^{-1}(V_\lambda) : \lambda \in L_0\}$ (see Proposition 3.2.14); according to the discussion about the definition of subdivision, one can take h to be the identity. Now choose, for any vertex $\boldsymbol{\kappa} \in K'_0$, a vertex $\lambda \in L_0$ such that $|\mathrm{st}_{K'}\boldsymbol{\kappa}| \subset g^{-1}(V_\lambda)$. The assignment $\boldsymbol{\kappa} \mapsto \lambda$ gives a function $f : K'_0 \to L_0$. The objective is to prove that this function f is a simplicial map with the desired properties. To this end, consider a point $x \in |K'|$; let $\boldsymbol{\kappa}_0, \ldots, \boldsymbol{\kappa}_q$ denote the vertices of its carrier. Then,

$$g(x) \in \bigcap g(|\mathrm{st}_{K'}\boldsymbol{\kappa}_i|) \subset \bigcap V_{f(\kappa_i)},$$

and so the intersection on the right-hand side is non-empty, implying that the vertices $f(\boldsymbol{\kappa}_0), \ldots, f(\boldsymbol{\kappa}_q)$ form a simplex of L, and hence the function f is a simplicial map. Moreover, the simplex $\{f(\boldsymbol{\kappa}_0), \ldots, f(\boldsymbol{\kappa}_q)\}$ is a face of the carrier of $g(x)$ and contains $|f|(x)$. Thus, the homotopy $H : |L| \times I \to \mathbf{R}^n$, $(\mathbf{x}, t) \mapsto t \cdot g(\mathbf{x}) + (1-t) \cdot |f|(x)$ factors through $|L|$, thus completing the proof. □

Exercises

1. Show that the forgetful functor *SiCo* → *Sets*, which assigns to every simplicial complex its vertex set, has a left as well as a right adjoint.
2. Show that the operation of 'taking the vertex scheme of a Euclidean complex' commutes with 'barycentric subdivision'!
3. Show that the category *SiCo* has products, but that the geometric realization does not commute with products!
4. Let X be a space and let the family $\{u_\lambda : \lambda \in \Lambda\}$ be a locally finite partition of unity on X. Let $\{U_\lambda : \lambda \in \Lambda\}$ denote the induced open covering of X and form its nerve $K(\Lambda)$. Let L be a subdivision of $K(\Lambda)$ with vertex set Λ and let $h : |L| \to |K(\Lambda)|$ be a homeomorphism mapping each simplex of $|L|$ linearly into a simplex of $|K(\Lambda)|$.

   (a) Show that the family $(p_\lambda \circ h^{-1} \circ \psi_\Lambda : \lambda \in \Lambda)$ is a locally finite partition of unity on X! (Here, as in the main text, p_λ denotes the restriction of the coordinate function $I^\Lambda \to I$ to $|L|$.)

   (b) Let $\{U_\gamma : \gamma \in \Gamma\}$ denote the covering of X which is induced by the partition of unity in (a). Show that its nerve $K(\Gamma)$ can be considered as a subcomplex of L, and that

   $$h|\,|K(\Gamma)| \circ \psi_\Gamma = \psi_\Lambda!$$

5. Let X be a space and let the families $\{u_\lambda : \lambda \in \Lambda\}$, $\{u_\gamma : \gamma \in \Gamma\}$ be locally finite partitions of unity on X. Let $\{U_\lambda : \lambda \in \Lambda\}$ and $\{U_\gamma : \gamma \in \Gamma\}$ denote

the respective induced open coverings and $\psi_\Lambda : X \to |K(\Lambda)|$, $\psi_\Gamma : X \to |K(\Gamma)|$ the canonical maps (Exercise 4).

(a) Assume $\{U_\gamma : \gamma \in \Gamma\}$ to be a refinement of $\{U_\lambda : \lambda \in \Lambda\}$. Show that there is a simplicial map $f : K(\Gamma) \to K(\Lambda)$ such that $|f| \circ \psi_\Gamma \simeq \psi_\Lambda$!

(b) Show that the nerve of the covering $\Lambda \tilde{\times} \Gamma$, which is induced by the product of the given partitions of unity, has a canonical embedding $j : K(\Lambda \tilde{\times} \Gamma) \to K(\Lambda) \times K(\Lambda) \times K(\Gamma)$ into the product $K(\Lambda) \times K(\Gamma)$ (in the sense of Exercise 3), such that $|p_\Lambda \circ j| \circ \psi_{\Lambda \tilde{\times} \Gamma} = \psi_\Lambda$ and $|p_\Gamma \circ j| \circ \psi_{\Lambda \tilde{\times} \Gamma} = \psi_\Gamma$! (Here p_Λ and p_Γ denote the projections from the product $K(\Lambda) \times K(\Gamma)$ onto the respective factors.)

6. For any space X, the *proper diagram of nerves* is the diagram containing all nerves of locally finite partitions of unity on X as objects, and having as maps either the embeddings $h| \, |K(\Gamma)|$ of Exercise 4 or the 'projections' $|p_\Lambda| K(\Lambda \tilde{\times} \Gamma)|$, $|p_\Gamma| K(\Lambda \tilde{\times} \Gamma)|$ of Exercise 5(b). Show that any paracompact space is the (inverse) limit of its proper diagram of nerves. (Alder, 1974)
7. A simplicial complex is said to be *full* if any finite set of vertices that pairwise form 1-simplices is a simplex itself. Let K be any simplicial complex and let L be a full simplicial complex. Show that a function $f : K_0 \to L_0$ is a simplicial map $K \to L$ iff it is a simplicial map $K^1 \to L$.
8. Show for simplicial complexes K, L:

$$K \cong L \Leftrightarrow K' \cong L'!$$

Here the isomorphism on the right-hand side is not assumed to be order preserving; but this property automatically holds if K (and therefore also L) is not the vertex scheme of the boundary $\dot{\Delta}$ of a geometric simplex Δ. Prove that in this case every isomorphism g on the right-hand side is of the form $g = f'$ for some isomorphism on the left-hand side. (Finney, 1965)
9. Show for simplicial complexes K, L :

$$K \cong L \Leftrightarrow (K')^1 \cong (L')^1.$$

(Segal, 1965)
10. In some textbooks, the proof of Theorem 3.3.15 is based on the assumption that the sequence $\{\mathbf{v}_j, j \in \mathbf{N}\}$, besides being of general position, only satisfies the condition of not having a cluster point in the ambient space $\mathbf{R}^{2n+1}$. Show by a counterexample that the given construction does not then necessarily yield a Euclidean complex. (Hint : Take $K = \mathbf{N} \cup \{\{2n, 2n+1\} : n \in \mathbf{N}\}$ and choose the sequence $\{\mathbf{v}_j, j \in \mathbf{N}\}$ in such a way that the sequence of barycentres of the 1-simplices obtained converges to one of them.)
11. Prove that any ELCX-space is an LEC-space.

12. Prove the *relative simplicial approximation theorem*: let K, L be finite simplicial complexes, and D a subcomplex of K. Let $f : |K| \to L|$ be a map such that $f \mid |D| = |g|$ for a simplicial map $g : D \to L$. Then, there exist a subdivision K' of K containing D as a subcomplex and a simplicial map $k : K' \to L$ such that $k|D = g$ and $|k| \simeq f$ rel. $|D|$. (Zeeman, 1964)

For the fatidic number 13, a bad property:

13. Show by an example that for relative simplicial approximation one cannot require the homotopy to move every point only on the carrier of its image. (Zeeman, 1964)

3.4 Triangulations

Simplicial complexes will also be used in connection with general spaces. If X is a space, a pair (K, h) consisting of a simplicial complex K and a homeomorphism $h : |K| \to X$ is called a *triangulation* of X. A space X is said to be *triangulable* if it possesses a triangulation. Clearly, simplicial complexes are triangulable, but this does not hold true for all CW-complexes.

Example Intuitively, the CW-complex to be constructed is obtained by taking a sheet of paper and folding it infinitely many times with one edge pressed into one line segment. To render this precise, first define an auxiliary map $f : \boldsymbol{I} \to \mathbf{R}$ by taking $f(0) = 0$ and $f(t) = t \cdot \sin(\pi/2t)$ for $t > 0$. This function has the absolute maximum 1, and, furthermore, has an infinite sequence $t'_1 > t'_2 > \cdots t'_n > \cdots$ of relative maxima $(1 > t'_1)$. Denote by t'' the absolute minimum of f.

Now take the space X to be the image of the square $\boldsymbol{I}^2$ under the map $g : \boldsymbol{I}^2 \to \mathbf{R}^3$, $(s, t) \mapsto (s, s \cdot t, f(t))$. The following filtration is evidently a CW-structure for X:

$$X^0 = \{\mathbf{0}, \boldsymbol{e}_0, \boldsymbol{e}_2, \boldsymbol{e}_0 + \boldsymbol{e}_1 + \boldsymbol{e}_2, t''\boldsymbol{e}_2\},$$

$$X^1 = \{(s, s \cdot t, t) | s \in \boldsymbol{I}, t \in \{0, 1\}\} \cup \{t\boldsymbol{e}_2 | t \in [t'', 1]\} \cup \{(1, t, f(t)) | t \in \boldsymbol{I}\},$$

$$X^2 = X.$$

The corresponding cell decomposition of X contains five 0-cells, five 1-cells and one 2-cell (see Figure 11).

The space X is compact. Thus, if (K, h) were a triangulation of X, $|K|$ would be compact and therefore K would be a finite simplicial complex (see Proposition 3.3.2 and Proposition 1.5.8). But it will be shown that the infinitely many points $t'_1\boldsymbol{e}_2, t'_2\boldsymbol{e}_2, \ldots, t'_n\boldsymbol{e}_2, \ldots$ must correspond to vertices of any triangulation of X!

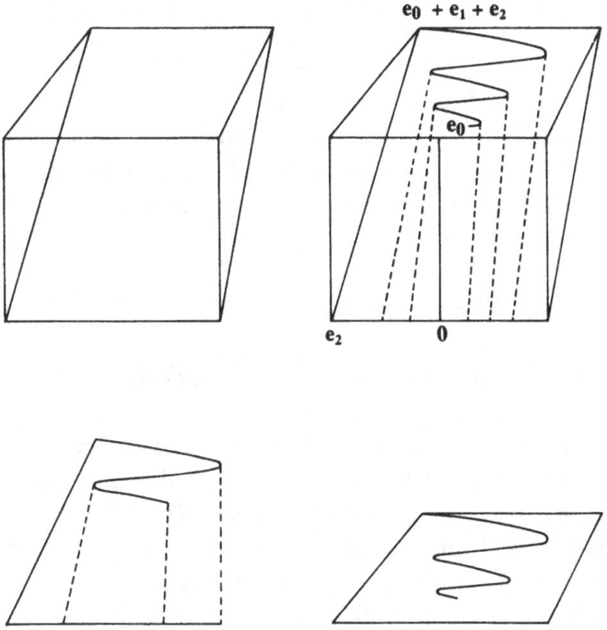

Figure 11

Because dim $X=2$ (see Proposition 1.5.14), a triangulation of X could contain only simplices of dimension $\leqslant 2$. An interior point of a 2-simplex has neighbourhoods homeomorphic to the interior of the 2-ball. Thus, none of the points $t\boldsymbol{e}_2$, $t \in \boldsymbol{I}$ can correspond to an interior point of a 2-simplex. Now assume, for some $n, t'_n\boldsymbol{e}_2$ corresponding to an interior point of some 1-simplex. Take a point $\boldsymbol{x} \in X$, $\boldsymbol{x} \neq t'_n\boldsymbol{e}_2$, belonging to the same 1-simplex as an interior point. Then, every sufficiently small neighbourhood of $t'_n\boldsymbol{e}_2$ has to be homeomorphic to an open neighbourhood of $\boldsymbol{x}$. But this is impossible, as one can see from the shape of the following typical neighbourhoods of $t'_n\boldsymbol{e}_2$ and $\boldsymbol{x}$.

There is a local base (base for the neighbourhood system) at $t'_n\boldsymbol{e}_2$ in X consisting of subspaces homeomorphic to the space V constructed below via a homeomorphism mapping $t'_n\boldsymbol{e}_2$ to the centre $\boldsymbol{v}$ of V. The basic bricks of V are half discs

$$D^{1/2}=\{(s,t):\ -1<s<1, 0\leqslant t<\sqrt{1-s^2}\}.$$

First, take $2n-1$ copies of $D^{1/2}$ and patch them together along their bounding diameters. Denote the resulting space by V_n. Secondly, take a further copy of $D^{1/2}$ and identify $(s,0)$ with $(-s,0)$, thus obtaining a space V'. Thirdly, take the canonical embeddings $[0,1)\to V_n$, $[0,1)\to V'$, both describable by $s \mapsto (s,0)$, and define V to be the union space

$$V = V' \bigsqcup_{[0,1)} V_n.$$

On the other hand, considering x close to $t'_n e_2$, one can find three different types of local bases, consisting of spaces of the following forms: (i) open 2-cells, (ii) V_n as described before, (iii) V_{n+1} in the same sense. But all these bases are incompatible with the local base at $t'_n e_2$ described.

□

However, regular CW-complexes are well behaved.

Theorem 3.4.1 *A regular CW-complex is triangulable.*

Proof Let X be a regular CW-complex. Construct inductively triangulations $(K(n), h_n)$ of the skeleta X^n, such that every closed cell of dimension $\leqslant n$ corresponds to a subcomplex of $K(n)$. Clearly, one can take $K(0) = \{\{x\} : x \in X^0\}$ and $h_0 : |K(0)| \to X^0$ induced by the identity.

Now, suppose $(K(n-1), h_{n-1})$ is already given. Take an n-cell e of X. Its boundary $\bar{e} \backslash e$ is a subcomplex of X (see Theorem 1.4.10), and thus triangulated by the inductive hypothesis; let (K_e, h_e) denote this triangulation of $\bar{e} \backslash e$. Moreover, let L_e be a Euclidean realization of K_e (Theorem 3.3.15). Choose the dimension of the ambient Euclidean space high enough to be able to form a cone $(C, |L_e|, \boldsymbol{p})$. C has a canonical simplicial decomposition (Proposition 3.2.13) whose vertex scheme $\tilde{K}_e$ may be viewed as containing K_e as a subcomplex. Extend the homeomorphism $h_e : |K_e| \to \bar{e} \backslash e$ to a homeomorphism $\tilde{h}_e : |\tilde{K}_e| \to \bar{e}$.

The homeomorphisms h_{n-1} and $\tilde{h}_e$, for each n-cell of X, fit together into a homeomorphism

$$h_n : |K(n)| = |K(n-1) \cup \bigcup \tilde{K}_e| \to X^n.$$

This finishes the induction.

Now define the simplicial complex $K = \bigcup_{n \in \mathbf{N}} K(n)$; since the covering of $|K|$ by its simplices refines the covering by the family $\{|K(n)| : n \in \mathbf{N}\}$, the space $|K|$ is the union space of the expanding sequence $\{|K(n)| : n \in \mathbf{N}\}$ (see Proposition A.2.1). Thus the desired homeomorphism $h : |K| \to X$ is obtained from the fact that X is the union space of the expanding sequence $\{X^n : n \in \mathbf{N}\}$. □

Remark A careful analysis of the preceding proof shows that, for every cell e of X, there is exactly one vertex $p_e \in K^0$ with $h(p_e) \in e$. Thus there is a one-to-one correspondence between the vertices of K and the cells of X. Moreover, the inclusion relation between the closed cells of X provides

K with a distinguished local vertex ordering. In particular, if X is (the geometric realization of) a simplicial complex L, then the simplicial complex K is nothing but the barycentric subdivision of $L, K = L'$. □

Notes to Chapter 3

Plane triangles have been the subject of mathematical research already at prehistorical times; they naturally evolved to arbitrary dimensions in the form of geometric simplices. The topological invariance of simplices under simplicial retractions (see Lemma 3.1.1), which is crucial for the proof of the triangulability of simplicial sets (see Corollary 4.6.12), is due to Barratt (1956); an alternative, but not simpler, approach can be found in Lundell & Weingram (1969). The intersection property (see Theorem 3.1.5) was conjectured by A.N. Kolmogorov and proved in Borovikov (1952); our presentation follows Winkler (1985), where one also can find applications to probability theory, in particular Markov chains.

The history of polyhedra, the objects which form the main theme of Section 3.2, is nearly as old as that of the plane triangles. The outline given here is similar to that in Rourke & Sanderson (1972), where the local characterization of polyhedra in Theorem 3.2.20 was originally developped. The first counterexample to the 'Hauptvermutung' (see Remark following Example 2 in Section 3.3) was exhibited in Milnor (1961).

The abstract notion of a simplicial complex (see Section 3.3) appeared for the first time with full clarity in Alexandroff (1925). The nerve of a covering (see Section 3.3, Example 1) was also introduced by Alexandroff; this notion proved to be essential for the development of certain cohomology theories – Eilenberg & Steenrod (1952) and Dold (1972). Different topologies for a simplicial complex were compared in Dowker (1952). In particular, it is there shown that the topology determined by the simplices and the metric topology lead to spaces of the same homotopy type (see Proposition 3.3.7); the proof given here is based on ideas in Mather ((1964). The class of LEC-spaces was first studied in Fox (1943) and Serre (1951); the subclass of ELCX-spaces and its relationship to simplicial complexes with strong topology (see Proposition 3.3.8, Corollary 3.3.9) are due to Milnor (1959). In Hanner (1951), it is proved that those complexes are also ANRs. The embedding theorem for finite-dimensional, countable and locally finite simplicial complexes (see Theorem 3.3.15) can be found in Seifert & Threlfall (1934), but might be much older. The absolute simplicial approximation theorem (see Theorem 3.3.17) is due to Alexander (1915).

The example of a non-triangulable CW-complex (see Section 3.4, Example) presented here in a purely combinatorial fashion was first given by Metzler (1967) who proved its crucial property by means of local homology; it was the first example of that sort.

4

Simplicial sets

4.1 The category Δ of finite ordinals

In the most general form of the theory of combinatorial complexes, the interplay between the different simplices of a structure is ruled by an action of the category Δ of finite ordinals, whose morphisms serve as operators. Therefore it is necessary to describe this category in minute detail. The main part of the abstract material in this section can be better understood if one looks at the geometric interpretation of the objects and morphisms of the category Δ. Hence the necessary geometric considerations are included here.

For every natural number n, let $[n]$ denote the corresponding *ordinal*, i.e., the set $\{0, 1, \ldots, n\}$ of natural numbers equipped with its natural ordering. Geometrically one should view $[n]$ as the standard-n-simplex Δ^n of $\mathbf{R}^{n+1}$ defined in Section 3.1. In this context, it is often convenient to label the vertices e_i of Δ^n just by the natural number i; e_i is said to be the ith vertex of Δ^n. The (small) category Δ of finite ordinals has as objects the ordered sets $[n]$, for all $n \in \mathbf{N}$, and, as morphisms – which already now will be called *operators* – all order-preserving, i.e., weakly increasing, functions between such ordered sets. Geometrically, an operator $\alpha : [m] \to [n]$ describes the map $\Delta^\alpha : \Delta^m \to \Delta^n$ induced by the linear map $\mathbf{R}^{m+1} \to \mathbf{R}^{n+1}$, $e_i \mapsto e_{\alpha(i)}$; the natural number m is called the *dimension* of α, notation:

$$m = \dim \alpha;$$

thus, in accordance with the intuition,

$$\dim \alpha = \dim(\operatorname{dom} \Delta^\alpha).$$

Note that all the maps Δ^α are closed maps because the geometric simplices are compact spaces.

In abstract language the assignments $[n] \mapsto \Delta^n$, $\alpha \mapsto \Delta^\alpha$ form a covariant functor $\Delta^- : \Delta \to Top$. But one may also think of a left action of the category Δ on the (disjoint) union of the spaces Δ^n; according to this, it is customary to use the short notation

$$\alpha t = \Delta^\alpha(t),$$

for all $t \in \Delta^{\dim \alpha}$. The coordinates of αt can be computed by

$$(\alpha t)_i = \sum_{\alpha(j)=i} t_j.$$

There is a tricky, but useful, application of the geometric interpretation of the category $\boldsymbol{\Delta}$.

Lemma 4.1.1 *For fixed $m, n \in \mathbf{N}$, an operator $\alpha : [m] \to [n]$ is uniquely determined by the value αt for one fixed interior point $t \in (\Delta^m)^\circ$.*

Proof Let $\alpha, \beta : [m] \to [n]$ be different operators. Then there is a smallest $i \in [m]$ such that $\alpha(i) \neq \beta(i)$; assume $\alpha(i) < \beta(i)$. Take $t \in (\Delta^m)^\circ$ and look at the $\alpha(i)$th (barycentric) coordinates of αt and βt. The first is the sum of all coordinates t_j of t, with $\alpha(j) = \alpha(i)$; the second is the sum of the t_j's, with $\beta(j) = \alpha(i)$. Thus

$$(\alpha t)_{\alpha(i)} \geqslant (\beta t)_{\alpha(i)} + t_i > (\beta t)_{\alpha(i)}. \qquad \square$$

A more formal feature of the category $\boldsymbol{\Delta}$ – without a convenient geometric interpretation – is the order structure on the sets $\boldsymbol{\Delta}([m],[n])$ of operators from $[m]$ to $[n]$. It is a partial order, inherited from the fact that operators are number-valued functions. Define for operators $\alpha, \beta : [m] \to [n]$:

$$\alpha \leqslant \beta \Leftrightarrow \alpha(i) \leqslant \beta(i) \quad \text{for all } i \in [m].$$

Clearly, this order is compatible with composition; thus $\boldsymbol{\Delta}$ has the structure of a 2-category.

The subcategory of $\boldsymbol{\Delta}$ formed by its monomorphisms will be denoted by $\boldsymbol{M}$. A monomorphism – in general, denoted by μ or ν – is an injective operator, i.e., a strictly increasing function. If $\mu : [k] \to [n]$ is an injective operator, then $k \leqslant n$; geometrically, there is an embedding of the standard-k-simplex Δ^k as a k-dimensional face, the μth *face*, into the standard-n-simplex Δ^n. Therefore the morphisms of $\boldsymbol{M}$, i.e., the monomorphisms in $\boldsymbol{\Delta}$, are called *face operators*. Every point $s \in \Delta^n$ determines a unique face operator $s^\#$ and a unique interior point $s^\flat \in \Delta^k$, with $k = \dim s^\#$ such that

$$s = s^\# s^\flat.$$

Special face operators are:

(1) the *identity* operators

$$i^n : [n] \to [n]$$

mapping every element onto itself;

(2) the *elementary* face operators

$$\delta_i^n : [n-1] \to [n],$$

which are injective and omit the index i in the image; geometrically, they represent the embedding of Δ^{n-1} into Δ^n as the ith face, which is the $(n-1)$-dimensional face opposite to the vertex i;

(3) the *vertex* operators

$$\varepsilon_i^n : [0] \to [n]$$

mapping the unique element 0 in the domain onto the element i of the codomain; geometrically, these operators exhibit the vertex i, with $0 \leqslant i \leqslant n$;

if no confusion arises, the (upper) index n will be suppressed from the notation.

The composition of the elementary face operators is subject to the rule

$$\delta_i \circ \delta_j = \delta_j \circ \delta_{i-1}, \qquad j < i.$$

The category $\boldsymbol{M}$ is generated by the elementary face operators; every *proper*, i.e., non-identity, face operator $\mu : [k] \to [n]$ has a unique decomposition of the form

$$\mu = \delta_{i_r} \circ \cdots \circ \delta_{i_1},$$

with $0 \leqslant i_1 < \cdots i_r \leqslant n$; the indices i_j are those elements of $[n]$ that are not in the image of α, i.e., those vertices of Δ^n that do not belong to the μth face.

The assignment $\mu \mapsto \operatorname{im} \mu$ induces a bijection between the set of the face operators with the same fixed codomain $[n]$ and the set of non-empty subsets of $[n]$, and thus the set of the face operators with codomain $[n]$ inherits another order, the structure of a (partially) ordered set with suprema. This leads to the following notations:

(1) $\mu \subset \nu$, if image $\mu \subset$ image ν, i.e. if the μth face of Δ^n is contained in its νth face, and

(2) $\mu \cup \nu$ for the unique face operator whose image is image $\mu \cup$ image ν, i.e., which corresponds to the convex hull of the union of the μth face and the νth face,

whenever μ and ν are face operators with the same codomain $[n]$.

Viewing the face operators as sets, one can easily describe a functor $\wedge : \boldsymbol{M} \to \boldsymbol{M}$ useful in the context of simplicial subdivisions (see Lemma 4.6.14). It associates to every face operator $\mu : [k] \to [n]$ the unique face operator $\hat{\mu} : [k+1] \to [n+1]$, satisfying

$$\text{image}\, \hat{\mu} = \text{image}\, \mu \cup \{n+1\}.$$

The function $\hat{\mu}$ is given explicitly by

$$i \mapsto \begin{cases} \mu(i), & i \in [k] \\ n+1, & i = k+1, \end{cases}$$

showing the functoriality of the assignment $\hat{}$. Notice that, up to the dimension indices, the face operators μ and $\hat{\mu}$ have the same decomposition into elementary face operators; more precisely, if

$$\mu = \delta^n_{i_r} \circ \dots \circ \delta^{k+1}_{i_1},$$

then

$$\hat{\mu} = \delta^{n+1}_{i_r} \circ \dots \circ \delta^{k+2}_{i_1}.$$

Dually to the category $\boldsymbol{M}$, one considers the subcategory $\boldsymbol{E}$ of $\boldsymbol{\Delta}$ formed by the epimorphisms in Δ. An epimorphism – in general denoted by ρ or τ – is just a surjective operator. If $\rho : [m] \to [k]$ is a surjective operator, then $m \geqslant k$; if it is *proper*, i.e., if $m > k$, then geometrically the standard-m-simplex Δ^m 'degenerates' to the standard-k-simplex Δ^k. Therefore, the morphisms of $\boldsymbol{E}$, i.e., the epimorphisms in $\boldsymbol{\Delta}$, are called *degeneracy operators.* Special degeneracy operators, besides the identity operators, are

(1) the *elementary* degeneracy operators

$$\sigma^n_i : [n+1] \to [n],$$

$0 \leqslant i \leqslant n$, which are surjective and map the index i, as well as its successor $i+1$, to i, i.e., which degenerate the line segments parallel to the edge connecting the vertices i and $i+1$ in Δ^{n+1} to a point;

(2) the *preterminal* operators

$$\omega^m_i : [m] \to [1],$$

$0 \leqslant i \leqslant m-1$, characterized by $\omega^m_i(i) = 0$ and $\omega^m_i(i+1) = 1$, i.e., the maps which degenerate the i-dimensional face of Δ^m spanned by the vertices $0, \dots, i$ to the vertex 0 of Δ^1 and the opposite face of Δ^m to the vertex 1 of Δ^1;

(3) the *terminal* operators

$$\omega^m : [m] \to [0],$$

mapping all elements of the domain onto the unique element 0 in the codomain, i.e., the constant functions

$$\Delta^m \to \Delta^0.$$

Again, if no confusion arises, the (upper) indices n, m will be suppressed from the notation.

The composition of the elementary degeneracy operators is subject to the following rule:

$$\sigma_j \circ \sigma_i = \sigma_{i-1} \circ \sigma_j, \qquad j < i.$$

This implies that every proper degeneracy operator $\rho : [m] \to [k]$ has an unique decomposition of the form

$$\rho = \sigma_{j_1} \circ \dots \circ \sigma_{j_s},$$

with $0 \leqslant j_1 < \cdots j_s < m$, where the indices j are the element of $[m]$ that have the same image as their successor under α. The following geometric preservation property of degeneracy operators holds true.

Lemma 4.1.2 *Degeneracy operators preserve interiors; i.e., if $\rho : [m] \to [k]$ is a degeneracy operator and $\boldsymbol{t}$ is an interior point of Δ^m then ρt is an interior point of Δ^k.*

Proof It is to be shown that all the barycentric coordinates

$$(\rho \boldsymbol{t})_i = \sum_{\rho(j)=i} t_j$$

of $\rho \boldsymbol{t}$ are positive. Because ρ is surjective there is, for every $i \in [k]$, at least one summand t_j on the right-hand side of the above equation and because $\boldsymbol{t}$ is an interior point, each of the t_j's is positive. □

The categories $\boldsymbol{E}$ and $\boldsymbol{M}$ are related to each other in several ways. To begin with, observe that the assignment $\sigma_i^n \mapsto \delta_i^{n+1}$ induces a contravariant functor

$$-^{\perp} : \boldsymbol{E} \to \boldsymbol{M}, \qquad \rho \mapsto \rho^{\perp},$$

which is an embedding and can be explicitly described by

$$\rho^{\perp}(j) = \max \rho^{-1}(j)$$

for all $j \in \operatorname{cod} \rho$; hence,

$$\rho\rho^{\perp} = \iota.$$

This implies, for every degeneracy operator ρ,

$$\rho^{\perp} = \max(\mu \in \boldsymbol{M} : \rho\mu = \iota),$$

i.e., $\rho^{\perp}$ is the maximum of the set of the *sections* of ρ with respect to the order '$\leqslant$'. Similarly, one obtains a contravariant functor $-_{\perp} : \boldsymbol{E} \to \boldsymbol{M}$, $\rho \mapsto \rho_{\perp}$, which assigns to each degeneracy operator its minimal section and is also an embedding. Conversely, one has the functors $-^{\perp}, -_{\perp} : \boldsymbol{M} \to \boldsymbol{E}$, $\mu \mapsto \mu^{\perp}$, $\mu_{\perp}$, which assign to each face operator its maximal and minimal *retraction* respectively. These are not embeddings: two face operators $\mu, \nu : [k] \to [n]$ have the same maximal retraction if they differ only at the kth place, i.e., if $\mu(i) = \nu(i)$ for $0 \leqslant i < k$; they have the same minimal retraction if they only differ at the 0th place.

The composition of elementary face and elementary degeneracy operators is subject to the rules

$$\sigma_i \circ \delta_j = \delta_j \circ \sigma_{i-1}, \qquad j < i,$$
$$\sigma_i \circ \delta_j = \iota, \qquad i \leqslant j \leqslant i+1,$$
$$\sigma_i \circ \delta_j = \delta_{j-1} \circ \sigma_i, \qquad j > i+1;$$

a useful crib for these and the former interchanging laws is that the smaller index is maintained, with only one exception. Noteworthy also are the following composition laws involving vertex operators:

$$\sigma \circ \varepsilon_i = \varepsilon_{\alpha(i)}, \qquad i \in \operatorname{dom} \alpha,$$
$$\omega_j \circ \varepsilon_i = \delta_1^1, \qquad i \leqslant j,$$
$$\omega_j \circ \varepsilon_i = \delta_0^1, \qquad i > j.$$

There are some technical characterizations of face and degeneracy operators that will prove useful in the sequel.

Lemma 4.1.3 (i) *Any degeneracy operator is uniquely determined by the set of its sections;*

(ii) *any face operator of dimension >0 is uniquely determined by the set of its retractions;*

(iii) *all vertex operators with a fixed codomain have the same set of retractions.*

Proof (i) Let ρ, τ be degeneracy operators with the same set of sections. Then, the maxima of these sets coincide, i.e. $\rho^{\perp} = \tau^{\perp}$, implying that $\rho = \tau$.

(ii) Let $\mu, \nu : [k] \to [n]$, $k > 0$, be face operators with the same set of retractions. Then, $\mu^{\perp} = \nu^{\perp}$; this implies that $\mu(i) = \nu(i)$ for $0 \leqslant i < k$. But also $\mu_{\perp} = \nu_{\perp}$; because $k > 0$, one finally obtains $\mu(k) = \nu(k)$.

(iii) The set of retractions of a vertex operator contains just one element, namely the corresponding terminal operator. □

Thus, to distinguish between vertex operators, another criterion is needed.

Lemma 4.1.4 *Let $\varepsilon_i, \varepsilon_j$ be vertex operators with the same codomain. Then $i > j$ iff there is a degeneracy operator τ such that $\tau\varepsilon_i = \delta_0$ and $\tau\varepsilon_j = \delta_1$.*

Proof '$\Rightarrow$': Take $\tau = \omega_j$.
'$\Leftarrow$': Since τ is non-decreasing, the equations $\tau\varepsilon_i(0) = \delta_0(0) = 1$ and $\tau\varepsilon_j(0) = \delta_1(0) = 0$ together imply $i = \varepsilon_i(0) > \varepsilon(0) = j$. □

An arbitrary operator $\alpha : [m] \to [n]$ has a unique decomposition into a degeneracy operator $\alpha^{\flat}$, followed by a face operator $\alpha^{\#}$:

$$\alpha = \alpha^{\#} \circ \alpha^{\flat}.$$

With respect to composition, one has the evident rules

$$(\alpha\beta)^{\#} = \alpha^{\#}(\alpha^{\flat}\beta^{\#})^{\#}, \qquad (\alpha\beta)^{\flat} = (\alpha^{\flat}\beta^{\#})^{\flat}\beta^{\flat}.$$

The images under a map Δ^{α} of two faces of Δ^m, one of which is contained in the other, are two faces of Δ^n, one of which is again contained in the other; this consideration proves that

$$\mu \subset \nu \Rightarrow (\alpha\mu)^{\#} \subset (\alpha\nu)^{\#}$$

for face operators μ, ν and any operator α such that $\operatorname{dom} \alpha = \operatorname{cod} \mu = \operatorname{cod} \nu$.

For some purposes, it is convenient to use another coordinate description for the points of the standard-simplex Δ^n. The *sum coordinates* of the point $\boldsymbol{t} = (t_0, t_1, \ldots, t_n)$ are the $n+2$ numbers $s_{-1}, s_0, \ldots, s_n$ given by

$$s_j = \sum_{i=0}^{j} t_i$$

for $-1 \leqslant j \leqslant n$; note

$$s_{-1} = 0 \leqslant s_0 \leqslant \cdots \leqslant s_{n-1} \leqslant s_n = 1.$$

A point is an interior point of Δ^n iff it has $n+2$ different sum coordinates. The effect of an operator to the sum coordinates consists in omitting some of them and repeating some others. More precisely,

$$\sigma_i(s_{-1}, \ldots, s_{i-1}, s_i, \ldots, s_{n+1}) = (s_{-1}, \ldots, s_{i-1}, s_{i+1}, \ldots, s_{n+1}),$$
$$\delta_i(s_{-1}, \ldots, s_{i-1}, s_i, \ldots, s_{n-1}) = (s_{-1}, \ldots, s_{i-1}, s_{i-1}, s_i, \ldots, s_{n-1})$$

for $0 \leqslant i \leqslant n$. The following rather technical fact will be needed.

Lemma 4.1.5 *Let $\boldsymbol{t}' \in \Delta^m$ be an interior point and let $\boldsymbol{t} \in \Delta^n$ be a point whose sum coordinates are among those of $\boldsymbol{t}'$. Then there is a unique operator $\alpha : [m] \to [n]$ with $\alpha\boldsymbol{t}' = \boldsymbol{t}$.*

Proof Compose an operator α of the degeneracy operators that are necessary to kill the superfluous sum coordinates of $\boldsymbol{t}'$ and the face operators that force the desired repetitions. Because $\boldsymbol{t}'$ is assumed to be an interior point of Δ^m, the result is uniquely determined (see Lemma 4.1.1). □

The abstract framework for the development of the next sections is given by the notions of:

- *simplicial object* in a category $\mathscr{C}$; that is, a contravariant functor $\Delta \to \mathscr{C}$;
- *cosimplicial object* in a category $\mathscr{C}$; that is, a covariant functor $\Delta \to \mathscr{C}$; and
- *presimplicial object* in a category $\mathscr{C}$; that is, a contravariant functor $\boldsymbol{M} \to \mathscr{C}$.

These objects, together with the corresponding natural transformations, form functor categories, which will be denoted by $Si\mathscr{C}$, $CSi\mathscr{C}$ and $PSi\mathscr{C}$ respectively. Moreover, the forgetful functor that assigns to a functor with domain Δ its restriction on the subcategory $\boldsymbol{M}$ will be denoted by $P : Si\mathscr{C} \to PSi\mathscr{C}$.

Exercise

If $\mu, \nu : [k] \to [n]$ are such that

$$\mu = \delta_{i_r} \circ \cdots \circ \delta_{i_1}, \qquad \nu = \delta_{j_s} \circ \cdots \circ \delta_{j_1}$$

with $r = s, i_1 < \cdots < i_r, j_1 < \cdots < j_s$, prove the equivalence

$$\mu \leqslant \nu \Leftrightarrow i_k \geqslant j_k \quad \text{for } k = 1, \ldots, r = s.$$

4.2 Simplicial and cosimplicial sets

A simplicial object in the category Sets of sets, i.e., an object of the category *SiSets*, is called a *simplicial set*. In dealing with simplicial sets, it is more convenient to think of a simplicial set X as an $\mathbb{N}$-graded set $X = \sqcup X_n$ with the small category $\boldsymbol{\Delta}$ operating on the right; more precisely, X is considered to be the disjoint union of the sequence $X_0 = X([0]), \ldots, X_n = X([n]), \ldots$ of sets together with given set maps

$$\alpha^* = X(\alpha) : X_n \to X_m, \qquad x \mapsto x\alpha$$

for each operator $\alpha : (m) \to [n]$, such that

$$\iota^{n*} = \mathbf{1}_{X_n}$$

and

$$(\alpha\beta)^* = \beta^* \alpha^*$$

for every pair α, β of operators whose corresponding composition is defined, i.e.,

$$x(\alpha\beta) = (x\alpha)\beta$$

for all x, α, β for which $x\alpha$ and $\alpha\beta$ are defined. The elements of X are called *simplices* of X and the elements of a single X_n are called, more specifically, *n-simplices* of X; a pair (x, α) consisting of a simplex $x \in X$ and an operator $\alpha \in \Delta$ such that $x\alpha$ is defined will be called *composable*. If $x \in X_n$, then, in the terminology of $\mathbb{N}$-graded sets, the natural number n is the degree of x; however, here one opts for the denomination *dimension* of x – notation:

$$n = \dim x$$

– in view of the geometric intuition behind the concept of simplicial set: one should think of every (abstract) simplex $x \in X_n$ as a copy Δ_x of the geometric standard-simplex Δ^n, and these are glued together by means of the maps $\Delta^\alpha : \Delta_{x\alpha} \to \Delta_x$; more precisely, the *geometric realization* $|X|$ of the simplicial set X is defined to be the quotient space of $\bigsqcup X_n \times \Delta^n$ – all X_n endowed with the discrete topology – with respect to the relation

$$(x\alpha, \boldsymbol{t}) \sim (x, \alpha\boldsymbol{t})$$

for any simplex $x \in X_n$, any operator $\alpha : [m] \to [n]$ and any point $\boldsymbol{t} \in \Delta^m$. The class of a pair $(x, \boldsymbol{t}) \in X_n \times \Delta^n$ with respect to the induced equivalence relation, which is a point of $|X|$, will be denoted by $[x, \boldsymbol{t}]$.

Remark Since the codomain of an identification map whose domain is a k-space is automatically a k-space, the geometric realization $|X|$ of a simplicial set X as defined here is always a k-space. The fact that $|X|$ also belongs to the class of weak Hausdorff k-spaces, the favourite spaces of this book, remains unproved until it will be shown that $|X|$ has an intrinsic CW-structure (Theorem 4.3.5).

Example 1 The (*simplicial*) *standard-p-simplex* $\Delta[p]$ is the contravariant hom-functor $\Delta \to Sets$ represented by the ordinal $[p]$; its n-simplices are all the operators $\gamma : [n] \to [p]$. Notice that an operator $\alpha : [m] \to [n]$ acts by composition:

$$\alpha^*(\gamma) = \gamma\alpha = \gamma \circ \alpha.$$

There is a geometric justification for the terminology. The geometric realization of a simplicial standard-simplex is a geometric standard-simplex, up to the natural homeomorphisms which are induced by the assignments

$$[\gamma, t] \mapsto \gamma t, \qquad t \mapsto [\iota, t]. \qquad \square$$

The morphisms of the category *SiSets* are called *simplicial maps*. Taking the operational point of view, a simplicial map $f : Y \to X$ from a simplicial set Y to a simplicial set X is considered to be a function $Y \to X$, which preserves the grading and is compatible with the operators, i.e., which satisfies

$$f(y\alpha) = (f(y))\alpha,$$

for all composable pairs (y, α). Clearly, the monomorphisms and epimorphisms in the category *SiSets* of simplicial sets are just the simplicial maps which are given by injective and surjective functions, respectively. The *geometric realization* of the simplicial map f is the well-defined map $|f|:|Y| \to |X|$ given by $[y, t] \mapsto [f(y), t]$. Thus, one has defined a *geometric realization functor* $|-|: SiSets \to Top$, which will be discussed in great detail in the next section.

Example 1^{m} An operator $\varphi : [p] \to [q]$ gives rise to a simplicial map $\Delta\varphi : \Delta[p] \to \Delta[q]$ by composition

$$\Delta\varphi(\gamma) = \varphi \circ \gamma.$$

Identifying $|\Delta[p]|$ and $|\Delta[q]|$ with Δ^p and Δ^q, respectively, via the homeomorphisms described in Example 1, one obtains the geometric realization of the simplicial map $\Delta\varphi$ to be $|\Delta\varphi| = \Delta^\varphi$. $\square$

Examples 1 and 1^{m} may be summed up by some categorical terms. The assignments $[p] \mapsto \Delta[p]$, $\varphi \mapsto \Delta\varphi$ yield a functor $\Delta - : \Delta \to SiSets$, which satisfies the equation

$$|\Delta -| = \Delta^-,$$

and is a full embedding, a so-called *Yoneda embedding* in category theory (it may also be viewed as a cosimplicial object in *SiSets*). To formulate the corresponding *Yoneda lemma*, another rather abstract but quite useful construction is helpful. To each simplicial set X, one associates a small category C_X, the *category of simplices of* X:

the objects are the simplices of X;
the morphisms are the composable pairs (x, α);
the domain of (x, α) is $x\alpha$, the codomain is x;
$\mathbf{1}_x = (x, \iota)$;
if the domain of (x, α) is equal to the codomain of (x', α'), i.e., $x\alpha = x'$, then $(x, \alpha) \circ (x', \alpha') = (x, \alpha\alpha')$.

The category of simplices is connected to the category of finite ordinals by the forgetful functor $D_X : C_X \to \Delta$, $x \mapsto [\dim x]$, $(x, \alpha) \mapsto \alpha$; its composition with the Yoneda embedding will be denoted by ΔX. Any simplicial map $f: Y \to X$ gives rise to the functor $C_f : C_Y \to C_X$, $y \mapsto f(y)$, $(y, \alpha) \mapsto (f(y), \alpha)$, which satisfies the equation $D_X \circ C_f = D_Y$; thus, one has the *category of simplices functor*

$$C_- : SiSets \to Cat,$$

where *Cat* denotes the category of small categories.

Lemma 4.2.1 (*Yoneda lemma*) *Let X be a simplicial set.*

(i) *The assignment $f \mapsto f(\iota^p)$, where $f: \Delta[p] \to X$ is a simplicial map, describes a natural one-to-one correspondence between the set of all simplicial maps $\Delta[p] \to X$ and the set X_p.*

(ii) $X = \operatorname{colim} \Delta X$.

Instead of an explicit proof, done in general form in category theory, only the inverse assignment needed for proving (i) is indicated: make the simplicial map

$$\varphi_x : \Delta[p] \to X, \qquad \gamma \mapsto x\gamma$$

correspond to each p-simplex x. Moreover, note that in the language of category theory, statement (ii) expresses the fact that every set-valued functor is a colimit of representable functors. □

As a set-valued functor category, the category *SiSets* has all kinds of limits and colimits. They are computed *pointwise*; i.e., the set of n-simplices of a desired (co)limit is taken as the (co)limit of the involved sets of n-simplices in the category *Sets*. Thus, a simplicial map is a monomorphism iff it is injective and an epimorphism iff it is surjective.

Some kinds of limits have to be discussed in detail. A simplicial set Y is a *simplicial subset* of a simplicial set X if Y is a subset of X such that the inclusion $Y \to X$ is a simplicial map. A subset Y of a simplicial set X *forms* or *is* a simplicial subset of X if it is closed under the operations. For example, the image of a simplicial map (in the set theoretical sense) is a simplicial subset of its codomain; conversely, given a simplicial subset of the codomain of a simplicial map, its inverse image (again in the set theoretical sense) is a simplicial subset of its codomain. Clearly, arbitrary intersections and unions of simplicial subsets of a fixed simplicial set X again form a simplicial subset of X. Thus, every subset Y of a simplicial set X *generates* a simplicial subset $\tilde{Y}$, the intersection of all simplicial subsets of X that contain the set Y; $\tilde{Y}$ consists of all simplices $x \in X$ that have a representation of the form $x = y\alpha$ for some simplex $y \in Y$ and some operator $\alpha \in \boldsymbol{\Delta}$. A special application of this generation process yields the skeletal decomposition of a simplicial set X; its *n-skeleton* X^n is the simplicial subset of X that is generated by the set of all simplices of dimension at most n. (Another specific type of simplicial subset is the following. Let a simplicial map $p : X \to Z$ be given; then, a simplicial subset Y of X is a *retract of* X *over* Z if there is a simplicial map $r: X \to Y$ such that $r|Y = 1_Y$ and $p|Y \circ r = p$.) A simplicial map $f : Y \to X$ is called *constant* (*with value* $\tilde{x}$) if its image is generated by a 0-simplex (by the simplex $\tilde{x} \in X_0$).

Example 2 The proper face operators with codomain $[p]$ generate a simplicial subset of $\Delta[p]$, namely, its *boundary* $\delta\Delta[p]$; it is the $(p-1)$-skeleton of $\Delta[p]$. The geometric realization of $\delta\Delta[p]$ is clearly homeomorphic to the boundary of the geometric standard-p-simplex (in the sense of Section 3.1). □

The *product* of the simplicial sets Y and X is the simplicial set $Y \times X$ given by $(Y \times X)_n = Y_n \times X_n$, for all $n \in \mathbf{N}$, and $(y, x)\alpha = (y\alpha, x\alpha)$ whenever $\operatorname{cod}\alpha = [\dim y] = [\dim x]$. The simplicial maps $pr_Y: Y \times X \to Y$, $(y, x) \mapsto y$ and $pr_X: Y \times X \to X$, $(y, x) \mapsto x$ are the *projections of* $Y \times X$ *onto* Y and X respectively. It will be shown in the next section that geometric realization commutes with (finite) products.

Now, pullbacks in *SiSets* can be explicitly described. Given simplicial maps $p: Z \to X$, $f: Y \to X$, one takes the simplicial subset

$$W = \{(y, z): f(y) = p(z)\}$$

of the product $Y \times Z$. Then, the map $\bar{p}: W \to Y$, $(y, z) \mapsto y$ is *induced from p by f* and the map $\bar{f} : W \to Z$, $(y,z) \mapsto z$ is *induced from f by p*. A specific form of pullback is hidden in the following notion. Given any simplicial map $p : Z \to X$ and a simplex $\tilde{x} \in X_0$, the inverse image of the simplicial subset generated by $\tilde{x}$ is called the *fibre of p over x*; it is (up to isomorphism) the domain of the map induced from p by the simplicial map $\Delta[0] \to X$, $\iota \mapsto \tilde{x}$.

Forming the product with the standard-simplex $\Delta[1]$, one can transfer the basic notions of homotopy to the combinatorial theory. Let $f, g : Y \to X$ be simplicial maps. A *simplicial homotopy from f to g* is a simplicial map $H : Y \times \Delta[1] \to X$ such that $H(y, \delta_1 \omega) = f(y)$ and $H(y, \delta_0 \omega) = g(y)$, for all $y \in Y$. If D is a simplicial subset of Y, such that the restriction of H to $D \times \Delta[1]$ factors through the projection of $D \times \Delta[1]$ onto D, one has a *simplicial homotopy rel. D*; if $p: X \to Z$ is a simplicial map such that the composition of H with p factors through the projection of $Y \times \Delta[1]$ onto Y one has a *simplicial homotopy over Z*. Clearly, these definitions induce relations called *simplicial homotopies* (rel. D, over Z) on the set of simplicial maps from Y to X; in general, these relations are reflexive, but neither symmetric nor transitive. Nevertheless, they allow to define *simplicial homotopy equivalences, simplicial deformation retracts* and *simplicial contractibility*; all these notions are based on the induced symmetric relation. This will be explained for contractibility: a simplicial set X is *simplicially contractible to the* 0*-simplex* $x_0 \in X$ if there is either a simplicial homotopy from $\mathbf{1}_X$ to the constant simplicial map with value x_0, *or* a simplicial homotopy from the constant map to $\mathbf{1}_X$.

Example 3 The standard-simplex $\Delta[p]$ is simplicially contractible to ε_0, as well as to ε_p. A simplicial homotopy H from the constant map with value ε_0 to $\mathbf{1}_{\Delta[p]}$ is given by taking, for $\alpha : [n] \to [p]$, $j \in [n-1]$, $H(\alpha, \omega_j) = \alpha'$, with $\alpha'(k) = 0$, for $0 \leqslant k \leqslant j$, and $\alpha'(k) = \alpha(k)$ otherwise (note that the simplicial set $\Delta[p] \times \Delta[1]$ is generated by the pairs (α, ω_j)). □

There is another geometric notion whose simplicial analogue can be defined by means of products. A simplicial map $p : Z \to X$ is *locally trivial*, if, for every simplicial map $f : \Delta[n] \to X$, there is an isomorphism $h : W \to \Delta[n] \times F$ with $pr_{\Delta[n]} \circ h = \bar{p}$, where $\bar{p} : W \to \Delta[n]$ denotes the simplicial map which is induced from p by f, and F denotes the fibre of p over $f(\varepsilon_0)$. A locally trivial simplicial map does not have many really different fibres.

Proposition 4.2.2 *If* $p : Y \to X$ *is a locally trivial simplicial map and* x, $\tilde{x}$ *are simplices in* X *with* $\tilde{x} = x\alpha$, *for some operator* α, *then the fibres over* $x\varepsilon_0$ *and* $\tilde{x}\varepsilon_0$ *are isomorphic.*

Proof Assume $x \in X_n$, $\tilde{x} \in X_m$ and define $f : \Delta[n] \to X$ by taking $f(\iota) = x$. By assumption, the simplicial map $\bar{p}$, which is induced from p by f, can be chosen as the projection of the product $\Delta[n] \times F$ onto $\Delta[n]$, where F denotes the fibre of p over $f(\varepsilon_0) = x\varepsilon_0$. Then, the simplicial map $\bar{\bar{p}}$, which is induced from $\bar{p}$ by $\Delta(\alpha\varepsilon_0)$ can be chosen as the unique simplicial map $F \to \Delta[0]$. Since $f \circ \Delta(\alpha\varepsilon_0)(\iota) = f(\alpha\varepsilon_0) = x\alpha\varepsilon_0 = \tilde{x}\varepsilon_0$ the domain F of $\bar{\bar{p}}$ is isomorphic to the fibre of p over $\tilde{x}\varepsilon_0$. □

The existence of colimits in the category *SiSets* implies that simplicial sets have an intrinsically algebraic nature. One consequence of this fact is that simplicial sets may be described in terms of generators and relations. A subset Y of a simplicial set X is said to be a *set of generators for* X if it generates the whole simplicial set X itself. Any **N**-graded set Y generates the *free* simplicial set FY, consisting of all formal expressions $y\alpha$ with $y \in Y$, $\alpha \in \mathbf{\Delta}$ and $\operatorname{cod} \alpha = [\dim y]$ (here, '$\dim y$' clearly means the degree of y) and the evident operations; for convenience, one shortens the notation to 'y' instead of writing '$y\iota$'. A *relation* in a free simplicial set is an equation of the form

$$y\alpha = z\beta,$$

with y, α and z, β as above, and $\operatorname{dom} \alpha = \operatorname{dom} \beta$. Any set of relations R induces an equivalence relation on the set FY which is compatible with the grading. Thus, the resulting set of equivalence classes has a canonical grading and allows an induced operation of the category $\mathbf{\Delta}$. This is the simplicial set *generated by the set* Y, *subject to the relations* R.

A special colimit construction is the simplicial analogue of attachings. Given a simplicial subset D of a simplicial set Y, a simplicial set A and a simplicial map $f : D \to A$, one has a *partial simplicial map* $f : Y \text{-/}\!\!\to A$ *with domain* D and forms the simplicial set X by taking

$$X_n = A_n \sqcup (Y_n \backslash D_n)$$

with suitably defined operations. This simplicial set X is said to be *obtained from* A *by (simplicially) attaching* Y *via* f; the canonical simplicial map $\bar{f} : Y \to X$ is called – as in the continuous case – a *characteristic map* of the simplicial attaching.

A simplex x of a simplicial set X is called *degenerate* if x splits off a degeneracy operator, i.e., if it can be represented in the form $x = y\sigma_i$ with some $y \in X_{\dim x - 1}$ and some $i \in [\dim y]$; otherwise, one has a *non-degenerate* simplex. Clearly, all 0-simplices are non-degenerate. The non-degenerate simplices in the standard-simplex $\Delta[p]$ are the face operators with codomain $[p]$. If X is a simplicial set, denote by $X^{\#}$ the set of its non-degenerate simplices, and by $X^{\flat}$ the set of its degenerate simplices;

specifying dimension n, one also writes $X_n^{\#}$ and $X_n^{\flat}$ respectively. With this terminology, one proves a basic fact in the simplicial theory, namely, the 'Eilenberg–Zilber lemma'.

Theorem 4.2.3 *Any simplex x of a simplicial set X has a unique decomposition in the form*

$$x = x^{\#}x^{\flat},$$

with a non-degenerate simplex $x^{\#} \in X$ and a degeneracy operator $x^{\flat}$.

Proof Splitting off a degeneracy operator decreases the dimension. Since the dimension numbers are bounded below, this cannot be done infinitely many times (after starting with a certain simplex x). Taking the remaining non-degenerate simplex, and composing the split degeneracy operators, one obtains a representation of the desired form.

To prove uniqueness, assume that

$$x\rho = y\tau$$

with x, y non-degenerate simplices and ρ, τ degeneracy operators. Application of a section μ of ρ to this equation yields $x = y\tau\mu$. Since x is non-degenerate, the operator $\tau\mu$ cannot contain a proper degeneracy operator; thus it is a face operator, and therefore $\dim x \leqslant \dim y$. The opposite inequality is obtained by symmetry, and thus $\dim x = \dim y$. But then, $\tau\mu$ is a face operator, whose domain and codomain coincide, and, consequently, an identity operator. This implies that $y\tau\mu = y$, and so $x = y$. Moreover, $\tau\mu = \iota$ shows that every section of ρ is also a section of τ, and vice versa. But degeneracy operators with the same set of sections are equal (see Lemma 4.1.3(i)). □

As an application, one can describe the n-skeleton X^n of a simplicial set X in terms of a simplicial attaching.

Corollary 4.2.4 (i) *A simplicial set is generated by the set of its non-degenerate simplices.*

(ii) *The n-skeleton X^n of a simplicial set X is obtained from its $(n-1)$-skeleton X^{n-1} by attaching the non-degenerate n-simplices; more precisely: if Δ_x denotes a copy of the standard-simplex $\Delta[n]$ and $\delta\varphi_x : \delta\Delta_x \to X^{n-1}$ is the simplicial map given by $\delta\varphi_x(\alpha) = x\alpha$, for each $x \in X_n^{\#}$, then X is obtained from X^{n-1} by attaching $\sqcup\Delta_x$ via $\{\delta\varphi_x\}$.* □

Example 4 Take $p \in \mathbf{N}$ and let f denote the unique simplicial map $\delta\Delta[p] \to \Delta[0]$. The *simplicial p-sphere* $S[p]$ is obtained from $\Delta[0]$ by attaching $\Delta[p]$ via f; $S[p]$ contains exactly two non-degenerate simplices, one in dimension 0 and a further one in dimension p (thus two 0-simplices in case $p = 0$). Moreover, $S[p]$ can easily be described in terms of generators and relations; indeed, it can be obtained by taking one generator x in dimension p subject to the relations

$$x\delta_0^n = \cdots = x\delta_n^n = x\varepsilon_0^n \omega^{n-1}.$$

In the next section, it will be explained that the geometric realization of $S(p)$ is actually a sphere (see the Example in Section 4.3). □

Given a simplicial set X and simplices $x, y \in X$ such that $x = y\alpha$, for some operator α, then x is a *(proper) face* or *degeneracy* of y if α is a (proper) face operator or degeneracy operator respectively. The simplicial subset of X which is generated by the proper faces of a simplex x is the *boundary* δx of x.

The following facts are evident.

Lemma 4.2.5 (i) *A proper degeneracy of a simplex is degenerate.*

(ii) *If two degenerate simplices have the same faces they are equal.*

(iii) *The non-degenerate part of a simplex is a face of this simplex.*

(iv) *If Y is a simplicial subset of the simplicial set X then a simplex of X belongs to Y iff its non-degenerate part belongs to Y.*

(v) *An injective simplicial map transforms non-degenerate simplices into non-degenerate simplices.* □

A non-empty simplicial set X always has simplices of arbitrary high dimensions, which can be exhibited by applying suitable degeneracy operators; however, the dimensions of the non-degenerate simplices of X may be bounded. In this case, the simplicial set X is said to *have finite dimension* and its *dimension* – notation: $\dim X$ – is defined by taking

$$\dim X = \max\{\dim x : x \in X^{\#}\}.$$

Now turn to *cosimplicial sets*, i.e. cosimplicial objects over the category *Sets*. Again, it is convenient to view cosimplicial sets as **N**-graded sets with the category $\boldsymbol{\Delta}$ operating on them, but now by a left action. In this sense, a *cosimplicial map* between cosimplicial sets is clearly a function between the corresponding sets which respects the grading and is compatible with the left action. The functor $\Delta^{-}:\boldsymbol{\Delta} \to Top$ described in Section 4.1 composed with the forgetful functor $V : Top \to Sets$ provides an illuminating example for a cosimplicial set and explains the following terminology. If Y is a cosimplicial set then the elements of $Y = \sqcup Y_n$ are called *points* of Y. Dually to the notion of a non-degenerate simplex, one has *interior* points: these are the points $y \in Y$ which cannot be written in the form $y = \delta_i z$ for some suitable point z and some suitable operator δ_i, i.e., which do not allow the extraction of a face operator. Now one may be interested in dualizing the Eilenberg–Zilber lemma. Because of the formal differences between face and degeneracy operators (see Lemma 4.1.3), this is not possible in general.

Proposition 4.2.6 *For a cosimplicial set Y, the following conditions are equivalent:*

(i) *every point $y \in Y$ has a unique decomposition of the form*

$$y = y^{\#} y^{\flat}$$

with $y^{\#}$ a face operator and $y^{\flat}$ an interior point;

(ii) *for all points $y \in Y_0$,*

$$\delta_0 y \neq \delta_1 y.$$

Proof (ii) is a special case of (i), so it suffices to show that (ii) also implies (i). This is done by dualizing the proof of the Eilenberg–Zilber lemma (see Theorem 4.2.3). The only problem lies in the fact that vertex operators are not determined by their set of retractions (see lemma 4.1.3(ii) and (iii)). Thus an extra argument is necessary for an equation of the form

$$\varepsilon_i y = \varepsilon_j y.$$

Assume $i > j$; then applying ω_j to the previous equation would yield the equation $\delta_0 y = \delta_1 y$ contradicting condition (ii). □

A cosimplicial set is said to *have the Eilenberg–Zilber property* if it satisfies the equivalent conditions of Proposition 4.2.6. Similarly, a cosimplicial space, i.e., a cosimplicial object in *Top*, has the Eilenberg–Zilber property if its composition with the underlying set functor has the Eilenberg–Zilber property. As pointed out in Section 4.1, the cosimplicial space Δ^- is an example of a cosimplicial space with the Eilenberg–Zilber property. Also, a cosimplicial object in a set-valued functor category has the Eilenberg–Zilber property if it has this property pointwise. More down to earth: a covariant functor $\Phi : \Delta \to SiSets$ has the Eilenberg–Zilber property if, for all $n \in \mathbf{N}$, the functors

$$\Phi_n : \Delta \to Sets, \qquad [p] \mapsto (\Phi[p])_n$$

have the Eilenberg–Zilber property.

Example 5 Normal subdivision of standard-simplices is a cosimplicial object in *SiSets* with the Eilenberg–Zilber property (see Lemma 4.6.2). □

If, within an algebraic context, there are given a set of operators acting on one object on the right and on another object on the left, then the familiar procedure is to form a tensor product. Similarly, a *tensor product* $X \otimes Y$ of a simplicial set X and a cosimplicial set Y can be defined. One takes the disjoint union $\sqcup X_n \times Y_n$ and generates on this set an equivalence relation $\sim$ by

$$(x\alpha, y) \sim (x, \alpha y);$$

then $X \otimes Y$ is just the set of the corresponding equivalence classes. Clearly, with respect to this definition, the underlying set of the geometric realization of a simplicial set X is nothing but the tensor product of X and $V\Delta^-$ where $V : Top \to Sets$ again denotes the underlying set functor. This also justifies the notation $[x, y]$ for the equivalence class of the pair $(x, y) \in X_n \times Y_n$, in the general case. Evidently this concept is bifunctorial, i.e., a simplicial map $f : X'$ and a cosimplicial map $g : Y \to Y'$ yield a well-defined function $f \otimes g : X \otimes Y \to X' \otimes Y'$ by taking $f \otimes g([x, y]) = [fx, gy]$; the interchanging law $f \otimes \mathbf{1}_{Y'} \circ \mathbf{1}_X \otimes g = \mathbf{1}_{X'} \otimes g \circ f \otimes \mathbf{1}_Y$ holds true.

As in algebra, the tensor product of a right action and a left action is just a set. But if the object with the left action has also a right action, then the tensor product inherits also a right action. More precisely, if X is a simplicial set and $\Phi : \Delta \to SiSets$ is a covariant functor, then the *tensor product* $X \otimes \Phi$ is the simplicial set given by $(X \otimes \Phi)_n = X \otimes \Phi_n$, $[x, y]\alpha = [x, y\alpha]$. Again, this construction is bifunctorial, i.e., a simplicial map $f : X \to X'$ and a natural transformation $g : \Phi \to \Phi'$ yield a well-defined simplicial map $f \otimes g : X \otimes \Phi \to X' \otimes \Phi'$ by taking $f \otimes g([x, y]) = [fx, gy]$ and the corresponding interchanging law holds.

Example 6 For any simplicial set $X, X \otimes \Delta - \cong X$ holds true. Isomorphisms are provided by the assignments $[x, y] \mapsto xy$ and $x \mapsto [x, \iota]$. Using functoriality in the first variable, one obtains that the functor $- \otimes \Delta - : SiSets \to SiSets$, i.e., the tensor product with the Yoneda embedding, is naturally equivalent to the identity functor Id on the category *SiSets*. □

Example 7 For any simplicial set X, the tensor product with the normal subdivision of simplices

$$\operatorname{Sd} X = X \otimes \Delta'$$

yields the *normal subdivision* of X (see Section 4.6). □

An agreeable feature of the algebraic part of this theory is that in many cases there are canonical representatives for the elements of a tensor product. To describe them, a further notion should be introduced. If X is a simplicial set and Y is a cosimplicial set, a pair $(x, y) \in X_n \times Y_n$ is called *minimal* if x is a non-degenerate simplex and y is an interior point. Moreover, given an arbitrary pair $(x, y) \in X_n \times Y_n$, it is convenient to say that one has a pair of *dimension* n – notation:

$$\dim(x, y) = n.$$

Proposition 4.2.7 *Let X be a simplicial set and let Y be a cosimplicial set with the Eilenberg–Zilber property. Then, any element of the tensor product $X \otimes Y$ can be represented by a unique minimal pair.*

Proof Take $Z = \sqcup X_n \times Y_n$ and define functions $t_1, t_r, t : Z \to Z$ by setting

$$t_1(x, y) = (x^\#, x^\flat y), t_r(x, y) = (x y^\#, y^\flat)$$

and

$$t = t_1 \circ t_r.$$

Then the following facts are evident:

(i) $(x, y) \sim t_1(x, y) \sim t_r(x, y) \sim t(x, y)$,
(ii) $t(x, y) \neq (x, y) \Rightarrow \dim t(x, y) < \dim (x, y)$,
(iii) $t_1(x, y) = (x, y) \Leftrightarrow x$ is a non-degenerate simplex,
(iv) $t_r(x, y) = (x, y) \Leftrightarrow y$ is an interior point and
(v) $t(x, y) = (x, y) \Leftrightarrow (x, y)$ is a minimal pair.

From (iii) and (iv), it follows that the functions t_1 and t_r are *idempotent*, i.e., $t_1^2 = t_1$, and $t_r^2 = t_r$, respectively. Moreover, since the set of dimensions is bounded below, it follows from (ii) that, for any pair (x, y), the sequence $(t^n(x, y))$ becomes stationary. Thus, by (v), it contains a minimal pair which, by (i), is equivalent to the initial pair (x, y). This proves the existence of a minimal pair in every class. This part of the assertion does not depend on the Eilenberg Zilber property for Y.

Now assume simplices $x, x' \in X$, points $y, y' \in Y$ and an operator α to be given such that $x = x'\alpha$ and $\alpha y = y'$. Then, the pairs (x, y) and (x', y') are equivalent and one says that the pair (x, y) *is directly equivalent to* the pair (x', y') (*via* α). In this situation, it follows that the pair $t_r(x', y')$ is directly equivalent to the pair $t(x, y)$ via

$$\alpha' = (x'(\alpha y^\#)^\#)^\flat((\alpha y^\#)^\flat y^\flat)^\#.$$

Observe that the idempotency of the function t_r implies $tt_r = t$. Thus, by iteration, it follows that, for all $n \in \mathbf{N}$, the pair $t_r t^n(x, y)$ is directly equivalent to the pair $t^n(x', y')$. For sufficiently large n, these pairs are the minimal pairs associated to the original pairs (x, y) and (x', y') respectively, given in the first part of the proof; but minimal directly equivalent pairs are equal. Since the equivalence relation under consideration is generated by the (non-symmetric) direct equivalence, this finishes the proof. □

A technical detail of the given proof deserves special attention.

Addendum 4.2.8 *Let X be a simplicial set and let Y be a cosimplicial set with the Eilenberg–Zilber property. Then, if $(x, y) \in X_n \times Y_n$ is an arbitrary pair and $(x_m, y_m) \in X_m \times Y_m$ is the minimal pair representing the same element of the tensor product $X \otimes Y$ as (x, y), there are a (not necessarily unique) face operator $\mu : [m] \to [n]$ and a (not necessarily unique) degeneracy operator $\rho : [n] \to [m]$ such that $x_m = x\mu$ and $y_m = \rho y$.*

Proof Inspect the functions $t_l, t_r, t : Z \to Z$ of the proof just given, and observe that $x^{\#}$ can be obtained from x by applying the face operator $(x^{\flat})^{\perp}$ and that $y^{\flat}$ can be obtained from y by applying the degeneracy operator $(y^{\#})^{\perp}$. □

Remark In many applications of the proposition, the cosimplicial sets under consideration have the property that interior points are mapped onto interior points by degeneracy operators (see Lemma 4.1.2 and Exercise 2). In this case, the pair $t(x, y)$ is already minimal, for any pair (x, y). But that this is a special property becomes quite clear if one looks at the dual situation. It would say that any face of a non-degenerate simplex should be non-degenerate, giving a presimplicial set (to be discussed in Section 4.4). The geometric appeal of simplicial sets is derived from the fact that non-degenerate simplices may have degenerate faces. □

Corollary 4.2.9 *If Y is a cosimplicial set with the Eilenberg–Zilber property, then the tensor product $-\otimes Y : SiSets \to Sets$ preserves and reflects monomorphisms.*

Proof Let $f : Z \to Y$ be an injective simplicial map. Assume $(z, y), (z', y')$ to be minimal pairs such that

$$f \otimes \mathbf{1}([(z, y)]) = f \otimes \mathbf{1}([(z', y')]),$$

i.e.,

$$[f(z), y] = [f(z'), y'].$$

Since f is injective, the pairs $(f(z), y)$ and $(f(z'), y')$ are still minimal (see Lemma 4.2.5 (v)). Thus, the uniqueness of the representation by minimal pairs implies $f(z) = f(z')$ and $y = y'$. Again using the injectivity of f, one obtains $z = z'$ and, finally, $[(z, y)] = [(z', y')]$.

Let a simplicial map $f : Z \to X$ be given such that $f \otimes \mathbf{1}$ is injective and take $z_1, z_2 \in Z$ with $f(z_1) = f(z_2)$; without loss of generality, one may assume z_1 to be non-degenerate. Define $n = \dim z_1 = \dim z_2$ and choose an interior point y in Y_n. The injectivity of $f \otimes \mathbf{1}$ implies $[z_1, y] = [z_2, y]$. Since the pair (z_1, y) is minimal, the simplex z_1 is a face of the simplex z_2 (see Addendum 4.2.8) and since $\dim z_1 = \dim z_2$ it follows that $z_1 = z_2$. □

Another property of these tensor products follows from the adjoint functor generating principle (see Section A.10).

Proposition 4.2.10 *If Y is a cosimplicial set, then the tensor product $-\otimes Y$: SiSets → Sets is left adjoint to the functor S_Y : Sets → SiSets given by*

$$(S_Y T)_n = \text{set of all functions } Y_n \to T$$

for all sets T, all $n \in \mathbf{N}$.

$$x\alpha = x \circ Y(\alpha)$$

for all elements $x \in (S_Y T)_n$, all operators α with $\operatorname{cod} \alpha = [n]$,

$$S_Y f(x) = f \circ x$$

for any function f with domain T and any $x \in S_Y T$. □

This fact has an essential consequence.

Corollary 4.2.11 *If Y is a cosimplicial set, then the tensor product $-\otimes Y$: SiSets → Sets preserves all colimits.* □

Taking a cosimplicial object in the category *SiSets*, one obtains similarly.

Proposition 4.2.12 *If $\Phi : \Delta \to$ SiSets is a covariant functor, then the tensor product $-\otimes \Phi$ is a left adjoint functor and preserves all colimits. Moreover, if Φ has the Eilenberg–Zilber property, then $-\otimes \Phi$ transforms simplicial attachings into simplicial attachings.*

Proof See the adjoint functor generating principle (Section A.10) and Corollary 4.2.9. □

Finally, note a very general statement which nevertheless is sometimes useful.

Proposition 4.2.13 *Let $\mathscr{C}$ either be Sets or SiSets and let $\Phi : \Delta \to \mathscr{C}$ be a cosimplicial object in $\mathscr{C}$. Then, for any simplicial set X,*

$$X \otimes \Phi = \operatorname{colim} \Phi \circ D_X$$

where $D_X : C_X \to \Delta$ denotes the forgetful functor. □

In view of this fact, one extends the terminology 'tensor product' to a more general situation. A cosimplicial object Φ in an arbitrary cocomplete category $\mathscr{C}$ induces, by the adjoint functor generating principle (see Section A.10), a cocontinuous functor $SiSets \to \mathscr{C}$ which is given on the objects by the right-hand side of the equation in Proposition 4.2.13. Thus, this functor will be called again *tensor product with* Φ.

Exercises

1. Show that the standard-simplex $\Delta[p]$ is not contractible to any ε_k with $0<k<p$.
2. Give explicit descriptions of the *simplicial dunce hat*, i.e., the simplicial set generated by one 2-simplex x subject to the relations $x\delta_0 = x\delta_1 = x\delta_2$, and the *simplicial projective plane*, i.e., the simplicial set generated by one 2-simplex y subject to the relations $y\delta_0 = y\delta_2$, $y\delta_1 = y\varepsilon_0\sigma_0$.
3. Construct examples for cosimplicial sets

   which do not have the Eilenberg–Zilber property, and/or
   such that interiors are not preserved by degeneracy operators.
4. Let (K, R) be an ordered simplicial complex. Its *associated* simplicial set $(K, R)_s$ is generated by the set K and subject to the relations
$$x\delta_i = x_i$$
   with $x \in K, i \in [\dim x]$ and x_i the ith face of x, i.e., the face obtained by omitting the ith vertex. Extend this definition to an embedding of the category *OSiCo* into the category *SiSets* as a coreflective subcategory. (*Coreflective* means that the embedding is a right adjoint functor, i.e., the embedding has a coadjoint, a so-called *coreflector*, and that the restriction of the coreflector to the subcategory obtained is equivalent to the identity.)
5. Construct simplicial mapping sets; i.e., show that, for any simplicial set Z, the product functor $-\times Z$ has a right adjoint $(-)^Z$. (*Hint*: According to the Yoneda lemma (Lemma 4.2.1), the n-simplices of a simplicial mapping set X^Z have to be in one-to-one correspondence with the simplicial maps $\Delta[n] \to X^Z$, which in turn – by adjointness – should correspond to the simplicial maps $\Delta[n] \times Z \to X$; take the set of these maps as $(X^Z)_n$, define suitable operations and show functoriality.)

4.3 Properties of the geometric realization functor

In the preceding section, the geometric realization $|X|$ of a simplicial set X was defined as the quotient space of $\sqcup X_n \times \Delta^n$ – all X_n endowed with the discrete topology – with respect to the relation
$$(x\alpha, t) \sim (x, \alpha t)$$

for any simplex $x \in X_n$, any operator $\alpha : [m] \to [n]$ and any point $t \in \Delta^m$; recall that the class of a pair $(x, t) \in X_n \times \Delta^n$ with respect to the induced equivalence relation, which is a point of $|X|$, is denoted by $[x, t]$.

Moreover, the geometric realization of a simplicial map $f: Y \to X$ has been defined as the map $|f|: |Y| \to |X|$, taking $[y, t] \mapsto [f(y), t]$. This completed the definition of the geometric realization functor $|-|: SiSets \to Top$.

There is another description of the geometric realization $|X|$ of a simplicial set X which is useful for some purposes. First – as exhibited in the previous section – the underlying set of the space $|X|$ can be considered as the tensor product of the simplicial set X and the cosimplicial set $V\Delta^-$; second, the topology of $|X|$ is the final topology with respect to the family of maps

$$\bar{c}_x : \Delta^{\dim x} \to |X|, \qquad t \mapsto [x, t],$$

for all $x \in X$; note that $\bar{c}_x$ can be considered as the geometric realization of the simplicial map $\Delta[n] \to X, \alpha \mapsto x\alpha$, up to the natural homeomorphism $|\Delta[\dim x]| \to \Delta^{\dim x}$. Because

$$\bar{c}_{x\alpha} = \bar{c}_x \circ \Delta^\alpha$$

whenever $x\alpha$ is defined, this point of view immediately yields the following fact.

Proposition 4.3.1 *If the set Y generates the simplicial set X, then the geometric realization $|X|$ of X is a quotient space of the subspace $\sqcup (Y \cap X_n) \times \Delta^n$ of $\sqcup X_n \times \Delta^n$.* □

Corollary 4.3.2 *The geometric realization $|X|$ of a simplicial set X is a quotient space of the subspace $\sqcup X_n^\# \times \Delta^n$ of $\sqcup X_n \times \Delta^n$.*

(Recall that $X_n^\#$ denotes the set of non-degenerate n-simplices of X.)

Proof A simplicial set is generated by its non-degenerate simplices (see Corollary 4.2.4 (i)). □

Next, apply the Eilenberg–Zilber property of the cosimplicial space $\Delta^- : \Delta \to Top$.

Proposition 4.3.3 (i) *If X is a simplicial set, then any point of its geometric realization $|X|$ has a unique representation by a pair (x, t), with x a non-degenerate simplex of X and t an interior point of $\Delta^{\dim x}$.*

(ii) *The geometric realization of a simplicial map is injective iff the simplicial map itself is injective.*

Proof (i) In the presence of the Eilenberg–Zilber property, the representation of the elements of a tensor product by minimal pairs is unique (see Proposition 4.2.7).

(ii) The tensor product with a cosimplicial set satisfying the Eilenberg–Zilber property preserves and reflects monomorphisms (see Corollary 4.2.9). □

Later on, it will be shown that the geometric realization of an injective simplicial map is even a closed cofibration. A part of this fact is proved immediately below.

Lemma 4.3.4 *If the simplicial set Y is a simplicial subset of the simplicial set X, then the geometric realization $|Y|$ of Y is a closed subspace of the geometric realization $|X|$ of X.*

Proof To begin with, observe that a point $[x, t]$ of $|X|$ belongs to $|Y|$ iff $xt^{\#}$ is an element of Y (see Lemma 4.2.5 (iv)). Now, let C be a closed subset of $|Y|$ and take an n-simplex x of X. One has to show that $\bar{c}_x^{-1}(C)$, the inverse image of C in Δ^n with respect to the map $\bar{c}_x : \Delta^n \to |X|$ given at the beginning of this section, is closed in Δ^n. The set $\bar{c}_x^{-1}(C)$ is the union of the sets $\Delta^\mu(\bar{c}_{x\mu}^{-1}(C))$ taken over the finitely many face operators μ with $x\mu \in Y$. Since C is closed in $|Y|$, the sets $\bar{c}_{x\mu}^{-1}(C)$ are closed in $\operatorname{dom} \Delta^\mu$. The desired result now follows from the fact that all the maps Δ^μ are closed. □

Now it is possible to establish the connection between simplicial sets and CW-complexes.

Theorem 4.3.5 *If X is a simplicial set then the sequence $\{|X^n| : n \in \mathbf{N}\}$ is a CW-structure for its geometric realization $|X|$. The corresponding cell decomposition of $|X|$ is given by the subsets*

$$e_x = \{[x, t] : t \in (\Delta^{\dim x})^\circ\},$$

of $|X|$, x running through $X^{\#}$, the set of non-degenerate simplices of the simplicial set X; the closed cells are the subsets $\bar{e}_x = \bar{c}_x(\Delta^{\dim x})$, for all $x \in X^{\#}$. Consequently, if X has a finite dimension, then $|X|$ is finite-dimensional and it holds true

$$\dim |X| = \dim X.$$

Proof According to the previous result, all the $|X^n|$ are closed subspaces of $|X|$. Thus, the sequence $\{|X^n| : n \in \mathbf{N}\}$ is a filtration of $|X|$. Since $|X^0| = X_0 \times \Delta^0$ (see Corollary 4.3.2), $|X^0|$ is a discrete space.

Next, one has to show that every pair $(|X^n|, |X^{n-1}|)$ is an adjunction of n-cells. For every non-degenerate n-simplex x, the map $\bar{c}_x|\delta(\Delta^n)$ factors through a unique map $c_x : \delta(\Delta^n) \to |X^{n-1}|$. All these maps c_x fit together to define a partial map $c : X_n^\# \times \Delta^n -/\!\to |X^{n-1}|$. It follows from the unique representation of the points of $|X^n|$ (see Proposition 4.3.3 (i)) that $|X^n| = |X^{n-1}| \sqcup_c X_n^\# \times \Delta^n$, at least at the set level. Now observe that the identification map $\bigsqcup_{k=1}^n X_k^\# \times \Delta^k \to |X^n|$ (induced by Corollary 4.3.2) factors through a unique identification map $|X^{n-1}| \sqcup X_n^\# \times \Delta^n \to |X^n|$. Thus, $|X^n|$ is endowed with the final topology with respect to the induced restrictions $|X^{n-1}| \to |X^n|$ and $X_n^\# \times \Delta^n \to |X^n|$. Because $X_n^\# \times \Delta^n \approx X_n^\# \times B^n$ and $X_n^\# \times \delta\Delta^n \approx X_n^\# \times S^{n-1}$, the pair $(|X^n|, |X^{n-1}|)$ is an adjunction of n-cells.

Finally, one must prove that the space $|X|$ has the topology determined by the family $(|X^n| : n \in \mathbf{N})$. For this, take a function $f : |X| \to Z$, Z any space, whose restrictions to all $|X^n|$ are continuous. Since each of the maps $\bar{c}_x$ factors through some $|X^n|$, for all $x \in X$, the compositions $f \circ \bar{c}_x$ are continuous, and thus f is continuous; this finishes the proof. □

Corollary 4.3.6 *If X is a simplicial set,*

$$\dim X = \dim |X|. \qquad \square$$

Corollary 4.3.7 (i) *The geometric realization of a simplicial map is a regular map;*

(ii) *the geometric realization of a constant simplicial map is a constant map.*

Proof (i) Since the degeneracy operators Δ^ρ preserve interiors (see Lemma 4.1.2), the image of an open cell under the geometric realization of a simplicial map is always an open cell.

(ii) A constant simplicial map factors through $\Delta[0]$; thus its geometric realization factors through Δ^0, which is a one-point space. □

Thus geometric realization can be viewed as a functor $SiSets \to CW^r$. The fact mentioned before Lemma 4.3.4 is now also evident:

Corollary 4.3.8 (i) *If Y is a simplicial subset of the simplicial set X, then $|Y|$ is a CW-subcomplex of $|X|$;*

(ii) *if $\hat{Y}$ is a (CW-) subcomplex of the geometric realization of a simplicial set X, then $\hat{Y}$ is the geometric realization of a simplicial subset Y* of *X;*

(iii) *if $f : Y \to X$ is an injective simplicial map, then $|f|$ is an embedding of a CW-subcomplex, and thus a closed cofibration.* □

Remark The statement of Theorem 4.3.5 also implies that there are canonical characteristic maps for the cells of the geometric realization of a simplicial set, at least after fixing a sequence of homeomorphisms $B_n \to \Delta_n$. Just combine those homeomorphisms with the corresponding maps $\bar{c}_x : \Delta_n \to |X|$. In this sense, the maps $\bar{c}_x$ themselves will be called characteristic maps, for all non-degenerate simplices $x \in X$. □

Some crucial properties of geometric realization follow from the fact that it has a right adjoint $S : Top \to SiSets$, called the *singular* functor. Again (see Propositions 4.2.10 and 4.2.12), its existence is a consequence of the adjoint functor generating principle (see Section A.10), which gives still another interpretation of the geometric realization of a simplicial set.

Lemma 4.3.9 *If X is a simplicial set, then*

$$|X| = \operatorname{colim} \Delta_X,$$

where $\Delta_X = \Delta^- \circ DX$ is the composition of Δ^- with the forgetful functor D_X. □

Historically, the singular functor gives the first simplicial sets that were considered and can be explicitly described as follows. The *singular set* of a space T is the simplicial set ST obtained by taking

$$(ST)_n = \text{set of all maps } \Delta^n \to T$$

for all $n \in \mathbf{N}$ and

$$x\alpha = x \circ \Delta^\alpha$$

for all elements $x \in (ST)_n$, all operators $\alpha \in \Delta[n]$. The elements of $(ST)_n$ are the *singular n-simplices* of the space T. A continuous map $f : T \to U$ gives rise to a simplicial map $Sf : ST \to SU$ by taking

$$Sf(x) = f \circ x;$$

this formula really yields a simplicial map as a consequence of the associativity law for compositions of maps. It is worthwhile to exhibit in detail the tools that are provided by this adjointness: for a simplicial set X, one has the unit $\eta_X : X \to S|X|$, which is the simplicial embedding associating to a simplex $x \in X_n$ the singular simplex

$$\eta_X(x) : \Delta^n \to |X|, \qquad t \mapsto [x, t].$$

For a space T, the co-unit $j_T : |ST| \to T$ is given by

$$j_T([x, t]) = x(t),$$

for every singular simplex $x:\Delta^n \to T$ and every point $t \in \Delta^n$. (Later on – see Theorem 4.5.30 – it will be shown that the map j_T is a weak homotopy equivalence, for every space T.) The main point of the adjointness is that it gives a bijective correspondence between maps $g : |X| \to T$ and simplicial maps $f:X \to ST$. The adjoint of the map g is the simplicial map $g' = Sg \circ \eta_X$; the adjoint of the simplicial map f is the map $j_T \circ |f|$. That these processes of forming adjoints are inverse to each other is due to the *fundamental* equations between unit and co-unit:

$$j_{|X|} \circ |\eta_X| = \mathbf{1}_{|X|}, \qquad Sj_T \circ \eta_{ST} = \mathbf{1}_{ST}.$$

As a left adjoint functor, geometric realization preserves all colimits. More specifically, geometric realization preserves pushouts and also has a somewhat stronger property.

Proposition 4.3.10 *Geometric realization transforms simplicial attachings into (topological) attachings.*

Proof Since geometric realization transforms injective simplicial maps into closed cofibrations (see Corollary 4.3.8 (iii)), partial simplicial maps go over into partial (continuous) maps; thus, the pushout obtained from a simplicial attaching is indeed an adjunction square. □

Example Take the simplicial p-sphere $S[p]$ (see Example 4 of Section 4.2). The proposition tells us that its geometric realization is nothing but the quotient space $\Delta^p/\delta\Delta^p$. A homeomorphism $\Delta^p \to B^p$ induces a homeomorphism $\Delta^p/\delta\Delta^p \to B^p/S^{p-1}$ whose codomain is a p-sphere via the standard map b^p (see Section 1.0). □

The adjointness also implies that geometric realization preserves epimorphisms. More precisely:

Proposition 4.3.11 *A simplicial map is surjective iff its geometric realization is an identification.*

Proof '⇒': Since epimorphisms are preserved under geometric realization, a surjective simplicial map is transformed into a surjective regular map (see Corollary 4.3.7 (i)) which is an identification (see Corollary 2.1.2).

'⇐': Let $f : Y \to X$ be a simplicial map such that $|f|$ is surjective. Since X is generated by its non-degenerate simplices, it suffices to exhibit an inverse image for each non-degenerate simplex of X. Take $x \in X^\#$ and an interior point $t \in \Delta^{\dim x}$. The surjectivity of $|f|$ implies the existence of a simplex $y \in Y$ and a point $s \in \Delta^{\dim y}$ such that

$$[x,t] = |f|([y,s]) = [f(y),s] = [(f(y))^\#, (f(y))^\flat s];$$

one may assume s to be an interior point (see Proposition 4.3.3 (i)). Then, $(f(y))^{\flat}s$ is also an interior point (see Lemma 4.1.2), and consequently the pair $((f(y))^{\#}, (f(y))^{\flat}s)$ is minimal; therefore, $x=(f(y))^{\#}$ (again by Proposition 4.3.3 (i)). Thus, x is a face of $f(y)$ (see Lemma 4.2.5 (iii)), and so it is the image of the corresponding face of y under the simplicial map f. □

There is still another type of colimit construction, which has yet to be mentioned.

Proposition 4.3.12 *Let $X_0 \subset X_1 \subset \cdots \subset X_j \subset \cdots$ be an increasing sequence of simplicial sets and $X = \cup X_r$. Then, $\{|X_r| : r \in \mathbf{N}\}$ is an expanding sequence with union space $|X|$.* □

Geometric realization also commutes with finite limits. To see this, one has to show that it commutes with equalizers and finite products. Indeed, geometric realization preserves and reflects equalizers:

Proposition 4.3.13 *Let $f, g : Y \to X$ be a pair of simplicial maps and let Z be a simplicial subset of Y. Then Z is the difference kernel of the pair (f, g) iff $|Z|$ is the difference kernel of the pair $(|f|, |g|)$.*

Proof The geometric realization $|Z|$ of Z is a closed subspace of the geometric realization $|Y|$ of the simplicial set Y (see Lemma 4.3.4).

'⇒': Let Z be the difference kernel of the pair (f, g), i.e., $Z = \{y \in Y : f(y) = g(y)\}$. Then its geometric realization $|Z|$ is clearly contained in the difference kernel of the pair $(|f|, |g|)$. Now consider a point $[y, t]$ in the difference kernel of $(|f|, |g|)$; assume t to be an interior point (see Proposition 4.3.3 (i)). Then,

$$\begin{aligned}[(f(y))^{\#}, (f(y))^{\flat}t] &= [f(y), t] = |f|([y, t]) = |g|([y, t]) \\ &= [g(y), t] = [(g(y))^{\#}, (g(y))^{\flat}t].\end{aligned}$$

Since $(f(y))^{\flat}$ and $(g(y))^{\flat}$ are degeneracy operators, $(f(y))^{\flat}t$ and $(g(y))^{\flat}t$ are interior points (see Lemma 4.1.2). The uniqueness of the representation by minimal pairs (see again Proposition 4.3.3 (i)) now gives

$$(f(y))^{\#} = (g(y))^{\#}, \qquad (f(y))^{\flat}t = (g(y))^{\flat}t.$$

The second of these equations implies $(f(y))^{\flat} = (g(y))^{\flat}$ (see Lemma 4.1.1); combining it with the first, we obtain

$$f(y) = (f(y))^{\#}(f(y))^{\flat} = (g(y))^{\#}(g(y))^{\flat} = g(y),$$

i.e., $y \in Z$; thus, $[y, t] \in |Z|$.

'$\Leftarrow$': Take $|Z|$ to be the difference kernel of the pair $(|f|, |g|)$. Consider a simplex $y \in Y$ and choose an interior point $t \in \Delta^{\dim y}$. Now, if $y \in Z$, the same argument as above shows $f(y) = g(y)$. On the other hand, if y is in the difference kernel of the pair (f, g), then

$$|f|([y, t]) = [f(y), t] = [g(y), t] = |g|([y, t]),$$

i.e., $[y, t] \in |Z|$; this implies $y \in Z$. □

Now turn to finite products. Consider the product $Y \times X$ of the simplicial sets Y and X. The projections $p_Y : Y \times X \to Y$ and $p_X : Y \times X \to X$ induce a natural map $h_{Y,X} = (|p_Y|, |p_X|) : |Y \times X| \to |Y| \times |X|$.

Lemma 4.3.14 *For $q, p \in \mathbf{N}$, the natural map*

$$h_{\Delta[q], \Delta[p]} : |\Delta[q] \times \Delta[p]| \to |\Delta[q]| \times |\Delta[p]|$$

is a homeomorphism.

Proof Identify $|\Delta[q]|$ and $|\Delta[p]|$ with Δ^q and Δ^p respectively (via the homeomorphisms described in Example 1 of Section 4.2). The aim is to construct a function $g : \Delta^q \times \Delta^p \to |\Delta[q] \times \Delta[p]|$ which is inverse to the map $h_{\Delta[q], \Delta[p]}$. To this end, consider $(\boldsymbol{t}, \boldsymbol{t}') \in \Delta^q \times \Delta^p$. Order all of the sum coordinates of $\boldsymbol{t}$ and $\boldsymbol{t}'$ to form an $(r+2)$-tuple $(u_{-1}, u_0, \ldots, u_r)$ such that

$$0 = u_{-1} < u_0 < \cdots < u_{r-1} < u_r = 1,$$

thus obtaining the sum coordinates of an interior point $\boldsymbol{v} \in \Delta^r$. Now there are unique operators $\alpha : [r] \to [q]$, $\beta : [r] \to [p]$ such that $\alpha\boldsymbol{v} = \boldsymbol{t}$ and $\beta\boldsymbol{v} = \boldsymbol{t}'$ respectively (see Lemma 4.1.5). The pair (α, β) is an r-simplex of $\Delta[q] \times \Delta[p]$; thus, it is possible to define $g(\boldsymbol{t}, \boldsymbol{t}') = [(\alpha, \beta), \boldsymbol{v}]$. But

$$\begin{aligned} h_{\Delta[q], \Delta[p]} \circ g(\boldsymbol{t}, \boldsymbol{t}') &= h_{\Delta[q], \Delta[p]}([(\alpha, \beta), \boldsymbol{v}]) \\ &= ([\alpha, \boldsymbol{v}], [\beta, \boldsymbol{v}]) = (\alpha\boldsymbol{v}, \beta\boldsymbol{v}) = (\boldsymbol{t}, \boldsymbol{t}'), \end{aligned}$$

and so

$$h_{\Delta[q], \Delta[p]} \circ g = \mathbf{1}_{\Delta^q \times \Delta^p}.$$

For the other composition, take $([(\alpha, \beta), \boldsymbol{v}]) \in |\Delta[q] \times \Delta[p]|$ and assume (α, β) to be a non-degenerate r-simplex of $\Delta[q] \times \Delta[p]$ and $\boldsymbol{v} \in (\Delta^r)^\circ$. One again obtains

$$h_{\Delta[q], \Delta[p]}([(\alpha, \beta), \boldsymbol{v}]) = (\alpha\boldsymbol{v}, \beta\boldsymbol{v}).$$

Now, if $g(\alpha\boldsymbol{v}, \beta\boldsymbol{v})$ were to be different from $([(\alpha, \beta), \boldsymbol{v}])$, there would be at least one sum coordinate u_i of $\boldsymbol{v}$ which would be neither a sum coordinate of $\alpha\boldsymbol{v}$ nor of $\beta\boldsymbol{v}$. But this would imply $(\alpha, \beta) = (\alpha'\sigma_i, \beta'\sigma_i) = (\alpha', \beta')\sigma_i$ to be degenerate, a contradiction!

Thus, the natural map $h_{\Delta[q], \Delta[p]} : |\Delta[q] \times \Delta[p]| \to |\Delta[q]| \times |\Delta[p]|$ is bijective. Now observe that every r-simplex $(\alpha, \beta) \in \Delta[q] \times \Delta[p]$ with $r > q + p$ is degenerated. Indeed, inspection of the unique decompositions

$$\alpha^{\flat} = \sigma_{j_1} \circ \cdots \circ \sigma_{j_s}, \qquad \beta^{\flat} = \sigma_{j'_1} \circ \cdots \circ \sigma_{j'_{s'}},$$

with $0 \leqslant j_1 < \cdots j_s < r, 0 \leqslant j'_1 < \cdots j'_{s'} < r, s \geqslant r - q$ and $s' \geqslant r - p$ leads to $s + s' \geqslant 2r - q - p > r$, which implies that the sets $\{j_1, \ldots, j_s\}$ and $\{j'_1, \ldots, j'_{s'}\}$ have at least one element in common. Thus, $\Delta[q] \times \Delta[p]$ has only finitely many non-degenerate simplices; consequently, $|\Delta[q] \times \Delta[p]|$ is a finite CW-complex (see Theorem 4.3.5), and therefore compact (Proposition 1.5.8). But a continuous bijection with a compact domain and a Hausdorff codomain is a homeomorphism. □

Remark The argument just given shows that the product of a simplicial set of dimension q and a simplicial set of dimension p is of dimension $q + p$. This fact is not rewarded with a special statement because it is obtained from the similar theorem for CW-complexes (see Theorem 2.2.2) via geometric realization. □

Proposition 4.3.15 *The natural map $h_{Y,X} : |Y \times X| \to |Y| \times |X|$ is a homeomorphism, for all simplicial sets Y, X.*

Proof An inverse function g to $h_{Y,X}$ can be given as follows:

$$g([y, \boldsymbol{t}], [x, \boldsymbol{t}']) = [(y\alpha, x\beta), \boldsymbol{v})]$$

for $y \in Y_q$, $x \in Y_p$ and with α, β, $\boldsymbol{v}$ constructed from $(\boldsymbol{t}, \boldsymbol{t}')$ as in the preceding proof. The closed cells of the product $|Y| \times |X|$ are the sets $\bar{e}_y \times \bar{e}_x$ with $y \in Y^{\#}_q, x \in X^{\#}_p, q, p \in \mathbf{N}$ (Theorem 4.3.5 and Theorem 2.2.2). Thus, for the continuity of the function g, it suffices to show that all the restrictions $g | \bar{e}_y \times \bar{e}_x$ are continuous. Moreover, the maps $\bar{c}_y \times \bar{c}_x$ induce identification maps $\Delta^q \times \Delta^p \to \bar{e}_y \times \bar{e}_x$; thus, it remains to show that the compositions $g \circ (\bar{c}_y \times \bar{c}_x)$ are continuous. These functions are – up to suitable identifications – nothing but the compositions of the homeomorphisms $(h_{\Delta[q], \Delta[p]})^{-1}$ with the geometric realizations $|\varphi_y \times \varphi_x|$ of the simplicial maps $\varphi_y \times \varphi_x : \Delta[q] \times \Delta[p] \to Y \times X$ which are given by the assignment $(\gamma, \gamma') \mapsto (y\gamma, x\gamma')$ (cf. the sketch of the proof of Lemma 4.2.1). □

Propositions 4.3.13 and 4.3.15 together imply, as announced:

Theorem 4.3.16 *Geometric realization preserves finite limits.* □

The fact that geometric realization commutes with products shows that from a geometric point of view the simplicial homotopy notions defined in the previous section actually do what they are supposed to do. Observe that $|Y \times \Delta[1]| \cong |Y| \times |\Delta[1]| \cong |Y| \times \boldsymbol{I}$, according to Proposition 4.3.15 and the homeomorphism $|\Delta[1]| \to \Delta^1 \to \boldsymbol{I}, [\iota, t] \mapsto t \mapsto t_1$. This proves

Proposition 4.3.17 *The geometric realizations of a pair of simplicially homotopic simplicial maps is a pair of homotopic maps. Consequently, simplicial homotopy equivalences are transformed into (geometric) homotopy equivalences by geometric realization.* □

Conversely, it is easy to construct a pair of simplicial maps whose geometric realizations are homotopic without the simplicial maps themselves being simplicially homotopic. Nevertheless, a sort of reciprocal statement to Proposition 4.3.17 is true.

Proposition 4.3.18 *If $f, g : T \to U$ are homotopic maps, the corresponding singular maps Sf, $Sg : SY \to SX$ are simplicially homotopic. Consequently, homotopy equivalences are transformed into simplicial homotopy equivalences by the singular functor.*

Proof Let h be a homotopy from f to g. Assume the domain of h to be $T \times |\Delta[1]|$. As a right adjoint functor, the singular functor commutes with products. Thus one can consider Sh as a simplicial map from $ST \times S|\Delta[1]|$ to SU. Now composition of Sh with the simplicial map $\mathbf{1}_{ST} \times \eta_{\Delta[1]}$ yields the desired simplicial homotopy. □

Corollary 4.3.19 *The composed functors $|S-|$ and $S|-|$ preserve homotopies, homotopy equivalences, deformation retractions and contractibility.* □

Moreover, note that the composed functor $|S-|$ transforms subspaces into (CW–)subcomplexes.

There is some further terminology in this context. A simplicial map f is a *weak homotopy equivalence* if its geometric realization $|f|$ is an honest homotopy equivalence; a simplicial set is *weakly contractible* if its geometric realization is contractible.

The following is an application of the Eilenberg–Zilber property for convariant functors $\boldsymbol{\Delta} \to \mathit{SiSets}$.

Theorem 4.3.20 (*Comparison theorem*) *Let Φ, $\Phi' : \boldsymbol{\Delta} \to \mathit{SiSets}$ be (covariant) functors with the Eilenberg–Zilber property and let $\varphi : \Phi \to \Phi'$ be a natural transformation such that $\varphi_{[n]} : \Phi[n] \to \Phi'[n]$ is a weak homotopy equivalence, for every $n \in \mathbf{N}$. Then, the induced simplicial map*

$$\varphi_X = \mathbf{1}_X \otimes \varphi : X \otimes \Phi \to X \otimes \Phi'$$

is a weak homotopy equivalence, for all simplicial sets X.

Proof The claim is evident if X is a coproduct of standard-simplices. The statement will be proved next for simplicial sets having finite dimension, using induction on the dimension. If $\dim X = 0$ then, X is a coproduct of 0-simplices. Now assume $\dim X = n > 0$ and observe that X may be obtained from X^{n-1} by means of a simplicial attaching of n-simplices, i.e.,

$$X = X^{n-1} \sqcup_f \bigsqcup \Delta[n]$$

for some simplicial map $f : \sqcup \delta\Delta[n] \to X^{n-1}$ (see Corollary 4.2.4 (ii)). Since the functors Φ and Φ' have the Eilenberg–Zilber property, the tensor products $-\otimes\Phi$ and $-\otimes\Phi'$ preserve simplicial attachings (see Proposition 4.2.12); thus,

$$X \otimes \Phi = X^{n-1} \otimes \Phi \sqcup_{f\otimes\Phi} \bigsqcup \Phi[n],$$

as well as

$$X \otimes \Phi' = X^{n-1} \otimes \sqcup_{f\otimes\Phi'} \bigsqcup \Phi'[n].$$

Geometric realization preserves attachings (see Proposition 4.3.10); consequently, one obtains $|X \otimes \Phi|$ and $|X \otimes \Phi'|$ by the induced attachings. The map $|\varphi_{\sqcup\Delta[n]}| : |\bigsqcup \Phi[n]| \to |\bigsqcup \Phi'[n]|$ is a homotopy equivalence as stated in the beginning; by the induction hyptothesis, the same holds true for the maps $|\varphi_{\sqcup\delta\Delta[n]}| : \bigsqcup \delta\Delta[n] \otimes \Phi| \to |\bigsqcup \delta\Delta[n] \otimes \Phi'|$ and $|\varphi_{X^{n-1}}|$. The result now follows from the gluing theorem (see Theorem A.4.12).

If X does not have finite dimension, one has expanding sequences

$$\{|X^n \otimes \Phi| : n \in \mathbf{N}\}, \qquad \{|X^n \otimes \Phi'| : n \in \mathbf{N}\}$$

and a commutative ladder of homotopy equivalences connecting them; thus, the induced map between the union spaces (which can be considered as φ_X) is also a homotopy equivalence (see Proposition A.5.11). □

This shows that replacing the simplices of a simplicial set by some more complicated objects does not alter the weak homotopy type as long as the new objects are weakly contractible.

Corollary 4.3.21 *Let $\Phi : \mathbf{\Delta} \to SiSets$ be a (covariant) functor with the Eilenberg–Zilber property and such that $\Phi([n])$ is weakly contractible, for all $n \in \mathbf{N}$. Furthermore, let a natural transformation $\varphi : \Phi \to \Delta$ be given. Then, the induced simplicial map $\varphi_X = 1_X \otimes \phi : X \otimes \Phi \to X$ is a weak homotopy equivalence, for all simplicial sets X.* □

Remark Clearly, the same statement is true if one is given a natural transformation $\Delta \to \Phi$ instead of that assumed in the statement of Corollary 4.3.21. In some sense, it is not even necessary to require the existence of any natural transformation (see Section 4.5, Exercise 1). □

Recall that a (continuous) map $p : Y \to X$ is *locally trivial* if every point $x \in X$ has a neighbourhood U such that the map induced from p by the inclusion $U \subset X$ can be chosen as the projection of the product $U \times p^{-1}(x)$ onto U.

Proposition 4.3.22 *The geometric realization of a locally trivial simplicial map is a locally trivial (continuous) map.*

Proof Let $p : Y \to X$ be a locally trivial simplicial map and consider a point $[\tilde{x}, \tilde{t}] \in |X|$; without loss generality, one may assume $\tilde{x}$ to be a non-degenerate n-simplex in X and $\tilde{t}$ an interior point of Δ^n (see Proposition 4.3.3 (i)). The open cell of the CW-complex $|X|$ corresponding to the simplex $\tilde{x}$ is an open neighbourhood U_n of $[\tilde{x}, \tilde{t}]$ in $|X^n| = |X|^n$. Forming the infinite collar of this open cell with respect to the canonical characteristic maps (see the Remark after Corollary 4.3.8), one obtains a neighbourhood U of $[\tilde{x}, \tilde{t}]$ in $|X|$ (see Proposition 1.3.1 (ii)); U is the union space of the expanding sequence of the intermediate collars $U_m, m > n$ (see Proposition 1.3.1 (iv)). Let $q : V \to U$ and $q_m : V_m \to U_m$ denote the maps which are induced from $|p|$ by the inclusions $U \subset |X|$ and $U_m \subset |X|$, for $m \geqslant n$, respectively; note that the maps q_m can also be thought as induced from q by the inclusions $U_m \subset U$. The spaces V and V_m can be considered as subspaces of $|Z|$, and V is determined by the family $\{V_m : m \geqslant n\}$ (see Corollary A.2.3, which will be used several times in the sequel without explicit reference). For any non-degenerate simplex $x \in X_{m+1}$, let e_x denote the corresponding open cell of $|X|$. Set $U_x = U_m \cup (U_{m+1} \cap e_x)$ and let $q_x : V_x \to U_x$ denote the map induced from q_{m+1} by the inclusion $U_x \subset U_{m+1}$. Then, U_{m+1} is determined by the family $\{U_m\} \cup \{U_x : x \in X^{\#}_{m+1}\}$ (see Proposition 1.1.3 (ii)) and V_{m+1} is determined by the family $\{V_m\} \cup \{V_x : x \in X^{\#}_{m+1}\}$.

Now, let F denote the fibre of p over $\tilde{x}\varepsilon_0$. It will be shown that there is a homeomorphism $h : V \to U \times |F|$ whose composition with the projection onto U is just q. Since $U \times |F|$ is determined by the family $\{U_m \times |F| : m \geqslant n\}$, this can be done by an inductive construction of suitable homeomorphisms $h_m : V_m \to U_m \times |F|$.

Start with $m = n$: Because the simplicial map p is assumed locally trivial, there is an isomorphism $h' : W \to \Delta[n] \times F$, with $pr_{\Delta[n]} \circ h = \bar{p}$, where $\bar{p} : W \to \Delta[n]$ denotes the simplicial map which is induced from p by $\tilde{f} : \Delta[n] \to X, \alpha \mapsto \tilde{x}\alpha$. Since geometric realization preserves finite limits (see Theorem 4.3.16), the projection $\Delta^n \times |F| \to \Delta^n$ is (up to the homeomorphism $|h|$) induced from $|p|$ by the characteristic map $|\tilde{f}| = \bar{c}_{\tilde{x}}$. The map q_n can be thought as induced from this projection by the inclusion

$U_n \subset \Delta^n, [x,t] \mapsto t$; this yields the homeomorphism $h_n : |p|^{-1}(U_n) \to U_n \times |F|$.

Assume h_m is constructed. Again, $U_{m+1} \times |F|$ is determined by the family $\{U_m \times |F|\} \cup \{U_x \times |F|\}$, and it suffices to construct suitable homeomorphisms $h_x : V_x \to U_x \times |F|$. Because different U_x intersect exactly in U_m, one can restrict the argument to a single non-degenerate $(m+1)$-simplex x.

It follows from the definition of the collaring process that the starting simplex $\tilde{x}$ is a face of x; thus, the fibers of p over $\tilde{x}\varepsilon_0$ and $x\varepsilon_0$ can be identified (see Proposition 4.2.2). Now, local triviality implies, as in the case $n = m$, that the projection $\Delta^{m+1} \times |F| \to \Delta^{m+1}$ is induced from $|p|$ by the characteristic map $\bar{c}_x$; let $\tilde{c} : \Delta^{m+1} \times |F| \to |Z|$ denote the map which is correspondingly induced from $\bar{c}_x$ by $|p|$. Take $U' = \bar{c}_x^{-1}(U_x), \delta U' = \bar{c}_x^{-1}(U_m)$ and observe that there is a retraction $r : U' \to \delta U'$. Moreover, U_x is obtained from U_m by attaching U' via a map with domain $\delta U'$ (see Lemma 1.1.8); the characteristic map c' and the attaching map $\delta c'$ of this attaching can be taken as induced from $\bar{c}_x$ by the inclusion of U_x and U_m respectively, into $|X|$. Consequently, V_x is obtained from V_m by attaching $U' \times |F|$ via a map with domain $\delta U' \times |F|$ (see Proposition A.2.2); the characteristic map $\hat{c}$ and the attaching map $\delta\hat{c}$ of this attaching can be taken as induced from $\tilde{c}$ by the inclusion of V_x and V_m respectively, into $|Z|$. Therefore, in order to obtain the homeomorphism h_x, one needs a suitable map $U' \times |F| \to U_x \times |F|$ whose components $h' : U' \times |F| \to U_x$ and $h'' : U' \times |F| \to |F|$ can be defined as follows. For h', take the composition of the projection onto U' with c'. For h'', compose the maps $r \times \mathbf{1}_{|F|}, \delta\hat{c}, h_m$ and the projection $U_m \times |F| \to |F|$. The resulting map h_x is bijective; it remains to show the continuity of the inverse function. Since the functor $- \times |F|$ preserves attachings (see Section A.1) $U_x \times |F|$ is obtained from $U_m \times |F|$ by attaching $U' \times |F|$. Thus, it suffices to establish the continuity of the functions $h_x{}^{-1}|U_m \times |F|$ and $h_x{}^{-1} \circ (c' \times \mathbf{1}_{|F|})$. The first case is trivial: $h_x{}^{-1}|U_m \times |F| = h_m{}^{-1}$. For the second, note that the map $\delta\hat{c}$ is induced from $\delta c'$ by q_m; thus, the universal property of pullbacks can be used to obtain a map $g : U' \times |F| \to \delta U' \times |F|$, with $\delta\hat{c} \circ g = h_m{}^{-1} \circ (\delta c' \circ r \times \mathbf{1}_{|F|})$ and $pr_{\delta U'} \circ g = r \circ pr_{U'}$, where $pr_{\delta U'}$ and $pr_{U'}$ denote the respective projections. Take $g' : U' \times |F| \to U' \times |F|$ whose components are the projection $pr_{U'}$ and the composition of g with the projection onto $|F|$. This yields that the composition $h_x{}^{-1} \circ (c' \times \mathbf{1}_{|F|}) = \hat{c} \circ g'$ is also continuous. □

Corollary 4.3.23 *The geometric realization of a locally trivial simplicial map is a (locally trivial) fibration.*

Proof Let $p : Z \to X$ be a locally trivial simplicial map. Since the codomain of the map $|f|$ is a CW-complex (see Theorem 4.3.5), and therefore paracompact (see Theorem 1.3.5), the locally trivial map $|f|$ is a fibration (see Theorem A.4.22). □

Exercises

1. Show that the geometric realizations of an ordered simplicial complex in the sense of Section 3.3 and its associated simplicial set (see Section 4.2, Exercise 4) are naturally equivalent.
2. Show that the geometric realization functor $OSiCo \to CW^r$ preserves products.
3. Construct two simplicial maps that are not simplicially homotopic, but whose geometric realizations are homotopic.
4. Construct a simplicial map that is a weak, but not a simplicial, homotopy equivalence.

4.4 Presimplicial sets

A presimplicial object in the category *Sets* of sets, i.e., an object of the category *PSiSets*, is called a *presimplicial set*. In dealing with presimplicial sets, it is again convenient to view a presimplicial set X as an $\mathbf{N}$-graded set $X = \sqcup X_n$, now with the small category $\boldsymbol{M}$ operating on the right; the elements of a presimplicial set are also called simplices with the same notion of dimension. The category *PSiSets* has similar formal properties as the category *SiSets*; it has all limits and colimits as well as suitable notions of generators and relations. The Yoneda embedding $\Delta : \boldsymbol{\Delta} \to$ *SiSets* is only replaced by the Yoneda embedding $M : \boldsymbol{M} \to$ *PSiSets*. But in a presimplicial set there is no difference between degenerate and non-degenerate simplices. It might also happen that almost all X_n are empty; it is reasonable to define the dimension of a presimplicial set X – notation: $\dim X$ – by taking

$$\dim \varnothing = -1$$

and, for $X \neq \varnothing$,

$$\dim X = \max \{n : X_n \neq \varnothing\}.$$

If X is a simplicial set viewed as a functor $\boldsymbol{\Delta} \to$ *Sets*, one may form its restriction $X|\boldsymbol{M}$ to the subcategory $\boldsymbol{M}$ of $\boldsymbol{\Delta}$ to obtain a presimplicial set; by evident reasons, this process will be referred to as *forgetting degeneracies*. It extends to the forgetful functor P : *SiSets* $\to$ *PSiSets* which is right adjoint to a (non-full) embedding E : *PSiSets* $\to$ *SiSets*, described explicitly as follows. To a presimplicial set X assign the simplicial set EX, generated by all the elements of X (in the corresponding dimensions) and subject to the relations $x\mu = y$ for every triple (x, μ, y) satisfying this equation in the presimplicial set X. Thus, the n-simplices of EX can be viewed as pairs (x, ρ) consisting of a simplex x of X and a degeneracy operator $\rho : [n] \to [\dim x]$; the operation of Δ is given by the formula

$$(x, \rho)\alpha = (x(\rho\alpha)^{\#}, (\rho\alpha)^{\flat}),$$

for all such pairs (x, ρ) and all operators α with $\operatorname{cod} \alpha = [n]$. If $f : Y \to X$ is a *presimplicial map*, i.e., a morphism in the category *PSiSets*, then the simplicial map $Ef : EY \to EX$ is given by the assignment $(y, \rho) \mapsto (f(y), \rho)$. The unit for the adjunction is the family $\{u_X : X \in \mathrm{Ob}\, PSiSets\}$ of the presimplicial maps $u_X : X \to PEX, x \mapsto (x, \iota)$; the counit is the family $\{p_X : X \in \mathrm{Ob}\, SiSets\}$ of the simplicial maps $p_X : EPX \to X, (x, \rho) \mapsto x\rho$. In particular, the embedding E commutes with the Yoneda embeddings, i.e., $\Delta | M = E \circ M$.

The geometric realization of presimplicial sets is defined via simplicial sets, just as the composed functor $|\text{—}| \circ E = |E-|$. The simplices of a presimplicial set X correspond bijectively to the non-degenerate simplices of the simplicial set EX, i.e., $X = (EX)^{\#}$. In particular, this implies that the introduced notion of dimension for a presimplicial set X is compatible with the geometric realization:

$$\dim X = \dim |EX|.$$

For a simplicial set X, the geometric realization $X = |EPX|$ is also called the *fat realization* of X; it is infinite-dimensional except for $X = \varnothing$.

The geometric realization functor $|E-| : PSiSets \to Top$ is left adjoint to the composite functor $PS : Top \to PSiSets$. The unit and the co-unit of this adjunction are given by the families $\{\eta'_X : X \in \mathrm{Ob}\, PSiSets\}$, $\{j'_T : T \in \mathrm{Ob}\, Top\}$ with $\eta'_X = P\eta_{EX} \circ u_X$, $j'_T = j_T \circ |p_{ST}|$ respectively.

By abuse of language, one says that a simplicial set (map) is a *presimplicial set* (*map*) if it is in the image of the embedding functor E; this is the case

for a simplicial set iff the faces of non-degenerate simplices are again always non-degenerate; and

for a simplicial map iff domain and codomain are presimplicial sets and non-degenerate simplices are mapped onto non-degenerate simplices.

On the other hand, pairs of presimplicial maps are *simplicially homotopic* if they are such as simplicial maps. Similarly, one carries over the notions *simplicial* or *weak homotopy equivalence*, *simplicially* or *weakly contractible* into considerations of presimplicial maps and sets respectively.

Remark It does not make sense to look for an analogue presimplicial homotopy within the category *PSiSets*. The formal reason is that the embedding E is left adjoint; thus, it does not commute with products in general. For instance, the product $M[1] \times M[1]$ does not yield a square (after geometric realization), but, rather, the disjoint union of an interval and two single points. However, there is a nice sufficient condition for simplicial contractibility within the category *PSiSets*. For this one needs a special construction. □

The *cone of* the presimplicial set X is the presimplicial set CX given by

$$(CX)_0 = X_0 \cup \{*\},$$

where $*$ is just one extra element which does not belong to X,

$$(CX)_n = X_n \sqcup X_{n-1},$$

for $n > 0$ – if an $(n-1)$-simplex x of X is considered as an n-simplex of CX it will be denoted by 'x_c' in the sequel –

$$CX(\delta_i)(x) = X(\delta_i)(x),$$

for $x \in X_n \subset (CX)_n, 0 \leqslant i \leqslant n, n > 0,$

$$CX(\delta_0)(x_c) = x,$$

for $x \in X_{n-1}, n > 0,$

$$CX(\delta_1)(x_c) = *,$$

for $x \in X_0 \subset (CX)_1,$

$$CX(\delta_i)(x_c) = X(\delta_{i-1})(x),$$

for $x \in X_{n-1} \subset (CX)_n, 0 < i \leqslant n, n > 0.$

If $X = \varnothing$, the cone CX consists of exactly one simplex (of dimension 0); otherwise, CX is generated by the set $\{x_c : x \in X\}$ subject to the relations

$$x_c \delta_i = (x\delta_i)_c$$

for $x \in X_n \subset (CX)_{n+1}, 0 < i \leqslant n+1, n \geqslant 0.$

This construction extends in the evident manner to presimplicial maps giving rise to the *cone functor* $C : PSiSets \to PSiSets$. It is connected to the identity functor via a natural transformation which is given by the embeddings $c_X : X \to CX, x \mapsto x_c \delta_0$.

Theorem 4.4.1 *The cone functor C commutes with the geometric realization; more precisely, there is a natural equivalence* $|EC-| \dot{\to} C|E-| : PSiSets \to CW^r$.

Proof Let X be a non-empty presimplicial set. The geometric realization $|ECX|$ of its cone CX is a quotient space of $\sqcup X_n \times \Delta^{n+1}$ (see

Proposition 4.3.1). Consider Δ^{n+1} as a cone with peak e_0, i.e., represent any $s \in \Delta^{n+1}$ in the form

$$s = (1-t) \cdot \delta_0 s' + t e_0,$$

with $t \in \boldsymbol{I}$ and $s' \in \Delta^n$ (see Section 3.1, Example 2). Then, define a natural homeomorphism $h : |ECX| \to CX$ by taking

$$h([x_c, s]) = [[x, s'], t]$$

for $x \in X_n$ and $s \in \Delta^{n+1}$. □

Remark Notice that the homeomorphism just constructed is a regular map between the corresponding CW-complexes; thus, it is cellular in both directions. □

The following is a technical lemma which is useful in dealing with the formal apparatus.

Lemma 4.4.2 *Let the simplicial set X be a presimplicial set. Then the simplicial set $X \times \Delta[1]$ is isomorphic to the simplicial set X_I which has the sets $X_n \times [n]$ as generators in dimension $n+1$ and is subject to the relations*

$$\begin{aligned} (x, i)\delta_j &= (x\delta_j, i-1), & j &< i, \\ (x, i)\delta_i &= (x, i-1)\delta_i, & 0 &< i \leqslant n, \\ (x, i)\delta_j &= (x\delta_{j-1}, i), & j &> i+1. \end{aligned}$$

Proof The assignment $(x, i) \mapsto (x\sigma_i, \omega_i)$ induces an isomorphism $X_I \to X \times \Delta[1]$. □

Now, the announced condition for simplicial contractibility can be discussed.

Proposition 4.4.3 *A presimplicial set X is simplicially contractible if there is a simplicial retraction $CX \to X$.*

Proof Define a simplicial homotopy $H : X_I \to ECX$ from the constant map $X \to CX$ with value $*$ to the inclusion c_X by taking

$$H(x, i) = ((x(\delta_0)^i)_c, (\sigma_0)^i).$$

If there is a simplicial retraction $r : ECX \to EX$, then the composition $r \circ H$ shows that EX is contractible to the 0-simplex $r(*)$. □

Now it can be shown that standard-simplices remain homotopically trivial under forgetting degeneracies.

Proposition 4.4.4 *The presimplicial set $P\Delta[n]$ is simplicially contractible, for every $n\in\mathbf{N}$.*

Proof For an operator $\alpha : [m]\to[n]$ define $\alpha' : [m+1]\to[n]$ by taking

$$\alpha'(0)=0, \qquad \alpha'(i+1)=\alpha(i), \qquad \text{for } i\in[m].$$

The assignment $\alpha_c\mapsto\alpha'$ induces a retraction $CP\Delta[n]\to P\Delta[n]$. □

This was the last tool to prove the following:

Theorem 4.4.5 *The simplicial map $p_X : EPX\to X$ is a weak homotopy equivalence, for every simplicial set X.*

Proof The functor EP can be viewed as a tensor product

$$EP=-\otimes EP\Delta,$$

the simplicial sets $EP\Delta[n]$ are simplicially – and therefore, weakly – contractible (see Proposition 4.4.4) and $p_{\Delta[-]} : EP|\Delta\to\Delta-$ is a natural transformation; thus, the simplicial map p_X is a weak homotopy equivalence, for every simplicial set X (see Corollary 4.3.21). □

The formalism of presimplicial sets allows an easy simplicial description of the 3-dimensional lens spaces:

Example Given $p\in\mathbf{N}\backslash\{0\}$ and $q\in\mathbf{Z}_p=\mathbf{Z}/p\mathbf{Z}$ the lens space $L(p,q)$ is the geometric realization of the following simplicial set. Take generators $x_i, i\in\mathbf{Z}_p$, in dimension 3 and require the relations

$$x_i\delta_3=x_{i+1}\delta_2, x_i\delta_0=x_{i+q}\delta_1.$$ □

Exercises

1. There is also a cone construction in the category *SiSets* which is given as follows. Let X be an arbitrary simplicial set and let $*$ be an extra element which does not belong to X. Set

   $$(CX)_n=\{(x,q)\in X\times\mathbf{N} : q+\dim x=n\}\cup\{(*,n)\},$$

   for all $n\in\mathbf{N}$,

   $$(x,q)\delta_i=\begin{cases}(x,q-1), & 0\leqslant i<q,\\ (*,q-1), & q=i, \dim x=0,\\ (x\delta_{i-q},q), & q\leqslant i\leqslant q+\dim x, \dim x>0,\end{cases}$$

   $$(x,q)\sigma_j=\begin{cases}(x,q+1), & 0\leqslant j<q,\\ (x\sigma_{j-q},q), & q\leqslant j\leqslant q+\dim x,\end{cases}$$

for all $(x, q) \in X \times \mathbf{N}$, and

$$(*, n)\alpha = (*, \dim \alpha),$$

for all $n \in \mathbf{N}$ and $\alpha \in \Delta[n]$. Show that these facts define a simplicial set CX and prove simplicial analogues to the presimplicial statements of Theorem 4.4.1 and Proposition 4.4.3. Compare CEX and ECX for a presimplicial set X.

2. Prove the following generalization of the Comparison Theorem (see Theorem 4.3.20): let each of the categories $\mathscr{C}$ and $\mathscr{C}'$ be either *SiSets* or *PSiSets*. Suppose $F, G : \mathscr{C}' \to \mathscr{C}$ are cocontinuous (i.e., compatible with all colimits) functors which preserve injections. In addition, assume $\varphi : F \to G$ is a natural transformation such that the (pre)simplicial maps $\varphi_{\Delta[n]}$ or $\varphi_{M[n]}$ are weak homotopy equivalences, for all $n \in \mathbf{N}$. Then, the (pre)simplicial maps φ_M are weak homotopy equivalences, for all (pre)simplicial sets X.
3. Show that the lens space $L(p, q)$ has fundamental group $\mathbf{Z}/p\mathbf{Z}$ and universal covering S^3. Derive from this that for a fixed p, but all possible q, the Lens spaces $L(p, q)$ have the same homotopy groups.

Remark Using cohomology rings – which are beyond the scope of this book – one can moreover show that $L(p, q)$ and $L(p, q')$ have the same homotopy type iff $q \cdot q'$ or $-q \cdot q'$ is a quadratic residue mod p (Hilton & Wylie 1960, Section 5.10). Thus, for example, $L(5, 1)$ and $L(5, 2)$ have isomorphic homotopy groups, but different type.

4.5 Kan fibrations and Kan sets

The categories *SiSets* and *PSiSets* admit combinatorial analogues of the geometric idea of fibration. The basic notions for these are the so-called *horns* and *anodyne extensions*. For $n \in \mathbf{N}$ and $k \in [n]$, take $\Lambda^k[n]$ as the simplicial subset of the simplicial standard-n-simplex $\Delta[n]$ which is generated by all the elementary face operators δ_i^n with $i \neq k$; in case $n = 0$, take $\Lambda^0[0] = \emptyset$. The simplicial set $\Lambda^k[n]$ is called the *k-th horn of* $\Delta[n]$; indeed, it is a presimplicial set (in the sense described in the previous section). Using the natural homeomorphism $|\Delta[n]| \to \Delta^n$ (described in Section 4.2, Example 1), identify the geometric realization of the horn $\Lambda^k[n]$ with the subspace of the geometric standard-n-simplex Δ^n which consists of the points $\boldsymbol{t} = (t_0, \ldots, t_n)$ with $t_i = 0$, for at least one index $i \neq k$; for $n > 0$, the homotopy $H : \Delta^n \times I \to \Delta^n$

$$(\boldsymbol{t}, s) \mapsto (t_0 - st', \ldots, t_k + nst', \ldots, t_n - st')$$

with $t' = \min\{t_i : i \neq k\}$ establishes $|\Lambda^k[n]|$ as a strong deformation retract of Δ^n.

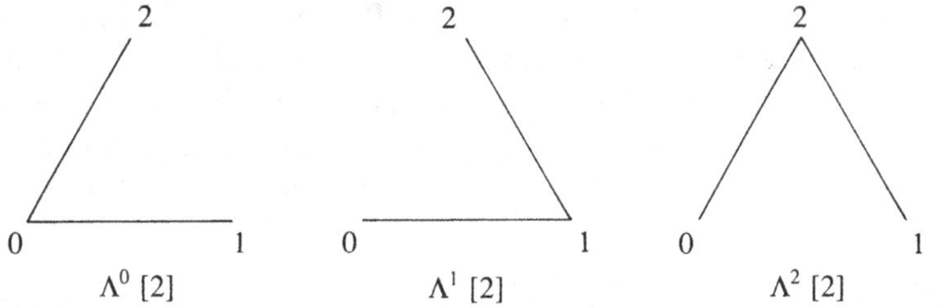

Figure 12

Examples 1 $\Lambda^0[1] \cong \Lambda^1[1] \cong \Delta[0]$; $\Lambda^i[2]$, $i=0,1,2$, see Figure 12.
Let Z be a simplicial set and let $f : \Lambda^k[n] \to Z$ be a simplicial map. The image of f is called an $(n-)$*horn in* Z; it is a simplicial subset of Z which is generated by a family $\{z_i : i \in [n], i \neq k\}$ of $(n-1)$-simplices, subject to the relations $z_i\delta_j = z_j\delta_{i-1}$ for $0 \leqslant j < i \leqslant n, j \neq k \neq i$; the empty set is the only 0-horn in Z. Each such family $\{z_i\}$ generates an n-horn in Z; by abuse of language, one says that the family itself *is an n-horn in Z*. An n-simplex $z \in Z$ is a *filling* of the horn $\{z_i\}$ if $z\delta_i = z_i$, for all $i \neq k$; each 0-simplex in Z is a filling of the 0-horn in Z. A horn can *be filled* if there exists a filling; this is the same as saying that the corresponding simplicial map $\Lambda^k[n] \to Z$ can be extended over the whole standard-n-simplex $\Delta[n]$.

Let $p : Z \to X$ be a simplicial map. If the family $\{z_i\}$ is a horn in Z, then the family $\{p(z_i)\}$ is a horn in X. The simplicial map p is called a *(Kan) fibration* with *base* X and *total set* Z, if, for every horn $\{z_i\}$ in Z and every filling x of the horn $\{p(z_i)\}$, there is a filling $z \in Z$ of the horn $\{z_i\}$ *over* x, i.e., such that $p(z) = x$. The following facts result immediately from the definition.

Proposition 4.5.1 (i) *The Kan fibrations form a subcategory of SiSets containing all isomorphisms, i.e., all identities and all other simplicial isomorphisms are Kan fibrations, and any composition of Kan fibrations is a Kan fibration.*

(ii) *If* $Z \to X$ *is a Kan fibration and* D *is a retract of* Z *over* X, *the restriction* $D \to X$ *is a Kan fibration.*

(iii) *If* $Z \to X$ *and* $Y \to W$ *are Kan fibrations, their product* $Z \times Y \to X \times W$ *is a Kan fibration.*

(iv) *If* $Z \to X$ *is induced from a Kan fibration (by means of a pullback construction), then* $Z \to X$ *is a Kan fibration.*

(v) *A Kan fibration with a non-empty base is surjective. In particular, every horn in the base is the image of a horn in the total set.* □

In contrast to the geometric situation (see Proposition A.4.7 (i)), it is not true in general that a terminal map in *SiSets* is a Kan fibration. Therefore, the simplicial set Z is called a *Kan set* if the unique simplicial map $Z \to \Delta[0]$ is a Kan fibration. Because every horn in $\Delta[0]$ has a unique filling, Kan sets can be characterized by the fact that all horns can be filled.

Corollary 4.5.2 (i) *A Kan set is non-empty.*
(ii) *The fibres of a Kan fibration are Kan sets.*
(iii) *The total set of a Kan fibration is a Kan set iff the base is a Kan set.*

Proof (i) If Z is a Kan set, then the unique simplicial map $Z \to \Delta[0]$ is surjective (see Proposition 4.5.1(v); thus, its domain cannot be empty.

(ii) A fibre of a Kan fibration may be thought out as the domain of a simplicial map with codomain $\Delta[0]$ induced from the Kan fibration.

(iii) '$\Rightarrow$': Assume the total set is a Kan set and consider a horn in the base. Take an inverse image in the total set (see Proposition 4.5.1 (v)). It has a filling whose image in the base is a filling of the given horn.

'$\Leftarrow$': If the base is a Kan set, then the image of every horn in the total set has a filling which, by the defining property of Kan fibrations, can be lifted to a filling in the total set. □

Example 2 Since $|\Lambda^k[n]|$ is a retract of $|\Delta[n]| = \Delta^n$, for all $n \in \mathbf{N}$ and all $k \in [n]$, the singular set ST of any topological space T is a Kan set. □

It is often tedious to reduce statements on Kan fibrations to fillings of horns. Therefore, it is worthwhile showing that Kan fibrations can be characterized by a more general extension property. To this end, one forms the subcategory $\boldsymbol{A} \subset$ *SiSets* of *anodyne extensions* whose objects are defined inductively as follows:

(0) all simplicial isomorphisms are anodyne extensions;
(1) all inclusions $\Lambda^k[n] \subset \Delta[n]$ are anodyne extensions;
(2) if the inclusion $\Lambda \subset D$ is an anodyne extension and D' is obtained from Λ' by attaching D via a partial simplicial map with domain Λ, the inclusion $\Lambda' \subset D'$ is an anodyne extension;
(3) if the inclusion $\Lambda \subset D$ is an anodyne extension and the inclusion $\Lambda' \subset D'$ is a *retract*, i.e., there are simplicial maps $g : D' \to D$, $f : D \to D'$ with $f \circ g = 1_{D'}$ and $g(\Lambda') \subset \Lambda \subset f^{-1}(\Lambda')$ the inclusion $\Lambda' \subset D'$ is an anodyne extension;
(4) if the inclusions $\Lambda(n) \subset \Lambda(n+1)$ are anodyne extensions, for all $n \in \mathbf{N}$, and $D = \cup \Lambda(n)$ the inclusion $\Lambda(0) \subset D$ is an anodyne extension;

(5) if the inclusions $\Lambda(j) \subset D(j)$ are anodyne extensions, for all j in some index set, the inclusion $\sqcup \Lambda(j) \subset \sqcup D(j)$ is an anodyne extension.

As a consequence of (0), one can assume, in most cases, that the anodyne extensions under consideration are inclusions of simplicial subsets.

Example 3 Let Λ be an *n-horn with r holes*; i.e., a simplicial subset of the standard simplex $\Delta[n]$ which is generated by $n+1-r$ elementary face operators, $0<r\leqslant n$; in this sense, the horn $\Lambda^k[n]$ is a horn with just one hole. Then, the inclusion $\Lambda \subset \Delta[n]$ is an anodyne extension. This follows by a double induction, first increasing on n, second on r. The key step lies in the fact that an n-horn with r holes, $r>1$, can be completed to an n-horn with $r-1$ holes by means of an attaching of an $(n-1)$ simplex where the attaching is defined on an $(n-1)$-horn with $r-1$ holes (see Condition (2) in the previous list). In particular, this shows that any embedding $\Delta\delta_i : \Delta[n-1] \to \Delta[n]$ is an anodyne extension. □

The description of anodyne extensions given before immediately implies the following:

Proposition 4.5.3 *A simplicial map $p : Z \to X$ is a Kan fibration iff for each anodyne extension $i : \Lambda \to D$ and each commutative square*

$$\begin{array}{ccc} \Lambda & \xrightarrow{\bar{f}} & Z \\ {\scriptstyle i}\downarrow & & \downarrow{\scriptstyle p} \\ D & \xrightarrow[f]{} & X \end{array}$$

there is a simplicial map $g : D \to Z$ such that $g \circ i = \bar{f}$ and $p \circ g = f$. □

Corollary 4.5.4 *A simplicial set Z is a Kan set iff for each anodyne extension $\Lambda \subset D$ each simplicial map $f : \Lambda \to Z$ can be extended over D.* □

Geometrically, anodyne extensions are more than just closed cofibrations.

Proposition 4.5.5 *The geometric realization of an anodyne extension $i : \Lambda \to D$ embeds $|\Lambda|$ as a strong deformation retract into $|D|$.*

Proof By induction. (0) The statement clearly holds true for simplicial isomorphisms.

(1) $|\Lambda^k[n]|$ is a strong deformation retract of $|\Delta[n]| = \Delta^n$, for all $n \in \mathbf{N}$ and all $k \in [n]$ (see the beginning of this section).

(2) Geometric realization transforms simplicial attachings into

attachings (see Proposition 4.3.10) and strong deformation retracts are preserved under attachings (see Proposition A.4.8(vi)).

(3) If there are simplicial maps $g : D' \to D$, $f : D \to D'$ with $f \circ g = \mathbf{1}_{D'}$ and $g(\Lambda') \subset \Lambda \subset f^{-1}(\Lambda')$ and a deformation retraction $H : |D| \times I \to |D|$, rel. $|\Lambda|$, then $\tilde{H} = |f| \circ H \circ (g \times \mathbf{1}_I)$ is a deformation retraction $|D'| \times I \to |D'|$ rel. $|\Lambda'|$.

(4) Given inclusions $\Lambda(n) \subset \Lambda(n+1)$, for all $n \in \mathbf{N}$, and $D = \cup \Lambda(n)$, the sequence $\{|\Lambda(n)| : n \in \mathbf{N}\}$ is an expanding sequence with union space $|D|$ (see Proposition 4.3.12). If, moreover, each $|\Lambda(n)|$ is a strong deformation retract of $|\Lambda(n+1)|$, then $|\Lambda(0)|$ is also a strong deformation retract of $|D|$ (see Corollary A.5.8).

(5) If each $|\Lambda(j)|$ is a strong deformation retract of $|D(j)|$, for all j in some index set, $|\sqcup \Lambda(j)| = \sqcup |\Lambda(j)|$ is a strong deformation retract of $|\sqcup D(j)| = \sqcup |D(j)|$. □

The analogy to geometry is more comprehensive. The following statement reflects the fact that any map can be decomposed into an injective homotopy equivalence followed by a fibration (see Proposition A.4.18).

Proposition 4.5.6 *Any simplicial map $f : Y \to X$ can be decomposed in the form $f = p \circ i$, where $p : Z \to X$ is a Kan fibration and $i : Y \to Z$ is an anodyne extension.*

Proof Define anodyne extensions $Y(n) \subset Y(n+1)$ and simplicial maps $f_{(n)} : Y(n) \to X$ with $f_{(n+1)} | Y(n) = f_{(n)}$, for all $n \in \mathbf{N}$, inductively as follows. First, take $Y(0) = Y$ and $f_0 = f$. Next, assume $Y(n)$ and $f_{(n)}$ are given. To obtain $Y(n+1)$, attach to $Y(n)$ fillings for all horns which do not have fillings but whose images under $f_{(n)}$ can be filled. Then, define $f_{(n+1)}$ on an attached filling by assigning to it some filling of the image of the generating horn under $f_{(n)}$.

Since each inclusion $Y(n) \subset Y(n+1)$ is an anodyne extension, so is the inclusion $Y = Y(0) \subset \cup Y(n) = Z$. The maps $f_{(n)}$ together define a map $p : Z \to X$. Since a horn is finitely generated, any horn in Z lives in some $Y(n)$. If the image of such a horn has a filling in Z, the horn itself has a suitable filling, at least in $Y(n+1)$, and therefore in Z. □

Corollary 4.5.7 *Any simplicial set can be embedded in a Kan set, by means of an anodyne extension.*

Proof Apply the proposition to the unique simplicial map from a given simplicial set to $\Delta[0]$. □

The main technical advantage of anodyne extensions comes from the following fact.

Proposition 4.5.8 *If A is a simplicial subset of the simplicial set X, and if the inclusion $\Lambda \subset D$ is an anodyne extension, then the inclusion*

$$X \times \Lambda \cup A \times D \subset X \times D$$

is also an anodyne extension.

Proof Again by induction. (1) *The inclusions*

$$\Delta[n] \times \Lambda^k[1] \cup \delta\Delta[n] \times \Delta[1] \subset \Delta[n] \times \Delta[1]$$

are anodyne extensions, for all $n \in \mathbf{N}$. First, consider the case $k = 0$. Then, $\Delta[n] \times \Delta[1]$ can be obtained from $\Delta[n] \times \Lambda^k[1] \cup \delta\Delta[n] \times \Delta[1]$ by successive attachings of the $(n+1)$-simplices $(\omega_n, \sigma_n), (\omega_{n-1}, \sigma_{n-1}), \ldots, (\omega_0, \sigma_0)$. Each single attaching is obtained via a partial simplicial map from $\Delta[n+1]$, with domain $\Lambda^m[n+1], m = n, n-1, \ldots, 0$, and, therefore, an anodyne extension. Then, the same holds true for the composition of these attachings. Second, in case $k = 1$, one has to attach the described $(n+1)$-simplices in the inverse order.

The inclusions

$$X \times \Lambda^k[1] \cup A \times \Delta[1] \subset X \times \Delta[1]$$

are anodyne extensions, for all simplicial sets X and all simplicial subsets $A \subset X$. Define a sequence of simplicial subsets of X by taking $X(-1) = A$, $X(n) = X^n \cup A$, for $n \in \mathbf{N}$, and observe that $X \times \Lambda^k[1] \cup X(n) \times \Delta[1]$ is obtained from $X \times \Lambda^k[1] \cup X(n-1) \times \Delta[1]$ by attaching a coproduct $\sqcup \Delta[n] \times \Delta[1]$ via a partial simplicial map with domain $\sqcup \Delta[n] \times \Lambda^k[1] \cup \delta\Delta[n] \times \Delta[1]$; since a coproduct of anodyne extensions is an anodyne extension, the inclusion

$$\sqcup \Delta[n] \times \Lambda^k[1] \cup \delta\Delta[n] \times \Delta[1] \subset \sqcup \Delta[n] \times \Delta[1]$$

is, according to the introductory step, an anodyne extension, and therefore the inclusion

$$X \times \Lambda^k[1] \cup X(n-1) \times \Delta[1] \subset X \times \Lambda^k[1] \cup X(n) \times \Delta[1]$$

is also an anodyne extension. Composing these anodyne extensions, for all n, one obtains that the inclusion of

$$X \times \Lambda^k[1] \cup X(-1) \times \Delta[1] = X \times \Lambda^k[1] \cup A \times \Delta[1]$$

into

$$\bigcup_{n=0}^{\infty} X \times \Lambda^k[1] \cup X(n) \times \Delta[1] = X \times \Delta[1]$$

is an anodyne extension.

The inclusions

$$X \times \Lambda^k[n] \cup A \times \Delta[n] \subset X \times \Delta[n]$$

are anodyne extensions, for all simplicial sets X, *all simplicial subsets* $A \subset X$, *all* $n \in \mathbf{N}$ *and all* $k \in [n]$. First, assume $k < n$. According to the previous step, the inclusion

$$X \times \Delta[n] \times \Lambda^0[1] \cup (X \times \Lambda^k[n] \cup A \times \Delta[n]) \times \Delta[1] \subset X \times \Delta[n] \times \Delta[1]$$

is an anodyne extension. It contains the simplicial map under consideration as a retract via the embedding

$$X \times \Delta[n] \to X \times \Delta[n] \times \Delta[1], (x, \alpha) \mapsto (x, \alpha, \delta_0 \omega)$$

and the retraction

$$X \times \Delta[n] \times \Delta[1] \to X \times \Delta[n], (x, \alpha, \beta) \mapsto (x, \tilde{\alpha}),$$

where the operator $\tilde{\alpha}$ is given by $\tilde{\alpha}(i) = k$, for $\beta(i) = 0$ and $\alpha(i) \geqslant k$, and $\tilde{\alpha}(i) = \alpha(i)$ otherwise. For $k = n$ – more generally, for $k > 0$ – replace $\Lambda^0[1]$ by $\Lambda^1[1], \delta_0$ by δ_1 and $\tilde{\alpha}$ by the operator $\hat{\alpha}$ given by $\hat{\alpha}(i) = k$, for $\beta(i) = 1$ and $\alpha(i) \leqslant k$, and $\hat{\alpha}(i) = \alpha(i)$ otherwise.

(2) *If* D' *is obtained from* Λ' *by attaching* D *via a partial simplicial map with domain* Λ *and the inclusion*

$$X \times \Lambda \cup A \times D \subset X \times D$$

is an anodyne extension, then the inclusion

$$X \times \Lambda' \cup A \times D' \subset X \times D'$$

is also an anodyne extension, because $X \times D'$ can be considered as obtained from $X \times \Lambda' \cup A \times D'$ by attaching $X \times D$ via a partial simplicial map with domain $X \times \Lambda \cup A \times D$.

(3) *If the inclusion* $\Lambda' \subset D'$ *is a retract of the inclusion* $\Lambda' \subset D'$ *and the inclusion*

$$X \times \Lambda \cup A \times D \subset X \times D$$

is an anodyne extension, then the inclusion

$$X \times \Lambda' \cup A \times D' \subset X \times D'$$

is also an anodyne extension because it is a retract of the given anodyne extension.

(4) *For every* $n \in \mathbf{N}$, *let* $\Lambda(n) \subset \Lambda(n+1)$ *be such that the inclusions*

$$X \times \Lambda(n) \cup A \times \Lambda(n+1) \subset X \times \Lambda(n+1)$$

are anodyne extensions; if $D = \cup \Lambda(n)$, *then the inclusion*

$$X \times \Lambda(0) \cup A \times D \subset X \times D$$

is an anodyne extension. Since $X \times \Lambda(n+1) \cup A \times D$ can be considered as obtained from $X \times \Lambda(n) \cup A \times D$, by attaching $X \times \Lambda(n+1)$ via a partial simplicial map with domain $X \times \Lambda(n) \cup A \times \Lambda(n+1)$, the inclusions

$$X \times \Lambda(n) \cup A \times D \subset X \times \Lambda(n+1) \cup A \times D$$

are anodyne extensions, for all $n \in \mathbf{N}$; but then, so is the inclusion

$$X \times \Lambda(0) \cup A \times D \subset \bigcup_{n=0}^{\infty} X \times \Lambda(n) \cup A \times D = X \times D.$$

(5) The claim for the transition to coproducts is obvious. □

Some tiny consequences of this proposition should be noted.

Corollary 4.5.9 (i) *If Z is a Kan set and some inclusion $\Delta[0] \subset Z$ is an anodyne extension, then Z is simplicially contractible.*

(ii) *Any Kan set is the base of a Kan fibration with a contractible total set.*

Proof (i) According to the proposition, the inclusion

$$Z \times \delta\Delta[1] \cup \Delta[0] \times \Delta[1] \subset Z \times \Delta[1]$$

is an anodyne extension; thus, there is a simplicial homotopy $H : Z \times \Delta[1] \to Z$ with $H(z, \varepsilon_0\omega) = z, H(z, \varepsilon_1\omega) = \omega$ and $H(\omega, \beta) = \omega$, for $z \in Z$ and $\beta \in \Delta[1]$.

(ii) A Kan set is non-empty (see Corollary 4.5.2 (i)); thus, it appears as codomain of simplicial maps with domain $\Delta[0]$. Apply the proposition to such a simplicial map. The total set Z of the resulting Kan fibration is again a Kan set (see Corollary 4.5.2 (iii)); as the codomain of an anodyne extension with domain $\Delta[0], Z$ is contractible by (i). □

Another application of anodyne extensions concerns homotopy over X.

Proposition 4.5.10 *Let Y be a simplicial set, let D be a simplicial subset of Y and let $p : Z \to X$ be a Kan fibration. Then, homotopy rel. D over X is an equivalence relation on the set of all simplicial maps $Y \to Z$.*

Proof All kinds of simplicial homotopies are reflexive relations. Now let $H : Y \times \Delta[1] \to Z$ and $\tilde{H} : Y \times \Delta[1] \to Z$ be simplicial homotopies rel. D over X from f to $g, \tilde{g}$ respectively. Together, they define a simplicial map

$$G : Y \times \Lambda^0[2] \cup D \times \Delta[2] \to Z.$$

The inclusion

$$j : Y \times \Lambda^0[2] \cup D \times \Delta[2] \to Y \times \Delta[2]$$

is an anodyne extension (see Proposition 4.5.8). Let $\bar{G} : Y \times \Delta[2] \to X$

denote the composition of $p \circ f$ and the projection from $Y \times \Delta[2]$ onto Y. Then, $\bar{G} \circ j = p \circ G$ and the extension property (see Proposition 4.5.3) ensures the existence of a simplicial map $\tilde{G} : Y \times \Delta[2] \to Z$ such that $p \circ \tilde{G} = \bar{G}$ and $\tilde{G} \circ j = G$. The composition $\tilde{G} \circ (1_Y \times \Delta\delta_0)$ is a homotopy between g and $\tilde{g}$ which proves the symmetry as well as the transitivity of the considered homotopy notion. □

Corollary 4.5.11 *If Y is an simplicial set, D is a simplicial subset of Y and Z is a Kan set, then homotopy rel. D is an equivalence relation on the set of all simplicial maps from Y to Z.* □

There are some specific types of simplicial fibrations to consider. A simplicial map $p : Z \to X$ is an *acyclic fibration* iff, for each commutative square

$$\begin{array}{ccc} \delta\Delta[n] & \xrightarrow{\bar{f}} & Z \\ {\scriptstyle i}\downarrow & & \downarrow{\scriptstyle p} \\ \Delta[n] & \xrightarrow[f]{} & X, \end{array}$$

where $i : \delta\Delta[n] \to \Delta[n]$ denotes the inclusion, there is a simplicial map $g : \Delta[n] \to Z$ such that $g|\delta\Delta[n] = \bar{f}$ and $p \circ g = f$.

Proposition 4.5.12 *An acyclic fibration is a Kan fibration.*

Proof Let $p : Z \to X$ be an acyclic fibration and let a horn $\{z_i\}$ in Z as well as a filling $x \in X$ of the horn $\{p(z_i)\}$ be given. Take the simplicial maps $\bar{f} : \Lambda^k[p] \to Z$, $\delta_i \mapsto z_i$ and $f : \Delta[p] \to X, \iota \mapsto x$; then, $p \circ \bar{f} = f | \Lambda^k[p]$. The inclusion $\Delta\delta_k : \Delta[p-1] \to \Delta[p]$ induces an inclusion $\delta : \delta\Delta[p-1] \to \Lambda^k[p]$ and by acyclicity there is a simplicial map $g : \Delta[p-1] \to Z$ such that $g|\delta\Delta[p-1] = \bar{f} \circ \delta$ and $p \circ g = f \circ \Delta\delta_k$. Now observe that $\delta\Delta[p]$ may be obtained from $\Lambda^k[p]$ by attaching $\Delta[p-1]$ via δ; thus, the simplicial maps $\bar{f}, g$ together define a simplicial map $\tilde{f} : \delta\Delta[p] \to Z$ such that $p \circ \tilde{f} = f | \delta\Delta[p]$. Using acyclicity again, one finds a simplicial map $\tilde{g} : \Delta[p] \to Z$ such that $\tilde{g}|\delta\Delta[p] = \tilde{f}$ and $p \circ \tilde{g} = f$; moreover, it follows that $\tilde{g}|\Lambda^k[p] = \bar{f}$. Therefore, the simplex $\tilde{g}(\iota)$ is a filling of the horn $\{z_i\}$ over x. □

There is a characterization of acyclic fibrations similar to that of ordinary Kan fibrations by means of anodyne extensions (see Proposition 4.5.3).

Proposition 4.5.13 *A simplicial map $p : Z \to X$ is an acyclic fibration iff for each injective simplicial map $i : A \to D$ and each commutative square*

$$\begin{array}{ccc} \Lambda & \xrightarrow{\bar{f}} & Z \\ {\scriptstyle i}\downarrow & & \downarrow{\scriptstyle p} \\ D & \xrightarrow[f]{} & X \end{array}$$

there is a simplicial map $g : D \to Z$ such that $g \circ i = \bar{f}$ and $p \circ g = f$.

Proof '⇒': Given a commutative square as described in the statement, one may assume that the injective simplicial map i is an inclusion. Define a sequence of simplicial subsets of D by taking $D(-1) = \Lambda$, $D(n) = D^n \cup \Lambda$, for $n \in \mathbf{N}$. Observe that $D = \cup D(n)$ and that $D(n)$ is obtained from $D(n-1)$ by attaching a coproduct $\sqcup \Delta[n]$ via a partial simplicial map d_n with domain $\sqcup \delta\Delta[n]$; let $\bar{d}_n : \sqcup \Delta[n] \to D(n)$ denote the corresponding characteristic map. Construct inductively simplicial maps $g_n : D(n) \to Z$ with $g_{-1} = \bar{f}, g_n | D(n-1) = g_{n-1}$ and $p \circ g_n = f$ as follows. Assume that g_{n-1} is constructed; using acyclicity, one finds a simplicial map $\tilde{g}_n : \sqcup \Delta[n] \to Z$ with $\tilde{g} | \sqcup \delta\Delta[n] = g_{n-1} \circ d_n$ and $p \circ \tilde{g}_n = f | D(n) \circ \bar{d}_n$. The simplicial maps g_{n-1} and $\tilde{g}_n$ together induce the desired simplicial map $g(n)$.

Finally, define $g : D \to Z$ by taking $g | D(n) = g_n$.

'⇐': Obvious. □

The ratio of Kan fibrations to acyclic fibrations can be described as follows.

Proposition 4.5.14 (i) *If the simplicial map $p : Z \to X$ is an acyclic fibration, there is a cross-section $s : X \to Z$ for p such that the composition $s \circ p$ is homotopic rel. $s(X)$ over X to $\mathbf{1}_Z$.*

(ii) *If the Kan fibration $p : Z \to X$ is a simplicial homotopy equivalence, then it is an acyclic fibration.*

Proof (i) Apply Proposition 4.5.13, with $\Lambda = \varnothing$ and $f = \mathbf{1}_X$. The result is a simplicial map $s : X \to Z$ with $p \circ s = \mathbf{1}_X$; i.e., a cross-section for p. Now, apply Proposition 4.5.13 once more, with

$$\bar{f} : Z \times \delta\Delta[1] \cup s(X) \times \Delta[1] \to Z,$$

defined by $(z, \varepsilon_1 \omega) \mapsto z, (z, \varepsilon_0 \omega) \mapsto s \circ p(z), (s(x), \beta) \mapsto s(x)$, and

$$f : Z \times \Delta[1] \to X,$$

defined by $(z, \delta_1 \omega) \mapsto p(z)$, which gives the desired homotopy.

(ii) First, from the assumption, one derives the existence of a cross-section $s : X \to Z$ for p such that the composition $s \circ p$ is homotopic to $\mathbf{1}_Z$ over X. To reach this goal, assume that there are given a simplicial map $q : X \to Z$ and simplicial homotopies $\bar{H} : Z \times \Delta[1] \to Z, H : X \times$

$\Delta[1] \to X$ from $q \circ p$ to $\mathbf{1}_Z$ and from $p \circ q$ to $\mathbf{1}_X$, respectively (for possibly other directions of these simplicial homotopies, the following proof can be suitably modified). Since p is a Kan fibration, there is a simplicial homotopy $\tilde{H} : X \times \Delta[1] \to Z$ from q to a cross-section s for p with $p \circ \tilde{H} = H$. Furthermore, there is a simplicial homotopy $G : Z \times \Delta[2] \to Z$ with $G(z, \delta_0 \beta) = \tilde{H}(p(z), \beta), G(z, \delta_1 \beta) = s \circ H(p(z), \beta)$ and $p \circ G(z, \gamma) = H(p(z), \sigma_0 \gamma)$. Since the inclusion

$$Z \times \Delta[1] \times \Lambda^0[1] \cup Z \times \delta\Delta[1] \times \Delta[1] \subset Z \times \Delta[1] \times \Delta[1]$$

is an anodyne extension, there is also a simplicial map

$$K : Z \times \Delta[1] \times \Delta[1] \to Z,$$

with $K(z, \beta, \varepsilon_0 \omega) = G(z, \delta_2 \beta)$, $K(z, \varepsilon_0 \omega, \beta') = s \circ p \circ \bar{H}(z, \beta')$, $K(z, \varepsilon_1 \omega, \beta') = \bar{H}(z, \beta')$ and $p \circ K(z, \beta, \beta') = p \circ \bar{H}(z, \beta')$. The assignment $(z, \beta) \mapsto K(z, \beta, \varepsilon_1 \omega)$ describes a homotopy from $s \circ p$ to $\mathbf{1}_Z$ over X.

Now, let $s : X \to Z$ be a cross-section for p and let $H : Z \times \Delta[1] \to Z$ be a simplicial homotopy over X from $s \circ p$ to $\mathbf{1}_Z$. Consider simplicial maps $\bar{f} : \delta\Delta[n] \to Z$, $f : \Delta[n] \to X$ such that $f | \delta\Delta[n] = p \circ \bar{f}$. Since p is a Kan fibration, there is a simplicial homotopy $G : \Delta[n] \times \Delta[1] \to Z$ with $G(\alpha, \varepsilon_0 \omega) = s \circ f(\alpha), G(\delta_i, \beta) = H(\bar{f}(\delta_i), \beta)$ and $p \circ G(\alpha, \beta) = f(\alpha)$. Then, the simplicial map $g : \Delta[n] \to Z, \alpha \mapsto G(\alpha, \varepsilon_1)$ satisfies the necessary equations for proving that p is an acyclic fibration. □

Corollary 4.5.15 *A Kan set Z is simplicially contractible iff any simplicial map $\delta\Delta[n] \to X$ can be extended over $\Delta[n]$.* □

It follows from Proposition 4.5.14 that the geometric realization of an acyclic fibration not only is a homotopy equivalence but also has another nice property:

Proposition 4.5.16 *The geometric realization of an acyclic fibration is a fibration.*

Proof Let $p : Z \to X$ be a acyclic fibration. Consider the commutative square

$$\begin{array}{ccc} Z & \xrightarrow{\mathbf{1}_Z} & Z \\ {\scriptstyle i}\downarrow & & \downarrow{\scriptstyle p} \\ Z \times X & \xrightarrow[f]{} & X \end{array}$$

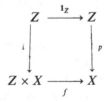

where f denotes the projection onto the second factor and i has the components $\mathbf{1}_Z, p$. Since p is an acyclic fibration, there is a simplicial map $g : Z \times X \to Z$ such that $g \circ i = \mathbf{1}_Z$ and $p \circ g = f$. Geometric realization preserves projections (see Proposition 4.3.15) and geometric projections are fibrations (see Section A.4, Example 5); thus, $|f|$ is a fibration. Moreover, the space $|Z|$ is a retract via $|g|$ of $|Z \times X|$ over the space $|X|$, and consequently the map $|p| = |f| \,|\, |Z|$ is a fibration (see Proposition A.4.7 (ii)). □

There is also a certain dual to Proposition 4.5.6. It corresponds to the geometric fact that any map can be decomposed into a cofibration followed by a homotopy equivalence (see Proposition A.4.10 (iv)).

Proposition 4.5.17 *Any simplicial map $f : Y \to X$ can be decomposed in the form $f = p \circ i$, where $p : Z \to X$ is an acyclic fibration and $i : Y \to Z$ is an inclusion.*

Proof Adapt the proof of Proposition 4.5.6. □

Let $p : Z \to X$ be a Kan fibration. Two simplices $z_0, z_1 \in Z_n$ are *fibre homotopic* if the corresponding simplicial maps $f_j : \Delta[n] \to Z, \iota \mapsto z_j, j \in [1]$ (see Lemma 4.2.1 (i)), are homotopic rel. $\delta\Delta[n]$ over X. Clearly, fibre homotopic simplices have the same boundary. The next property is also evident (see Proposition 4.5.10).

Proposition 4.5.18 *If $p : Z \to X$ is a Kan fibration, then fibre homotopy is an equivalence relation on Z_n, for all $n \in \mathbf{N}$.* □

The following are technical but useful criteria for recognizing fibre homotopic simplices.

Lemma 4.5.19 *Let $p : Z \to X$ be a Kan fibration. Two simplices $z_0, z_1 \in Z_n$ are fibre homotopic if*

(i) *there are homotopies $H_0, H_1 : \Delta[n] \times \Delta[1] \to Z$ such that*

(a) $H_0 | \delta\Delta[n] \times \Delta[1] = H_1 | \delta\Delta[n] \times \Delta[1]$,

(b) $p \circ H_0 = p \circ H_1$,

(c) $H_k(\iota, \varepsilon_0\omega) = z_k, k \in [1]$, *and* $H_0(\iota, \varepsilon_1\omega) = H_1(\iota, \varepsilon_1\omega)$

(*or*

(c′) $H_k(\iota, \varepsilon_1\omega) = z_k$ *and* $H_0(\iota, \varepsilon_0\omega) = H_1(\iota, \varepsilon_0\omega)$);

or

(ii) *there are simplices $z'_0, z'_1 \in Z_{n+1}$ with $p(z'_0) = p(z'_1)$ such that $z'_0\delta_j, z'_1\delta_j$ are fibre homotopic for some $j \in [n+1]$ and $z'_0\delta_r = z_0, z'_1\delta_r = z_1$ for some $r \in [n+1], r \neq j$.*

Proof (i): First assume (c). Since p is a Kan fibration, there is a simplicial map $\tilde{H} : \Delta[n] \times \Delta[2] \to Z$ such that $\tilde{H}(\alpha, \delta_k\beta) = H_k(\alpha, \beta), \tilde{H}(\delta_i, \gamma) = H_k(\delta_i, \sigma_0\gamma)$ and $p \circ \tilde{H}(\alpha, \gamma) = p \circ H_k(\alpha, \sigma_0\gamma)$, for $\alpha \in \Delta[n], \beta \in (\Delta[1])_{\dim \alpha}, \gamma \in (\Delta[2])_{\dim \alpha}, (\Delta[2])_{n-1}$, respectively, $k \in [1]$ and $i \in [n]$ (see Proposition 4.5.8 with $\Lambda = \Lambda^2[2], D = \Delta[2]$). The composition $\tilde{H} \circ (\mathbf{1}_{\Delta[n]} \times \Delta\delta_2)$ is a homotopy needed to show that z_0 and z_1 are fibre homotopic.

Second, if the conditions (c′) are satisfied the argument is similar. One has only to replace the operators δ_k, σ_0 by the operators δ_{k+1}, σ_1 respectively, the horn $\Lambda^2[2]$ by the horn $\Lambda^0[2]$ and the simplicial map $\Delta\delta_2$ by the simplicial map $\Delta\delta_0$.

(ii): The inclusion $\Delta\delta_j : \Delta[n] \subset \Delta[n+1]$ is an anodyne extension (see Example 3) and so is the induced inclusion

$$\Delta[n+1] \times \delta\Delta[1] \cup \Delta[n] \times \Delta[1] \subset \Delta[n+1] \times \Delta[1]$$

(see Proposition 4.5.8). Now, since p is a Kan fibration, there is a simplicial map $\tilde{H} : \Delta[n+1] \times \Delta[1] \to Z$ such that $\tilde{H}(\iota, \varepsilon_k\omega) = z'_k, \tilde{H}(\alpha, \beta) = H(\alpha, \beta)$ and $p \circ \tilde{H}(\alpha', \beta') = p(z'_k)\alpha'$, for $k \in [1], (\alpha, \beta) \in \Delta[n] \times \Delta[1], (\alpha', \beta') \in \Delta[n+1] \times \Delta[1]$, where $H : \Delta[n] \times \Delta[1] \to Z$ denotes a homotopy over X connecting $z'_0\delta_j$ and $z'_1\delta_j$. Then, $\tilde{H} \circ (\Delta\delta_r \times \mathbf{1}_{\Delta[1]})$ is a homotopy over X connecting z_0 and z_1. □

A simplicial map $p : Z \to X$ is a *minimal fibration* if it is a Kan fibration and fibre homotopic simplices are always equal. The simplicial set Z is a *minimal Kan set* if the unique simplicial map $Z \to \Delta[0]$ is a minimal fibration. Again, the following facts result immediately from the definitions.

Proposition 4.5.20 (i) *The minimal fibrations form a subcategory of SiSets containing all isomorphisms, i.e., all identities and all other simplicial isomorphisms are minimal fibrations, any composition of minimal fibrations is a minimal fibration.*

(ii) *If $Z \to X$ is a minimal fibration, and A is a retract of Z over X, the restriction $A \to X$ is a minimal fibration.*

(iii) *If $Z \to X$ is induced from a minimal fibration (by means of a pullback construction), it is a minimal fibration itself.*

(iv) *The fibres of a minimal fibration are minimal Kan sets.* □

One essential property of minimal fibrations is the following.

Proposition 4.5.21 *A minimal fibration is locally trivial.*

Proof Let $p : Z \to X$ be a minimal fibration and consider a simplicial map $f : \Delta[m] \to X$. The simplicial map $\bar{p} : Y \to \Delta[m]$ induced from p by f is a minimal fibration (see Proposition 4.5.20 (iii)). It is to be shown that the simplicial set Y is isomorphic to the product of $\Delta[m]$ and the fibre F of $\bar{p}$ over ε_0. To simplify the notation, assume that $\bar{p}=p$, and, consequently, $Y=Z$ and $X=\Delta[m]$.

Let $C : \Delta[m] \times \Delta[1] \to \Delta[m]$ denote the (unique) homotopy from the constant map with value ε_0 to $\mathbf{1}_{\Delta[m]}$ (see Section 4.2, Example 3). Since p is a Kan fibration, there is a simplicial deformation $\tilde{C} : Z \times \Delta[1] \to Z$ rel. F from a retraction $q : Z \to F$ to $\mathbf{1}_Z$ with $p \circ \tilde{C} = C \circ (p \times \mathbf{1}_{\Delta[1]})$ (see Propositions 4.5.8 and 4.5.3). The simplicial maps p and q together define the map $r : Z \to \Delta[m] \times F, z \mapsto (p(z), q(z))$. The existence of this simplicial map r depends only on the fact that p is a Kan fibration; minimality now will be used to show inductively that r is an isomorphism.

Injectivity Take two simplices $z_0, z_1 \in Z_n$ with $r(z_0) = r(z_1)$, i.e., $p(z_0) = p(z_1)$ and $\tilde{C}(z_0, \varepsilon_0\omega) = q(z_0) = q(z_1) = \tilde{C}(z_1, \varepsilon_0\omega)$. By the inductive hypothesis, they have the same boundary; thus, the homotopies

$$H_j : \Delta[n] \times \Delta[1] \to Z, \qquad (\alpha, \beta) \mapsto \tilde{C}(z_j\alpha, \beta),$$

$j \in [1]$, satisfy the necessary properties which assure that these simplices are fibre homotopic (see Lemma 4.5.19 (i)). By minimality, this implies $z_0 = z_1$.

Surjectivity Take $(\kappa, z) \in (\Delta[m] \times F)_n$. By the inductive hypothesis and the already proven injectivity of the simplicial map r, there is a simplicial map $h : \delta\Delta[n] \to Z$ such that $r \circ h(\delta_i) = (\kappa, z)\delta_i$, i.e., $p \circ h(\delta_i) = \kappa\delta_i$ and $\tilde{C}(h(\delta_i), \varepsilon_0\omega) = q \circ h(\delta_i) = z\delta_i$, for all $i \in [n]$. Since p is a Kan fibration, there is a simplicial homotopy $G : \Delta[n] \times \Delta[1] \to Z$ with $G(\alpha, \varepsilon_0\omega) = z\alpha$, $G(\delta_i, \beta) = \tilde{C}(h(\delta_i), \beta)$ and $p \circ G(\alpha, \beta) = C(\kappa\alpha, \beta)$ (see Propositions 4.5.8 and 4.5.3). Take $H_0 = G : \Delta[n] \times \Delta[1] \to Z$ and $H_1 : \Delta[n] \times \Delta[1] \to Z$, $(\alpha, \beta) \mapsto \tilde{C}(G(\alpha, \varepsilon_1\omega), \beta)$; these homotopies show that the simplices z and $q(\tilde{z})$ with $\tilde{z} = G(\iota, \varepsilon_1\omega)$ are fibre homotopic (see Lemma 4.5.19 (i)). By minimality, one obtains $z = q(\tilde{z})$, and, finally, $(\kappa, z) = r(\tilde{z})$. □

Corollary 4.5.22 *The geometric realization of a minimal fibration is a (locally trivial) fibration.*

Proof The geometric realization of a locally trivial simplicial map is a (locally trivial) fibration (see Corollary 4.3.23). □

Proposition 4.5.23 *Any Kan fibration contains a minimal fibration as a strong deformation retract.*

Proof Let $p : Z \to X$ be a Kan fibration. Choose a system Z' of representatives for the fibre homotopy classes of simplices of Z such that $Z^\flat \subset Z'$;

this is possible because each fibre homotopy class contains at most one degenerate simplex (see Corollary 4.2.4 (ii)). Using Zorn's lemma, find a simplicial subset $\hat{Z}$ of Z which is maximal with respect to the property $\hat{Z} \subset Z'$. The maximality of $\hat{Z}$ implies that any simplex in Z' whose faces all belong to $\hat{Z}$ also belongs to $\hat{Z}$.

It will be shown that $\hat{Z}$ is a strong deformation retract of Z, over X; i.e., that there is a simplicial homotopy rel. $\hat{Z}$ over X from a simplicial map whose image is contained in $\hat{Z}$ to $\mathbf{1}_Z$, by the method of the least criminal. Again, using Zorn's lemma, one finds a simplicial subset A of Z which is maximal with respect to the property that there is a homotopy $G : A \times \Delta[1] \to Z$ rel. $\hat{Z}$ over X from a simplicial map whose image is contained in $\hat{Z}$ to the inclusion of A into Z. Assume $A \neq Z$ and choose a simplex $z \in Z \setminus A$ of lowest dimension, say $\dim z = n$. The simplex z is non-degenerate (see Lemma 4.2.5 (iv)) and its boundary belongs to A. Let $G : A \times \Delta[1] \to Z$ denote a simplicial homotopy rel. $\hat{Z}$ over X, with $G(z, \varepsilon_0 \omega) \in \hat{Z}$ and $G(z, \varepsilon_1 \omega) = z$, for all $z \in A$, which exists by the hypothesis on A. The simplicial subset A' of Z generated by $A \cup \{z\}$ is strictly bigger than A. Since p is a Kan fibration, there is a simplicial homotopy $K : \Delta[n] \times \Delta[1] \to Z$ over X, with $K(\alpha, \varepsilon_1 \omega) = z\alpha$ and $K(\delta_i, \beta) = G(z\delta_i, \beta)$. The n-simplex $z'' = K(\iota, \varepsilon_0 \omega)$ has its boundary in $\hat{Z}$; the representative $z' \in Z'$ of the fibre homotopy class of z'' has the same boundary and therefore belongs also to $\hat{Z}$. Let $K' : \Delta[n] \times \Delta[1] \to Z$ denote a simplicial homotopy rel. $\delta\Delta[n]$ over X, with $K'(\iota, \varepsilon_0 \omega) = z'$ and $K'(\iota, \varepsilon_1 \omega) = z''$. Again using the fact that p is a Kan fibration, one obtains a simplicial map $\tilde{K} : \Delta[n] \times \Delta[1] \times \Delta[1] \to Z$ with $\tilde{K}(\alpha, \beta, \varepsilon_1 \omega) = K(\alpha, \beta)$, $\tilde{K}(\alpha, \varepsilon_0 \omega, \beta') = K'(\alpha, \beta')$, $\tilde{K}(\alpha, \varepsilon_1 \omega, \beta') = z\alpha$, $\tilde{K}(\delta_i, \beta, \beta') = G(z\delta_i, \beta)$ and $p \circ \tilde{K}(\alpha, \beta, \beta') = p(z\alpha)$. Now, the homotopy G can be extended to a homotopy $G' : A' \times \Delta[1] \to Z$ with the desired properties by taking $G'(z, \beta) = \tilde{K}(\iota, \beta, \varepsilon_0 \omega)$, contradicting the maximality of A. Thus, $A = Z$, and so $\hat{Z}$ is a strong deformation retract of Z over X.

It remains to show that the restriction $p|\hat{Z}$ is a minimal fibration. Since $\hat{Z}$ is a retract of Z over X, it is a Kan fibration (see Proposition 4.5.1 (ii)). Finally, observe that fibre homotopic simplices in $\hat{Z}$ are fibre homotopic in Z. Since $\hat{Z} \subset Z'$ contains at most one simplex of every fibre homotopy class in Z, fibre homotopic simplices in $\hat{Z}$ must be equal. □

Corollary 4.5.24 *Any Kan fibration may be factored into a composition of an acyclic fibration followed by a minimal fibration.*

Proof Let $p : Z \to X$ be a Kan fibration. Let $\hat{Z}$ be a simplicial subset of Z such that $p|\hat{Z}$ is a minimal fibration; take a homotopy $G : Z \times \Delta[1] \to Z$

rel. $\hat{Z}$ over X from a retraction $\hat{p} : Z \to \hat{Z}$ to $\mathbf{1}_Z$. It will be shown that $\hat{p}$ is an acyclic fibration. To this end, let a commutative square of the form

$$\begin{array}{ccc} \delta\Delta[n] & \xrightarrow{\bar{f}} & Z \\ {\scriptstyle i}\downarrow & & \downarrow{\scriptstyle \hat{p}} \\ \Delta[n] & \xrightarrow[f]{} & \hat{Z} \end{array}$$

be given. Since p is a Kan fibration, there is a simplicial homotopy $H_0 : \Delta[n] \times \Delta[1] \to Z$ with $H_0(\alpha, \varepsilon_0\omega) = f(\alpha)$, $H_0(\delta_i, \beta) = G(\bar{f}(\delta_i), \beta)$ and $p \circ H(\alpha, \beta) = p(f(\alpha))$, for $\alpha \in \Delta[n]$, $\beta \in (\Delta[1])_{n-1}, (\Delta[1])_{\dim \alpha}$ respectively, and $i \in [n]$. Take $z = H_0(\iota, \varepsilon_1)$ and $H_1 : \Delta[n] \times \Delta[1] \to Z, (\alpha, \beta) \mapsto G(z\alpha, \beta)$. Then, the simplices $f(\iota)$ and $\hat{p}(z)$ are fibre homotopic (see Lemma 4.5.19 (i)); since both belong to $\hat{Z}$, they are equal, by minimality. Thus, the simplicial map $g : \Delta[p] \to Z, \iota \mapsto z$, has the required properties. □

Now it is possible to prove the essential fact that fibrations are preserved under geometric realization.

Theorem 4.5.25 *The geometric realization of a Kan fibration is a fibration.*

Proof The geometric realization of an acyclic fibration is a fibration (see Proposition 4.5.16), the geometric realization of a minimal fibration is a fibration (see Corollary 4.5.22) and a composition of fibrations is a fibration (see Proposition A.4.7(i)). □

Remark This result states that the geometric realization of a Kan fibration has the homotopy lifting property for homotopies which are defined on weak Hausdorff k-spaces. The problem of deciding whether geometric realization transforms Kan fibrations into Hurewicz fibrations is still open, but is not a very interesting question. The difficulty arises from the fact that the product in the category *Top* used here is the cartesian product only in special cases. Thus, one surely obtains a Hurewicz fibration if the total set of the Kan fibration is transformed into a countable or locally finite CW-complex (see Proposition 2.2.3 and the remark preceding it). □

Conversely, one may ask which (continuous) maps are transformed into Kan fibrations by the singular functor.

Proposition 4.5.26 *If $p : Z \to X$ is any map, then the simplicial map Sp is a Kan fibration iff p is a Serre fibration.*

Proof '⇒': Let a map $g : \Delta^n \to Z$ and a homotopy $H : \Delta^n \times I \to X$ starting at $p \circ g$ be given; one has to look for a homotopy starting at g which is a lifting of H. Identify Δ^n with $|\Delta[n]|$ and $\Delta^n \times I$ with $|\Delta[n] \times \Delta(1)|$. Then, one can use the adjointness between geometric realization and the singular functor to obtain a simplicial map $g' : \Delta[n] \to SZ$ and a simplicial homotopy $H' : \Delta[n] \times \Delta[1] \to SX$ starting at $Sp \circ g'$. Since the inclusion

$$\Delta[n] = \Delta[n] \times \Lambda^0[1] \subset \Delta[n] \times \Delta[1]$$

is an anodyne extension, and Sp is assumed to be a Kan fibration, the simplicial homotopy H' lifts to a simplicial homotopy $G' : \Delta[n] \times \Delta[1] \to SZ$ starting at g'. The adjoint $G : \Delta^n \times I \to Z$ of G' is a homotopy of the desired kind.

'⇐': One has to fill n-horns in SZ whose images under Sp have fillings in SX. Thus consider simplicial maps $\bar{f} : \Lambda^k[n] \to SZ$ and $f : \Delta[n] \to SX$ such that $f | \Lambda^k[n]| = Sp \circ \bar{f}$. The respective adjoints are maps $\bar{f}' : |\Lambda^k[n]| \to Z$ and $f' : \Delta[n] \to X$ such that $f' | |\Lambda^k[p]| = p \circ \bar{f}'$. Since the subcomplex $|\Lambda^k[n]|$ is a strong deformation retract of the CW-complex Δ^n, and p is assumed to be a Serre fibration, there is a map $g' : \Delta[n] \to Z$ such that $g' | |\Lambda^k[p]| = \bar{f}'$ and $p \circ g' = f'$ (see Corollary 1.4.9). Its adjoint is a simplicial map $g : \Delta[n] \to SZ$ such that $g(\iota)$ is a filling of the horn given by $\bar{f}$ over the filling $f(\iota)$ of its image under p. □

Another nice property of Kan fibrations consists in the possibility of approximating certain maps by simplicial maps.

Theorem 4.5.27 *Let $p : Z \to X$ be a Kan fibration, let $i : \Lambda \to D$ be an injective simplicial map and let*

$$\begin{array}{ccc} \Lambda & \xrightarrow{\bar{f}} & Z \\ {\scriptstyle i}\downarrow & & \downarrow{\scriptstyle p} \\ D & \xrightarrow[f]{} & X \end{array}$$

be a commutative square in the category SiSets. Then, for each map $g' : |D| \to |Z|$ with $g' \circ |i| = |\bar{f}|$ and $|p| \circ g' = |f|$ there is a simplicial map $g : D \to Z$ with $g \circ i = \bar{f}$ and $p \circ g = f$ whose geometric realization is homotopic to g' rel. $|\Lambda|$ over $|X|$.

Proof If $D = \Delta[n]$ and $\Lambda = \delta\Delta[n]$, the claim can be reformulated in terms of fibre homotopic simplices: *A singular simplex z' of $|Z|$ with $\delta z' \subset Z$ and $x = |p| \circ z' \in X$ is fibre homotopic to a simplex $z \in Z$.* This will be proved, first by an induction on n and then by a further induction on $\tau = x^\flat$; the latter refers to the (partial) order given on the finite set of all degeneracy operators with dimension n by

$$\tau' \leqslant \tau \Leftrightarrow \tau'(i) \leqslant \tau(i), \qquad \text{for all } i \in [n]$$

which has ω as minimum and ι as maximum. Note that any simplicial set X can be considered as a simplicial subset of the singular set $S|X|$, via the injective simplicial map η_X.

If $n=0$, one has a point $z'(1)=[\tilde{z},t]\in|Z|$ such that $p([\tilde{z},t])=[p(\tilde{z}),t]=[x,1]$. This implies $p(\tilde{z})=x\omega$ and thus, $z=\tilde{z}\varepsilon_0$ is a 0-simplex over x; furthermore, the homotopy $I\to|Z|, s\mapsto[\tilde{z},(1-s)e_0+st]$ fulfils the required conditions.

Now suppose that $n>0$ and $\tau=\omega$. Then, $x^\#$ is a 0-simplex and the fibre F over $x^\#$ is a Kan set (see Corollary 4.5.2(ii)) containing the $(n-1)$-simplices $z'\delta_i$, for all $i\in[n]$. Since geometric realization commutes with pullbacks (see Theorem 4.3.16) the CW-complex $|F|$ can be identified with $|p|^{-1}([x^\#,1)]$ and in this way, it contains the image of the map z'. Choose a Kan fibration $q: W\to F$ with contractible total set W (see Corollary 4.5.9(ii)), an n-horn $\{w_i : i\in[n-1]\}$ in W over the n-horn $\{z'\delta_i : i\in[n-1]\}$ (see Proposition 4.5.1(v)) and a singular simplex $w'\in S|W|$ which fills this n-horn over z'; the latter choice is possible since $|q|$ is a fibration (see Theorem 4.5.25) and consequently $S|q|$ is a Kan fibration (see Proposition 4.5.26). The singular simplex $w'\delta_n$ is fibre homotopic to a simplex $w_n\in W_{n-1}$, by the inductive hypothesis; let $H: \Delta[n-1]\times\Delta[1]\to S|W|$ denote a simplicial homotopy rel. $\delta\Delta[n-1]$ over $S|F|$, with $H(\iota,\varepsilon_0)=w_n$ and $H(\iota,\varepsilon_1)=w'\delta_n$. Since W is contractible, there is a simplex $w\in W_n$ with $w\delta_i=w_i$, for all $i\in[n]$ (see Corollary 4.5.15). Next, consider the simplicial map

$$\tilde{H}: \Delta[n]\times\delta\Delta[1]\cup\delta\Delta[n]\times\Delta[1]\to S|W|$$

given by $\tilde{H}(\alpha,\varepsilon_0\omega)=w\alpha$, $\tilde{H}(\alpha,\varepsilon_1\omega)=w'\alpha$, $\tilde{H}(\delta_i,\beta)=w'\alpha$, for all $i\in[n-1]$, and $\tilde{H}(\delta_n,\beta)=H(\iota,\beta)$. The adjoint $\tilde{H}'$ of $\tilde{H}$ is a map from the boundary sphere $S=\Delta^n\times\{0,1\}\cup\delta\Delta^n\times I$ of the ball $|\Delta[n]\times\Delta[1]|=\Delta^n\times I$ to $|W|$. Since the CW-complex $|W|$ is contractible (see Proposition 4.3.17), $\tilde{H}'$ can be extended to a homotopy $\hat{H}: \Delta^n\times I\to|W|$ whose adjoint $\hat{H}'$, in turn, is a simplicial homotopy extending $\tilde{H}$. The composition $q\circ\hat{H}'$ shows that $z=q(w)\in Z$ is fibre homotopic to z'.

For the last step of the induction, assume that $\tau\neq\omega$; this implies $m=\dim x^\#>0$. Let k denote the smallest element of $[n]$ with $\tau(i)\neq\tau(i+1)$. Since p is a Kan fibration, there is a filling $\tilde{z}$ of the horn $\{z'\delta_i : i\in[n-1]\}$ in Z, over x. As before, since p is a Kan fibration, the simplicial map $S|p|$ also is a Kan fibration. Thus, there exists a singular simplex $\hat{z}\in S|Z|$ with $\hat{z}\delta_k=z_n, \hat{z}\delta_i=z'\sigma_k\delta_i$, for $i\in[n]\backslash\{k,k+1\}$, and $|p|\circ\hat{z}=x\sigma_k$. Compute

$(|p|\circ\hat{z})\delta_{n+1} = x\sigma_k\delta_{n+1}$ and $(x\sigma_k\delta_{n+1})^b < \tau$. Thus, by induction, the singular simplex $\hat{z}\delta_{n+1}$ is fibre homotopic to a simplex $z_{n+1}\in Z_n$; let $H : \Delta[n]\times\Delta[1]\to S|Z|$ denote a simplicial homotopy rel. $\delta\Delta[n]$ over $S|X|$ with $H(\iota,\varepsilon_0)=z_{n+1}$ and $H(\iota,\varepsilon_1)=\hat{z}\delta_{n+1}$. Using once more that p is a Kan fibration, one finds an $(n+1)$-simplex $w\in Z$ with $w\delta_k = z_n$, $w\delta_{n+1}=z_{n+1}$, $w\delta_i = z\sigma_k\delta_i$, for $i\in[n]\backslash\{k,k+1\}$, and $p(w)=x\sigma_k$. Since $S|p|$ is a Kan fibration and the inclusion

$$\Delta[n+1]\times\delta\Delta[1]\cup\Lambda^{k+1}[n+1]\times\Delta[1]\subset\Delta[n+1]\times\Delta[1]$$

is an anodyne extension (see Proposition 4.5.8), there is a simplicial homotopy $\tilde{H} : \Delta[n+1]\times\Delta[1]\to S|Z|$ with $\tilde{H}(\alpha,\varepsilon_0\omega)=w\alpha$, $\tilde{H}(\alpha,\varepsilon_1\omega)=\hat{z}\alpha$, $\tilde{H}(\delta_i,\beta)=\hat{z}\delta_i=w\delta_i$, for $k+1\neq i\in[n]$, $\tilde{H}(\delta_{n+1},\beta)=H(\iota,\beta)$ and $|p|\circ\tilde{H}(\alpha,\beta)=x\sigma_k\alpha$. The composition $\tilde{H}\circ(\Delta\delta_{k+1}\times\mathbf{1}_{\Delta[1]})$ shows that $z=w\delta_{k+1}\in Z$ is fibre homotopic to z'. This finishes the induction for the case in which one considers a single singular simplex.

The general case is dealt with using the method of the least criminal. Choose a simplicial subset $\tilde{D}\subset D$ containing Λ which is maximal with respect to the property that there are a simplicial map $\tilde{g} : \tilde{D}\to Z$ and a simplicial homotopy from $\eta_Z\circ\tilde{g}$ to $Sg'\circ\eta_D\big|\tilde{D}=(Sg'\big|S|\tilde{D}|)\circ\eta_{\tilde{D}}$ rel. Λ over $S|X|$. Assume that such a simplicial map $\tilde{g}$ and a corresponding homotopy $\tilde{H}$ are fixed. To simplify notation one may further assume that D is obtained from $\tilde{D}$ by attaching a simplicial standard-simplex $\Delta[n]$ via a simplicial map with domain $\delta\Delta[n]$; it is to be shown that $\tilde{g}$ can be extended over D such that the geometric realization of the extension is homotopic to g' rel. $|\Lambda|$ over $|X|$. Since $S|p|$ is a Kan fibration, one has a homotopy $\hat{H} : D\times\Delta[1]\to S|Z|$, over $S|X|$, from a simplicial map $g'' : D\to S|Z|$ to $Sg'\circ\eta_D$ which extends $\tilde{H}$. Let $a : \delta\Delta[n]\to\tilde{D}$ and $\bar{a} : \Delta[n]\to D$ denote the attaching and characteristic map respectively of the simplicial attaching which generates D out of $\tilde{D}$; take $d=d\bar{a}(\iota)$ and $z'=\hat{H}(d,\varepsilon_0\omega)$. Then, choose a simplex $z\in Z$ which is fibre homotopic to z' and a simplicial homotopy $G : \Delta[n]\times\Delta[1]\to S|Z|$ rel. $\delta\Delta[n]$ over $S|X|$. Extend $\tilde{g}$ to a simplicial map $g : D\to Z$ by taking $g(d)=z$. To obtain a suitable simplicial homotopy, first observe that $D\times\Delta[1]$ is obtained from $\tilde{D}\times\Delta[1]$ by attaching $\Delta[n]\times\Delta[1]$ via $a\times\mathbf{1}_{\Delta[1]}$. The universal property of this attaching implies the existence of a homotopy $\hat{G} : D\times\Delta[1]\to S|Z|$ rel. $\tilde{D}$ over $S|X|$, from $\eta_Z\circ g$ to g'', with $\hat{G}(d,\beta)=G(\iota,\beta)$. A homotopy rel. $\tilde{D}$ is also a homotopy rel. Λ; since $S|p|$ is a Kan fibration, homotopy rel. Λ over $S|X|$ is an equivalence relation (see Proposition 4.5.10). Therefore, the homotopies $\hat{H}$ and $\hat{G}$ show that the simplicial maps $\eta_D\circ g$ and $g'\circ\eta_Z$ are homotopic rel. Λ over $S|X|$; thus, by adjointness, the maps $|g|$ and g' are homotopic rel. $|\Lambda|$ over $|X|$ as desired. □

Corollary 4.5.28 *A Kan set is a strong deformation retract of the singular set of its geometric realization.*

Proof Let $p : Z \to \Delta[0]$ be a Kan fibration. In the diagram of Theorem 4.5.27, take

$i = \eta_Z : Z \to S|Z|$, the unit of the adjunction,
$\bar{f} = \mathbf{1}_Z : Z \to Z$,
$f : S|Z| \to \Delta[0]$, the unique possible simplicial map, and
$g' = j_{|Z|} : |S|Z|| \to |Z|$, the co-unit of the adjunction.

Then there is a homotopy $|S|Z| \times \Delta[1]| \to |Z|$, whose adjoint is a deformation of $S|Z|$ into Z (see Theorem 4.5.27). □

If the Kan set under consideration is the singular set of a space, then there is a distinguished retraction among those whose existence is assured in this statement; it arises from the co-unit of the adjointness between geometric realization and singular functor. Recall the fundamental equation

$$Sj_T \circ \eta_{ST} = \mathbf{1}_{ST}$$

and consider the converse composition of unit and co-unit. One cannot expect equality, but the best possible statement holds true.

Proposition 4.5.29 *If T is a space, the composition $\eta_{ST} \circ Sj_T$ of unit and co-unit is homotopic to $\mathbf{1}_{S|SY|}$ rel. ST.*

Proof The simplicial set ST is a Kan set (see Example 2); thus, one has a simplicial homotopy $H : S|ST| \times \Delta[1] \to S|ST|$ rel. ST from $\mathbf{1}_{S|ST|}$ to the composition $\eta_{ST} \circ r$ of the unit η_{ST} and some simplicial retraction $r : S|ST| \to ST$. The composition $\eta_{ST} \circ Sj_T \circ H$ is a simplicial homotopy rel. ST from $\eta_{ST} \circ Sj_T$ to $\eta_{ST} \circ Sj_T \circ \eta_{ST} \circ r = \eta_{ST} \circ r$. Since $S|ST|$ is also a Kan set, homotopy rel. ST is an equivalence relation on the set of all simplicial maps $S|ST| \to S|ST|$ (see Corollary 4.5.11) yielding the desired result. (That is, *cum grano salis*, the standard proof for showing that, given an invertible elements, any left inverse is also a right inverse.) □

Now it is possible to prove a deep result which has been already announced (see the discussion of the adjointness between geometric realization and singular functor in Section 4.3).

Theorem 4.5.30 *The co-unit $j_T : |ST| \to T$ is a weak homotopy equivalence, for any space T.*

Proof Let a space T be given; without loss of generality, one may assume T to be non-empty and path-connected. Choose a base point $x_0 \in T$ and let $\tilde{x}_0$ denote the unique point of the 0-cell in $|ST|$ which corresponds to the singular simplex $\Delta^0 \to T$, $1 \mapsto x_0$. One has to show that the functions

$$j_{T,n} : \pi_n(|ST|, \tilde{x}_0) \to \pi_n(T, x_0)$$

are bijective, for all $n \in \mathbf{N}$.

For $n = 0$, the assertion means that $|ST|$ should be path-connected, since T is assumed to be path-connected. Recall that every path component of the CW-complex $|ST|$ contains a 0-cell (see Proposition 1.4.15). It suffices to check that for every 0-cell $\tilde{x} \neq \tilde{x}_0$ in $|ST|$ there is a 1-cell in $|ST|$ with boundary $\{\tilde{x}, \tilde{x}_0\}$. To see this, take a path in T joining the points $j_T(\tilde{x})$ and x_0; it gives rise to a singular 1-simplex, and thus to a 1-cell in $|ST|$ of the desired kind.

Now assume $n > 0$ and take a representative $a : S^n \to T$ for an element of $\pi_n(T, x_0)$. Identify S^n with $|\delta\Delta[n+1]|$, such that $\boldsymbol{e}_0$ corresponds to some 0-cell of the CW-complex $|\delta\Delta[n+1]|$ and let $a' : \delta\Delta[n+1] \to ST$ denote the adjoint of a. Then, the geometric realization $|a'|$ of a' represents an element of $\pi_n(|ST|, \tilde{x}_0)$ which is mapped into the class of a. This shows the surjectivity of the function $j_{T,n}$.

To prove the injectivity, note that one deals with homomorphisms. Therefore, it is sufficient to verify that the appearing kernels are trivial, i.e., that any map $a : |\delta\Delta[n+1]| \to |ST|$ whose composition with j_T has an extension $\bar{a} : |\Delta[n+1]| \to T$ can be extended over $|\Delta[n+1]|$ itself. Let such maps $a, \bar{a}$ be given and let $a' : \delta\Delta[n+1] \to S|ST|, \bar{a}' : \Delta[n+1] \to ST$ denote their respective adjoints. Let $H : S|ST| \times \Delta[1] \to S|ST|$ denote a simplicial homotopy rel. ST from $\mathbf{1}_{S|ST|}$ to the composition $\eta_{ST} \circ Sj_T$ (see Proposition 4.5.29) and define a simplicial map $\tilde{a} : \Delta[n+1] \times \Lambda^1[1] \cup \delta\Delta[n+1] \times \Delta[1] \to S|ST|$ by taking $\tilde{a}(\alpha, \varepsilon_1 \omega) = \eta_{ST} \circ \bar{a}'(\alpha)$ and $\tilde{a}(\delta_i, \beta) = H(a'(\delta_i), \beta)$, for $\alpha \in \Delta[n+1], \beta \in (\Delta[1])_n$ and $i \in [n+1]$. Its adjoint has a domain which is homeomorphic to $|\Delta[n+1]|$ and because it agrees with the map a on the boundary it can be considered as the desired extension of a. □

Corollary 4.5.31 (i) *The co-unit $j_T : |ST| \to T$ is a homotopy equivalence, for any CW-complex T.*

(ii) *A map $f : U \to T$ is a weak homotopy equivalence iff $|Sf| : |SU| \to |ST|$ is a homotopy equivalence.*

(iii) *Let Y, X be Kan sets. Then, a simplicial map $f : Y \to X$ is a simplicial homotopy equivalence iff $|f| : |Y| \to |X|$ is a homotopy equivalence.*

(iv) *A map $f : U \to T$ is a weak homotopy equivalence iff $Sf : SU \to ST$ is a simplicial homotopy equivalence.*

(v) *A simplicial set X is weakly contractible iff the singular set $S|X|$ is simplicially contractible.*

(vi) *For any simplicial set X, the unit $\eta_X : X \to S|X|$ can be decomposed into an anodyne extension followed by an acyclic fibration.*

Proof (i) follows from Whitehead's realizability theorem (see Theorem 2.5.1).

For (ii), consider the equation

$$f \circ j_U = j_T \circ |Sf|$$

which arises from the naturality of the co-unit. Since the maps j_U, j_T are always weak homotopy equivalences, by the previous theorem it follows that f is a weak homotopy equivalence iff $|Sf|$ is a weak homotopy equivalence. But the latter holds true iff $|Sf|$ is a homotopy equivalence, again by Whitehead's realizability theorem.

(iii) '$\Rightarrow$': Geometric realization transforms simplicial homotopy equivalences into homotopy equivalences (see Proposition 4.3.17).

'$\Leftarrow$': Let $g : |X| \to |Y|$ be a homotopy inverse for $|f|$ and let $r : S|Y| \to Y$ be a simplicial homotopy inverse for η_Y. Then, the composition $r \circ Sg \circ \eta_X$ is a simplicial homotopy inverse for f. In fact, the composition $r \circ Sg \circ \eta_X \circ f = r \circ Sg \circ S|f| \circ \eta_Y$ is homotopic to $r \circ \eta_Y$, which in turn is homotopic to $\mathbf{1}_Y$; since homotopy between maps with a Kan set as codomain is transitive (see Corollary 4.5.11), $r \circ Sg \circ \eta_X \circ f$ is homotopic to $\mathbf{1}_Y$. The composition $f \circ r \circ Sg \circ \eta_X$ is homotopic to $r' \circ \eta_X \circ f \circ r \circ Sg \circ \eta_X = r' \circ S|f| \circ \eta_Y \circ r \circ Sg \circ \eta_X$, where r' denotes a simplicial homotopy inverse for η_X. The latter composition is homotopic to $r' \circ S|f| \circ Sg \circ \eta_X$ and two further homotopies lead to $\mathbf{1}_X$; again, one has to use the transitivity of the homotopy relation which is assured by the hypothesis that X is a Kan set.

The statement (iv) is an immediate consequence of (ii) and (iii); (v) results from application of (iv) to the unique simplicial map $X \to \Delta[0]$.

(vi) For any simplicial set X, the unit η_X can be decomposed in the form $p \circ i$ where p is a Kan fibration and i is an anodyne extension (see Proposition 4.5.6). Moreover, $|\eta_X|$ is a homotopy equivalence, since $j_{|X|}$ is a homotopy equivalence, by (i), and $j_{|X|} \circ |\eta_X| = \mathbf{1}_{|X|}$, by adjointness. But $|p| \circ |i| = |\eta_X|$, and $|i|$ is a homotopy equivalence (see Proposition 4.5.5); thus, $|p|$ is a homotopy equivalence. The base of p is a Kan set, and thus the total set of p is also a Kan set (see Corollary 4.5.2 (iii)), and, consequently, p is a simplicial homotopy equivalence, by statement (iii). Finally, a Kan fibration which is a homotopy equivalence is an acyclic fibration (see Proposition 4.5.14 (ii)). □

A special sort of Kan fibrations is given by the *simplicial resolutions of groups* defined as follows. Let G be a group with unit element denoted by

1. The *classifying set* of G is the simplicial set BG given by $(BG)_n = G^n$ for all $n \in \mathbf{N}$ together with the face operations

$$g\delta_j^1 = *$$

for $j = 0, 1$, where $*$ denotes the unique element of $(BG)_0 = G^0$,

$$(g_1, \ldots, g_n)\delta_0^n = (g_2, \ldots, g_n),$$
$$(g_1, \ldots, g_n)\delta_j^n = (g_1, \ldots, g_j \cdot g_{j+1}, \ldots, g_n),$$
$$(g_1, \ldots, g_n)\delta_n^n = (g_1, \ldots, g_{n-1}),$$

for $n > 1$ and $0 < j < n$, and the degeneracy operations

$$*\sigma_0^0 = \mathbf{1},$$
$$(g_1, \ldots, g_n)\sigma_i^n = (g_1, \ldots, g_i, \mathbf{1}, g_{i+1}, \ldots, g_n),$$

for $n > 0$ and $0 \leqslant i \leqslant n$. This will be the base of a Kan fibration, with total set EG given by $(EG)_n = G^{n+1}$ for all $n \in \mathbf{N}$, together with the face operations

$$(g_0, \ldots, g_n)\delta_j^n = (g_0, \ldots, g_j \cdot g_{j+1}, \ldots, g_n),$$
$$(g_1, \ldots, g_n)\delta_n^n = (g_1, \ldots, g_{n-1}),$$

for $n > 0$ and $0 \leqslant j < n$, and the degeneracy operations

$$(g_0, \ldots, g_n)\sigma_i^n = (g_0, \ldots, g_i, \mathbf{1}, g_{i+1}, \ldots, g_n),$$

for $n \leqslant 0$ and $0 \leqslant i \leqslant n$. Now, complete the construction of the simplicial resolution of G by defining the simplicial map

$$p_G : EG \to BG, \quad (g_0, \ldots, g_n) \mapsto (g_1, \ldots, g_n).$$

Lemma 4.5.32 *Let Y be a simplicial subset of the simplicial set X, let G be a group and let $f : Y \to EG$ be a simplicial map. Then there is a simplicial map $f' : X \to EG$ extending f.*

Proof The key to this fact lies in the observation that any simplex $g = (g_0, \ldots, g_n)$ is uniquely determined by its ordered set of vertices $(g\varepsilon_0, \ldots, g\varepsilon_n)$, since the following relations hold true:

$$g\varepsilon_i = g_0 \cdot \cdots \cdot g_i,$$
$$g_0 = g\varepsilon_0, \quad g_{i+1} = (g\varepsilon_i)^{-1} \cdot (g\varepsilon_{i+1});$$

moreover, the previous formulae show that every ordered set of $n+1$ 0-simplices of EG is the ordered set of vertices of a unique n-simplex of EG.

Now, in order to prove the statement, extend $f \mid Y^0$ arbitrarily over X^0. Then take an $x \in X_n$, with $n > 0$. The image of the vertices is already determined, and one just assigns to x the corresponding simplex of EG.

□

Proposition 4.5.33 *Let G be a group. Then:*

(i) *the simplicial map p_G is a Kan fibration with 0-dimensional fibre;*

(ii) *the total set EG is simplicially contractible;*

(iii) *the geometric realization $|BG|$ of BG is an Eilenberg–MacLane space of type $K(G,1)$.*

Proof (i) The simplicial map p_G is surjective; thus, any 0-horn in EG can be suitably filled. Considering the 0-simplex $\mathbf{g} = (g) \in EG$ as 0th 1-horn (respectively, 1st 1-horn) and the 1-simplex $\mathbf{g}' = (g') \in BG$ as the respective filling of the corresponding 1-horns in BG, then (g, g') (respectively, $(g \cdot (g')^{-1}, g')$) is a suitable filling of the horn in EG.

For any $n > 1$, observe first that any n-simplex in BG is determined by any two different faces; thus, any n-horn in BG has at most one filling. An n-horn in EG is uniquely determined by the ordered set of its vertices, and therefore has a unique filling. This filling lies over a prescribed simplex of BG because its image and the prescribed simplex have two different faces in common.

Since BG has only one 0-simplex, there is a unique fibre for p_G; it consists of the elements of the form $(g, \mathbf{1}, \ldots, \mathbf{1})$, and so it is 0-dimensional.

(ii) Extend the simplicial map $EG \times \delta\Delta[1] \to EG, (\mathbf{g}, \varepsilon_0 \omega) \mapsto \mathbf{g}, (\mathbf{g}, \varepsilon_1 \omega) \mapsto (\mathbf{1}, \ldots, \mathbf{1})$ over $EG \times \Delta[1]$ (see Lemma 4.5.32).

(iii) Since BG has only one 0-simplex, its geometric realization is, in a unique manner, a based CW-complex whose fundamental group is just G (see Theorem 2.6.8). The map $|p_G|$ is a fibration (see Theorem 4.5.25) with contractible total space (see (ii) and Proposition 4.3.17) and 0-dimensional fibre (see Theorems 4.3.5, 4.3.16). An inspection of the homotopy sequence of this fibration (see Proposition A.8.17) shows that the higher homotopy groups of $|BG|$ vanish. □

Remark Within the proof of part (i), it has been stated that every n-horn in BG (for $n > 1$) has at most one filling. In fact, such a filling always exists: since EG is a Kan set (also implicitly noted in the previous proof), so is BG as a consequence of statement (i) (see Corollary 4.5.2 (iii)). □

The previous proposition is particularly interesting for minimal Kan sets because their sets of 1-simplices have an intrinsic group structure which can be exploited to construct a useful Kan fibration.

Proposition 4.5.34 *Let X be a minimal Kan set with just one 0-simplex Then, the following hold true:*

(i) *X_1 has a canonical group structure isomorphic to the fundamental group of $|X|$;*

(ii) *there is a simplicial map $q_X : X \to BX_1$, which is a Kan fibration and induces an isomorphism between the fundamental groups of the respective geometric realizations;*

(iii) *the geometric realization of the fibre of q_X is simply connected.*

Proof (i): Given $x, y \in X_1$, there is a $z \in X_2$ such that $z\delta_0 = y$ and $z\delta_2 = x$; by minimality, the 1-simplex $z\delta_1$ is independent of the chosen z (see Lemma 4.5.19 (ii)), which allows one to define

$$x \cdot y = z\delta_1.$$

Clearly, the unique degenerate 1-simplex in X is the neutral element for this multiplication; the inverses can be found by filling of the corresponding horns. Since X has only one 0-simplex, its geometric realization can be viewed as a based CW-complex whose fundamental group is nothing but X_1 (see Theorem 2.6.8).

(ii): Define the simplicial map $q_X : X \to BX_1$ by the unique possible function in dimension 0, the identity in dimension 1 and the assignment

$$x \mapsto (x\delta_n \cdots \delta_2, \ldots, x\delta_n \cdots \delta_{i+1}\delta_{i-2} \cdots \delta_0, \ldots, x\delta_{n-2} \cdots \delta_0).$$

In order to show that this simplicial map is a Kan fibration, it suffices to consider n-horns for $n > 1$ only. Since X is a Kan set, every such horn has a filling; this lies over a prescribed simplex of BG, as in the proof of Proposition 4.5.33 (i). The second part of the statement follows from (i) (by means of Proposition 4.5.33 (iii)).

(iii): Since q_X is a Kan fibration, its geometric realization is a fibration (see Theorem 4.5.25), which moreover induces an isomorphism between the fundamental groups, by (ii). Inspection of the lowest terms of the homotopy sequence of $|q_X|$ (see Proposition A.8.17) shows that its fibre is simply connected. □

The concept of simplicial resolution of a group generalizes to the construction of simplicial universal coverings. To begin with, note that a simplicial set is called *connected* if its geometric realization is connected. Let X be a connected simplicial set and let π denote the fundamental group of its geometric realization $|X|$ with respect to a fixed base point corresponding to a 0-simplex $z_0 \in X_0$. Choose a *twisting function for* X, i.e., a function $\varphi : X \to \pi$ satisfying the following properties:

(1) $\varphi(x)$ depends only on the first edge of x, more precisely:

$$\varphi(x) = \varphi(x\delta_n \cdots \delta_2),$$

for any simplex x with $n = \dim x \geqslant 1$;

(2) for any degenerate 1-simplex x,

$$\varphi(x) = 1;$$

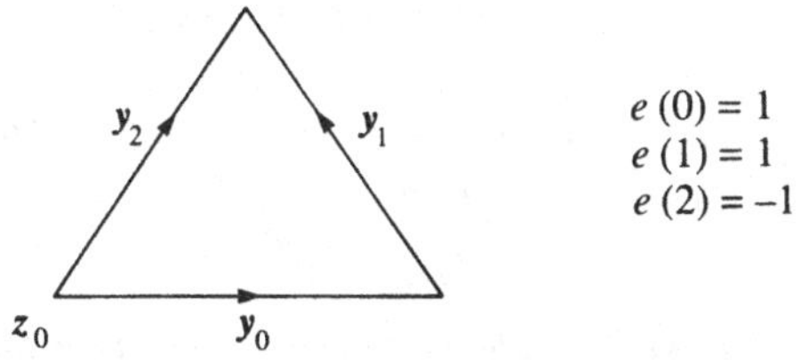

Figure 13

(3) for any 2-simplex x,

$$\varphi(x\delta_1) = \varphi(x\delta_2) \cdot \varphi(x\delta_0);$$

(4) any $\omega \in \pi$ has a representation of the form

$$\omega = \varphi(y_0)^{e(0)} \cdot \cdots \cdot \varphi(y_k)^{e(k)},$$

where $y_j \in X_1$, $e(j) \in \{-1, 1\}$, for $j \in [k]$, form a 'closed edge path' based at z_0 (Figure 13) i.e., satisfy the relations

$$y_0\varepsilon_f = y_k\varepsilon_{1-g} = z_0$$
$$\text{if } e(0) = (-1)^f,\ e(k) = (-1)^g,$$

$$y_j\varepsilon_{1-f} = y_{j+1}\varepsilon_g$$
$$\text{if } e(j) = (-1)^f,\ e(j+1) = (-1)^g,$$

for $f, g \in [1]$.

The values of φ on X_0 are meaningless and are included only in order to have a simple domain for φ. A possible way to obtain such a function is the following: choose a subset $S \subset X_1$ containing all degenerate simplices and such that the 1-cells corresponding to the non-degenerate 1-simplices in S span a maximal tree in $|X|$. To the elements of S, one associates the value $1 \in \pi$; to each other 1-simplex, assign the element of π that is generated by the corresponding 1-cell (in the direction given by the simplex; see Theorem 2.6.8). Condition (1) then forces the values on the higher-dimensional simplices; condition (3) is trivially satisfied for degenerate 2-simplices, while each non-degenerate 2-simplex gives a 2-cell, inducing the desired relation.

Now, the *simplicial universal covering* $\tilde{X}$ of the simplicial set X (with respect to the function φ) is given by $\tilde{X}_n = \pi \times X_n$ for all $n \in \mathbf{N}$, together with the face operations

$$(\omega, x)\delta_0 = (\omega \cdot \varphi(x), x\delta_0),$$
$$(\omega, x)\delta_j = (\omega, x\delta_j),$$
$$(\omega, x)\sigma_i = (\omega, x\sigma_i),$$

for all $\omega \in \pi$, $x \in X$, $0 < j \leqslant \dim x$ and $0 \leqslant i \leqslant \dim x$.†

Proposition 4.5.35 *Let X be a connected simplicial set and let φ be a twisting function for X. Then, the simplicial map*

$$p : \tilde{X} \to X, \qquad (\omega, x) \mapsto x$$

is a minimal fibration whose geometric realization is a universal covering projection.

Proof Let $\bar{f} : \Lambda^k[n] \to \tilde{X}$ be a simplicial map such that the composition $p \circ \bar{f}$ has an extension $f : \Delta[n] \to X$; one has to look for a simplicial map $g : \Delta[n] \to \tilde{X}$ such that $g | \Lambda^k[n] = \bar{f}$ and $p \circ g = f$. Consider the case $k \neq n$; take $\bar{f}(\delta_n) = (\omega, \tilde{x})$, $f(\iota) = x$ and define g by $\alpha \mapsto (\omega, x)\alpha$. If $k = n$ and $n > 1$, take $\bar{f}(\delta_{n-1})$ to find ω; if $k = n = 1$, assume $\bar{f}(\delta_0) = (\omega, \tilde{x})$, $f(\iota) = x$ and define $g(\alpha) = (\omega \cdot \varphi(x)^{-1}, x)\alpha$. This shows that p is a Kan fibration. The fibres p are 0-dimensional; hence, at least fibre homotopic 0-simplices have to be equal. Consequently, fibre homotopic simplices (ω, x), (ω', x) must have the same 0th vertex:

$$(\omega, x\varepsilon_0) = (\omega, x)\varepsilon_0 = (\omega', x)\varepsilon_0 = (\omega', x\varepsilon_0),$$

implying $\omega = \omega'$; therefore, p is a minimal fibration.

It follows that the geometric realization $|p|$ of p is a locally trivial fibration (see Corollary 4.5.22) with discrete fibre (see Theorem 4.3.16), and thus a covering projection. It remains to prove that $|\tilde{X}|$ is simply connected. To see that $|\tilde{X}|$ is path-connected, fix the base point $\hat{z}_0 \in |X|$ corresponding to the 0-simplex $z_0 \in X_0$ used in the definition of the twisting function φ. Now consider first a point $\tilde{x} \in |\tilde{X}|$ not belonging to the fibre over $\hat{z}$. Take a path in $|X|$ from $|p|(\tilde{x})$ to $\hat{z}$ and lift it to a path in $|\tilde{X}|$ starting at $\tilde{x}$. The lifted path ends in a point of the fibre over $\hat{z}$; thus, it suffices to check that any point in this fibre can be connected to the point corresponding to the 0-simplex $(1, z_0)$ by a path in $|\tilde{X}|$. Take $\omega \in \pi$ and represent it in the form described in condition (4) of the definition of twisting functions. Then, define elements $\varphi_r \in \pi$, for $r \in [k]$, by the formulae:

$$\varphi_0 = \begin{cases} 1, & e(0) = 1, \\ \varphi(y_0)^{-1}, & e(0) = -1, \end{cases}$$

$$\varphi_r = \begin{cases} \varphi(y_0)^{e(0)} \cdot \ldots \cdot \varphi(y_{r-1})^{e(r-1)}, & e(r) = 1, \\ \varphi(y_0)^{e(0)} \cdot \ldots \cdot \varphi(y_r)^{e(r)}, & e(r) = -1. \end{cases}$$

Now, the 1-simplices $(\varphi_0, y_0), \ldots, (\varphi_k, y_k)$ form a path connecting the points corresponding to the 0-simplices $(1, z_0)$ and (ω, z_0).

† The simplicial set obtained in this way is sometimes called the *twisted cartesian product* of X and π with respect to the twisting function φ.

The given argument also shows that π acts fixed-point-free on the fibre, implying thereby that $|\tilde{X}|$ is simply connected (see Corollary A.8.18).

□

Exercises

1. Let $\Phi : \Delta \to SiSets$ be a (covariant) functor satisfying the Eilenberg–Zilber property and such that the simplicial sets $\Phi([n])$ are weakly contractible, for all $n \in \mathbb{N}$. Then, the spaces $|X|$ and $|X \otimes \Phi|$ have the same homotopy type, for all simplicial sets X. (Fritsch & Latch, 1981, Lemma 4.7)
2. A Kan fibration is minimal iff any two n-simplices of the total set with the same image in the base have the same boundary whenever n of their $(n-1)$-dimensional faces coincide.
3. Show that a Kan set is connected iff each pair of 0-simplices generates the boundary of a 1-simplex. Show that a minimal Kan set is connected iff it contains exactly one 0-simplex.
4. Let Z be a connected minimal Kan set, and, for $n > 0$, let $\pi_n(Z)$ denote the set of n-simplices of Z whose boundary is generated by the unique 0-simplex $z_0 \in Z$. Define a binary operation on $\pi_n(Z)$ as follows. Given $z, z' \in \pi_n(Z)$ take a simplicial map $f : \Delta[n+1] \to Z$ with $f(\delta_0) = z$, $f(\delta_2) = z'$ and $f(\delta_i) = z_0\omega$, for all $i > 2$. Then, set $z \cdot z' = f(\delta_1)$. Show that this operation is a well-defined group structure on $\pi_n(Z)$ which is abelian for $n > 1$.
5. Let Z be a connected Kan set, choose a 0-simplex $z_0 \in Z$, and, for $n > 0$, let $\pi_n(Z, z_0)$ denote the set of fibre homotopy equivalence classes of n-simplices of Z whose boundary is generated by z_0. Define analogously to the previous exercise a binary operation on $\pi_n(Z, z_0)$, and show that it has the same properties. Moreover, show that the group obtained, the nth *homotopy group* of Z, depends on the choice of z_0 only up to isomorphism. (Kan, 1958c)
6. Extend the definition of homotopy groups for Kan sets to functors which are defined on the full subcategory of *SiSets* generated by the Kan sets. Show that these functors composed with the singular functor just yield the ordinary homotopy groups for based spaces.
7. Prove the simplicial analogue of Whitehead's realizability theorem (see Theorem 2.5.1): a simplicial map between Kan sets is a simplicial homotopy equivalence iff it induces isomorphisms for all homotopy groups. (Lamotke, 1968, VII,7.2 Folgerung)
8. Let p_G be the simplicial resolution of a group G. Show that the map $|p_G|$ is a universal covering projection.
9. Show that the construction of the simplicial universal covering

depends only up to simplicial isomorphism on the choice of the twisting function.

10. Let X be a connected simplicial set and let π be a subgroup of the fundamental group of $|X|$. Construct a Kan fibration $p : Z \to X$ whose geometric realization $|p|$ is a covering projection with image $|p|_1 = \pi$.
11. Show that the categories of fractions (see Gabriel & Zisman, 1967) $Top/\{j_T\}$ and $SiSets/\{\eta_X\}$ are equivalent to the homotopy category of CW-complexes, i.e., the category whose objects are the CW-complexes and whose morphism are the homotopy classes of maps between CW-complexes. (Ringel, 1970)
12. Degeneracy operators were not used for the definition of horns. So one can define presimplicial Kan sets by requiring that all horns can be filled. Show that any presimplicial Kan set alows the operation of degeneracy operators, i.e., belongs to the image of the forgetful functor $P : SiSets \to PSiSets$. (Kan, 1970; Fritsch, 1972)

4.6 Subdivision and triangulation of simplicial sets

For every $p \in \mathbf{N}$, define the simplicial set $\Delta'[p]$, the *normal subdivision* of $\Delta[p]$, as follows: take as n-simplices all weakly increasing (with respect to $\subset$) sequences $\boldsymbol{\mu} = (\mu_0, \ldots, \mu_n)$ of face operators with codomain $[p]$ and set $\boldsymbol{\mu}\alpha = (\mu_{\alpha(0)}, \ldots, \mu_{\alpha(m)})$, for all operators $\alpha : [m] \to [n]$.

Geometrically, one should view a single face operator μ_i of a sequence $\boldsymbol{\mu}$ as the barycentre $\boldsymbol{b}_i$ of the μ_ith face of Δ^p; the whole sequence $\boldsymbol{\mu}$ then corresponds to the simplex spanned by the vertices $\boldsymbol{b}_i$. The assignment

$$[\boldsymbol{\mu}, \boldsymbol{t}] \mapsto \sum_{i=0}^{n} t_i \boldsymbol{b}_i$$

describes a homeomorphism $\theta^p : |\Delta'[p]| \to \Delta^p$ (cf. the proof of Proposition 3.3.16). Thus, $|\Delta'[p]|$ can be viewed as barycentric subdivision of the Euclidean complex Δ^p (see Section 3.2, Example 3). Clearly, one speaks of an *interior point* in $|\Delta'[p]|$ if it is a point which is mapped onto an interior point of Δ^p by θ^p; the inverse image of the interior of Δ^p with respect to θ^p is the *interior* of $|\Delta'[p]|$. A nice combinatorial property of these subdivided standard-simplices is the following.

Lemma 4.6.1 *The simplicial set $\Delta'[p]$ is simplicially contractible to the vertex (ι).*

Proof A homotopy H from $\mathbf{1}_{\Delta'[p]}$ is given by taking for $\mu = (\mu_0, \ldots, \mu_n)$

and $j \in [n-1]$,

$$(H(\boldsymbol{\mu},\omega_j))_k = \begin{cases} \mu_k, & 0 \leqslant k \leqslant j, \\ \iota, & \text{otherwise.} \end{cases} \qquad \square$$

An operator $\varphi : [p] \to [q]$ gives rise to a simplicial map $\Delta'\varphi : \Delta'(p) \to \Delta'(q)$ by taking

$$\Delta'\varphi(\boldsymbol{\mu}) = ((\varphi\mu_0)^{\#}, \ldots, (\varphi\mu_n)^{\#});$$

thus, there is a covariant functor $\Delta' : \boldsymbol{\Delta} \to SiSets$ – the *normal subdivision of simplices* – which is a cosimplicial object in *SiSets*. Because $\boldsymbol{\Delta}$ operates on the left of the $\Delta'[p]$, instead of writing $\Delta'\varphi(\boldsymbol{\mu})$, one uses the shorter notation $\varphi\boldsymbol{\mu}$. An n-simplex $\boldsymbol{\mu}$ of $\Delta'[p]$ is an interior n-simplex iff $\mu_n = \iota^p$, i.e., if its highest vertex is an interior point of Δ^p, namely its barycentre.

Lemma 4.6.2 *The cosimplicial simplicial set Δ' has the Eilenberg–Zilber property.*

Proof The condition that any $\boldsymbol{\mu}$ is a non-decreasing sequence implies, for any pair of indices i,j with $i<j$, that there is a unique face operator $\mu_{j,i}$ with $\mu_i = \mu_j \mu_{j,i}$. This gives $\boldsymbol{\mu}$ the unique representation

$$\boldsymbol{\mu} = \mu_n(\mu_{n,0}, \ldots, \mu_{n,n} = \iota). \qquad \square$$

The cosimplicial simplicial sets $\Delta-$ and Δ' are related by a natural transformation $\mathrm{d}' : \Delta' \dot{\to} \Delta-$; it consists of the simplicial maps $\mathrm{d}'p : \Delta'[p] \to \Delta[p]$, which assign to each n-simplex $\boldsymbol{\mu} \in \Delta'[p]$ the operator $\mathrm{d}'\boldsymbol{\mu} : [n] \to [p]$ given by $\mathrm{d}'\boldsymbol{\mu}(i) = \mu_i(\dim \mu_i)$, the 'last element of μ_i'.

Remark In contrast to the simplicial maps $\mathrm{d}'p$, the homeomorphisms $\theta^p : |\Delta'[p]| \to \Delta^p$, although canonical, are not natural (see Section 3.3, Example 4, interpreting the simplicial maps f,g there as $\Delta^{\sigma_0}, \Delta^{\sigma_1}$ respectively. Moreover, this example shows that there cannot be any natural equivalence between the functors $|\Delta'-|$ and $|\Delta-| = \Delta^-$. In this sense, the natural transformation d' is the best available connection between the functors Δ' and $\Delta-$. $\square$

Reversing the order, i.e., replacing 'non-decreasing' by 'non-increasing' in the definition of $\Delta'[p]$, one obtains the *opnormal subdivision of simplices* ('op' derived from 'opposite'), again as a covariant functor $\Delta'' : \Delta \to SiSets$ satisfying the Eilenberg–Zilber property. The interior simplices are now those sequences $\boldsymbol{\mu}$ which start with $\mu_0 = \iota$. The simplicial sets $\Delta''[p]$ are also simplicially contractible to the vertex (ι). The corresponding natural transformation $\mathrm{d}'' : \Delta'' \dot{\to} \Delta-$ is given by $\mathrm{d}''\boldsymbol{\mu}(i) = \mu_i(0)$, the 'first element of μ_i'.

The cosimplicial sets discussed here suggest the formation of tensor products (see Section 4.2). The simplicial set $\mathrm{Sd}X = X \otimes \Delta'$ is called the *normal subdivision* of the simplicial set X. Functoriality in the first variable yields the functor *normal subdivision* $\mathrm{Sd} : \mathit{SiSets} \to \mathit{SiSets}$ and bifunctoriality with fixed second variable $\mathrm{d}' : \Delta' \dot{\to} \Delta -$ leads to a natural transformation $\mathrm{d}' : \mathrm{Sd} \dot{\to} \mathbf{1}$. In a similar way, the opnormal subdivision of simplices induces a functor *opnormal subdivision* $\mathrm{Sd}^{\mathrm{op}} : \mathit{SiSets} \to \mathit{SiSets}$ and a natural transformation $\mathrm{d}'' : \mathrm{Sd}^{\mathrm{op}} \dot{\to} \mathbf{1}$. The general theory of tensor products (see Section 4.2) yields the following properties of the normal subdivision.

Proposition 4.6.3 (i)*Any simplex of the normal (resp. opnormal) subdivision* $\mathrm{Sd}\,X$ *(resp.* $\mathrm{Sd}^{\mathrm{op}}\,X$*) of a simplicial set* X *has a unique representation by a pair* $(x, \boldsymbol{\mu})$, *with* x *a non-degenerate simplex of* X *and* $\boldsymbol{\mu}$ *an interior simplex of* $\Delta'[\dim x](\Delta''[\dim x])$.

(ii) *Normal subdivision preserves monomorphisms and simplicial attachings.*

(iii) *The simplicial maps* $\mathrm{d}'X$ *are weak homotopy equivalences, for all simplicial sets* X.

Proof For (i), see Proposition 4.2.7 and Lemma 4.6.2; for (ii), Corollary 4.2.9 and Proposition 4.2.12; for (iii), Corollary 4.3.21 and Lemma 4.6.1. □

A relative subdivision is necessary for some considerations. To describe it, let X be a simplicial set and let A be a simplicial subset of X. The normal subdivision $\mathrm{Sd}\,A$ of A can be viewed as a simplicial subset of the normal subdivision $\mathrm{Sd}\,X$ of X (see Proposition 4.6.3(ii)). Thus the natural simplicial map $\mathrm{d}'A : \mathrm{Sd}\,A \to A$ can be considered as a partial simplicial map $\mathrm{Sd}\,X - \backslash \to A$; forming the corresponding simplicial attaching, one obtains a simplicial set X' which is called the *normal subdivision of* X *rel.* A. Furthermore, one has a unique *canonical* simplicial map $\tilde{d} : X' \to X$ whose composition with the characteristic map $\bar{d} : \mathrm{Sd}X \to X'$ of this simplicial attaching is just the natural simplicial map $\mathrm{d}'X$ and whose existence is ensured by the naturality of the simplicial maps $\mathrm{d}'A, \mathrm{d}'X$. This process really yields a subdivision in the geometric sense, as one can deduce from the absolutely non-trivial content of the following statement.

Theorem 4.6.4 *The geometric realization* $|X'|$ *of the normal subdivision* X' *of a simplicial set* X *rel. to a simplicial subset* A *of* X *is homeomorphic to the geometric realization* $|X|$ *of the simplicial set* X *itself, via a homeomorphism which is homotopic rel.* $|A|$ *to the geometric realization* $|\tilde{d}|$

of the canonical simplicial map $\tilde{d} : X' \to X$, and which allows one to consider the CWcomplex $|X'|$ as a subdivision of the CW-complex $|X|$ (in the sense of Section 2.3).

Proof Since $\operatorname{Sd} X = \operatorname{colim} \Delta' \circ D_X$ (see Proposition 4.2.13) and geometric realization preserves colimits, one has $|\operatorname{Sd} X| = \operatorname{colim}(|\Delta' - | \circ D_X)$. Thus, in order to obtain a map $h : |\operatorname{Sd} X| \to |X|$, one needs a family $\{h_x : x \in X\}$ of maps $h_x : |\Delta'[\dim x]| \to \Delta^{\dim x}$ such that

(1) $\Delta^\alpha \circ h_{x\alpha} = h_x \circ |\Delta' \alpha|$, for all composable pairs (x, α).

Searching for a map defined on $|X'|$, one has to look for a map h decomposable in the form $h' \circ |\bar{d}|$, where $\bar{d} : \operatorname{Sd} X \to X'$ denotes the characteristic map of the attaching which produces X'. Moreover, this requires

(2) $h_x = |d'n|$, for each n-simplex $x \in A$.

The resulting map h' should be a homeomorphism; this is ensured by the condition:

(3) h_x maps the interior of $|\Delta'[\dim x]|$ bijectively onto the interior of $\Delta^{\dim x}$, for each non-degenerate simplex $x \in X \backslash A$.

The bulk of the proof consists in the construction of a family $\{h_x\}$ satisfying properties (1), (2) and (3). Once this is done, the remainder of the claim is nearly evident. The map $|\tilde{d}|$ may be thought of as obtained from a family $\{d_x\}$ of maps $d_x : |\Delta'[\dim x]| \to \Delta^{\dim x}$ satisfying similar conditions (1) and (2); in particular, observe that $d_x = h_x$, for all $x \in A$. Since the maps Δ^α are induced by linear maps, the family $\{H_x\}$ consisting of the homotopies

$$H_x : |\Delta'[\dim x]| \times I \to \Delta^{\dim x}, \qquad (\boldsymbol{t}, s) \mapsto (1 - s) \cdot h_x(\boldsymbol{t}) + s \cdot d_x(\boldsymbol{t})$$

again satisfies such conditions; for $x \in A$, these homotopies factor through the projection onto $|\Delta'[\dim x]|$. Since the product functor $- \times I$ preserves colimits, all homotopies H_x together define a homotopy $H : |X'| \times I \to |X|$ rel. $|A|$ from the homeomorphism h' to the canonical map $\tilde{d}$. Finally, each cell e of $|X'|$ is contained in the image of a map $|\bar{d}| \times \bar{c}'_x$, for some non-degenerate simplex $x \in X$ where $\bar{c}'_x : |\Delta'[\dim x]| \to |\operatorname{Sd} X|$ denotes the corresponding map from $|\Delta'[\dim x]|$ to the colimit; thus, $h'(e)$ is contained in the cell of $|X|$, corresponding to the simplex x (see Theorem 4.3.5). This proves that $|X'|$ can be considered as a (CW-)subdivision of $|X|$.

Now, fix $x \in X_n$. To define the map h_x, consider a pair $(\boldsymbol{\mu}, \boldsymbol{u})$, with $\boldsymbol{\mu} = (\mu_0, \ldots, \mu_q) \in \Delta'[n]$, $\boldsymbol{u} = (u_0, \ldots, u_q) \in \Delta^q$. Depending on $\boldsymbol{\mu}$, one has unique face operators μ_{kj}, and unique degeneracy operators ρ_k, ρ_{kj}, for

$0 \leqslant j \leqslant k \leqslant q$, such that

$$\mu_j = \mu_k u_{kj},$$
$$\rho_k = \omega^{\dim \mu_k}, \text{ for } x\mu_k \in A, \text{ and } \rho_k = (x\mu_k)^\flat, \text{ otherwise;}$$
$$\rho_{kj} = (\rho_k \mu_{kj})^\flat.$$

Recall that the maximal section of any degeneracy operator ρ is denoted by $\rho^\perp$ (see Section 4.1) and let $\boldsymbol{b}_{kj}$ denote the barycentre of the $\mu_j \rho_{kj}^\perp$th face of Δ^n. Now take

$$h_x([\boldsymbol{\mu}, \boldsymbol{u}]) = \Sigma u_k(1 - u_q - \cdots - u_{k+1})\boldsymbol{b}_{kk} + \Sigma u_j u_k \boldsymbol{b}_{kj},$$

where the first and the second sum run over all $k \in [q]$ and all pairs (j, k) with $0 \leqslant j < k \leqslant q$ respectively; for $k = q$, the expression in parentheses is taken to be 1. In order to verify that this formula yields a well-defined function $|\Delta'[n]| \to \Delta^n$, one has to check that the expression on the right-hand side depends only on the class $[\mu, \boldsymbol{u}]$, but not on the specific pair $(\boldsymbol{\mu}, \boldsymbol{u})$. This is like saying that the expression does not change if one replaces the given arbitrary pair by the minimal pair in the same class which can be obtained by an iteration of the following two steps:

skip μ_j and u_j if $u_j = 0$,
skip μ_{j+1} and combine u_j, u_{j+1} to $u_j + u_{j+1}$ if $\mu_{j+1} = \mu_j$

(see Proposition 4.2.7 and its proof). In both cases, this does not have any effect on the expression under consideration (The coefficients in both sums are all non-negative and their sum is equal to 1; thus, one has a convex combination of points in Δ^n, which, by the convexity of Δ^n, also belongs to Δ^n.) Since addition and multiplication of real numbers are continuous operations the function h_x depends continuously on $\boldsymbol{u}$ as long as $\boldsymbol{\mu}$ is fixed. But $|\Delta'[n]|$ is covered by the finitely many closed sets $\{[\boldsymbol{\mu},\boldsymbol{u}] : \boldsymbol{u} \in \Delta^{\dim \boldsymbol{\mu}}\}$, with non-degenerate $\boldsymbol{\mu}$; thus, the functions h_x are continuous, i.e., maps.

It remains to establish conditions (1), (2) and (3). For (1), take an operator $\alpha : [m] \to [n]$ and a pair $(\boldsymbol{v}, \boldsymbol{u})$, with $\boldsymbol{v} = (v_0, \ldots, v_q) \in (\Delta'[m])_q$, $\boldsymbol{u} \in \Delta^q$. Construct $v_{kj}, \tau_k, \tau_{kj}$ and $\boldsymbol{b}'_{jk}$ dependent on $x\alpha$ and $\boldsymbol{v}$, just as $\mu_{kj}, \rho_k, \rho_{kj}$ and $\boldsymbol{b}_{jk}$ respectively were derived from x and $\boldsymbol{\mu} = \alpha \boldsymbol{v}$. By the linearity of the map Δ^α, it is sufficient to verify $\alpha \boldsymbol{b}'_{jk} = \boldsymbol{b}_{jk}$, for all pairs (j, k) with $0 \leqslant j \leqslant k \leqslant q$. Notice that $\mu_j = (\alpha v_j)^\#$, $\mu_{kj} = ((\alpha v_k)^\flat v_{kj})^\#$, $\tau_k = \rho_k(\alpha v_k)^\flat$, $\tau_{kj} = \rho_{kj}(\alpha v_j)^\flat$, and, by the functoriality of $-^\perp$, $\alpha v_j \tau_{kj}^\perp = \mu_j \rho_{kj}^\perp$. Therefore the operator $\alpha v_j \tau_{kj}^\perp$ is injective, and consequently Δ^α maps the $v_j \tau_{kj}^\perp$th face of Δ^m isometrically onto the $\mu_j \rho_{kj}^\perp$th face of Δ^n; in particular, the barycentre $\boldsymbol{b}'_{jk}$ is transformed into the barycentre $\boldsymbol{b}_{jk}$.

Now, assume $x \in A$. In this case, all ρ_k, ρ_{kj} are terminal operators and all $\boldsymbol{b}_{kj}$ are vertices of Δ^n; more precisely, $\boldsymbol{b}_{kj}$ is the last 'last' vertex of the

μ_jth face of Δ^n. Thus, the defining formula for h_x is nothing but an explicit description of the geometric realization of d$'n$, which confirms condition (2).

For (3), first show that h_x transforms interior points into interior points if x is non-degenerate. To this end, note that the pair $(\boldsymbol{\mu},\boldsymbol{u})$ represents an interior point of $|\Delta'[n]|$ iff $\mu_q = \iota$ and $u_q > 0$. If x is non-degenerate, one has, moreover, $\rho_{qq}^{\perp} = \iota$ and the ith coordinate of $h_x([\boldsymbol{\mu},\boldsymbol{u}])$ can be estimated by

$$(h_x([\boldsymbol{\mu},\boldsymbol{u}]))_i \geqslant u_q(\boldsymbol{b}_{qq})_i = u_q/(n+1),$$

for all $i \in [n]$. Thus, all coordinates of $h_x([\boldsymbol{\mu},\boldsymbol{u}])$ are strictly positive, which characterizes an interior point of Δ^n.

Next, turn to injectivity. Every point in $|\Delta'[n]|$ can be represented by a pair $(\boldsymbol{\mu},\boldsymbol{u})$, with $\boldsymbol{\mu} = (\mu_0,\ldots,\mu_n) \in \Delta'[n]$, $\dim \mu_j = j$, for all $j \in [n]$, and $\boldsymbol{u} = (u_0,\ldots,u_n) \in \Delta^n$; in particular, that means that $\mu_n = \iota$. Moreover, one obtains a permutation φ of the set $\{0,1,\ldots,n\}$, such that image $\mu_j = \varphi(\{0,1,\ldots,j\})$, for all $j \in [n]$. If x is non-degenerate, then $\rho_{nj} = \rho_{nj}^{\perp} = \iota^j$, again for all $j \in [n]$. Now assume pairs $(\boldsymbol{\mu},\boldsymbol{u})$ and $(\boldsymbol{\mu}',\boldsymbol{u}')$ with the described properties are given such that

$$h_x([\boldsymbol{\mu},\boldsymbol{u}]) = h_x([\boldsymbol{\mu}',\boldsymbol{u}']) = (t_0,\ldots,t_n).$$

All the following constructions will be done simultaneously for both pairs and distinguished by attaching the prime $'$ to an object which comes from $(\boldsymbol{\mu}',\boldsymbol{u}')$. It has to be shown that $u_j = u_j'$, for all $j \in [n]$, and $\mu_j = \mu_j'$ if $u_j > 0$. This will be done by decreasing induction on j. Observe that $(\boldsymbol{b}_{kj})_i = 0$ if $i \notin \text{image}\,\mu_j$. This implies, in particular, that $u_n/(n+1) = t_{\varphi(n)} \geqslant u_n'/(n+1)$, i.e., $u_n \geqslant u_n'$; by symmetry, one obtains $u_n \leqslant u_n'$, and, so, $u_n = u_n'$, thereby starting the induction. Assume the claim is proved for all $j > l$. Then consider the point

$$\tilde{\boldsymbol{t}} = (\tilde{t}_0,\ldots,\tilde{t}_n) = \sum u_k(1 - u_n - \cdots - u_{k+1})\boldsymbol{b}_{kk} + \sum u_j u_k \boldsymbol{b}_{kj} \in \mathbf{R}^{n+1},$$

where the first and the second sum run over all $k \in [l]$ and all pairs (j,k) with $0 \leqslant j \leqslant l$, $j < k \leqslant n$ respectively; this point is obtained from the right-hand side of the defining equation for h_x by cancelling all summands depending only on indices $j,k > l$. By induction, $\tilde{\boldsymbol{t}}$ is equal to the corresponding point $\tilde{\boldsymbol{t}}'$. Now, if $\mu_l \neq \mu_l'$ there is an element $i \in \text{image}\,\mu_l$ which does not belong to image μ_l'. For such an i, one computes

$$u_l u_n/(l+1) = u_l u_n(\boldsymbol{b}_{nl})_i \leqslant \tilde{t}_i = \tilde{t}_i' = 0,$$

since the ith component of all summands forming $\tilde{\boldsymbol{t}}'$ vanishes. Thus $u_l = 0$, and, by symmetry, $u_l' = 0$. If $\mu_l = \mu_l'$, then, again by induction, the points

$$\hat{\boldsymbol{t}} = (\hat{t}_0,\ldots,\hat{t}_n) = (1 - u_n - \cdots - u_{l+1})\boldsymbol{b}_{kk} + \sum u_k \boldsymbol{b}_{kj} \in \mathbf{R}^{n+1}$$

the sum running over k with $l < k \leqslant n$ and $\hat{\boldsymbol{t}}'$ coincide. Moreover, for

$i = \varphi(l)$, one has

$$u_l \hat{t}_i = \tilde{t}_i = \tilde{t}'_i \geqslant u'_l \hat{t}'_i = u'_l \hat{t}_i.$$

Because $\hat{t}_i \geqslant u_n/(l+1) > 0$, this implies $u_l \geqslant u'_l$, and, once more by symmetry, $u_l \leqslant u'_l$. Thus, $u_l = u'_l$ completing the induction.

Finally, one has to check that h_x maps the interior $\mathring{\Delta}'$ of $|\Delta'[n]|$ onto the interior $\mathring{\Delta}$ of Δ^n if x is non-degenerate. By the theorem of the invariance of domain (see Theorem A.9.6), $h_x(\mathring{\Delta}')$ is open in $\mathring{\Delta}$. On the other hand, $h_x(\mathring{\Delta}) = \mathring{\Delta} \cap h_x(|\Delta'[n]|)$ is closed in $\mathring{\Delta}$. Thus, being non-empty, $h_x(\mathring{\Delta}')$ must be equal to $\mathring{\Delta}$. □

Taking $A = \varnothing$ in this theorem, one obtains as a special case:

Corollary 4.6.5 *The geometric realization of the normal subdivision of a simplicial set is homeomorphic to the geometric realization of the simplicial set itself. The homeomorphism is not natural but homotopic to a natural map. In particular, the natural map $|d'X| : |\mathrm{Sd}\, X| \to |X|$ is a homotopy equivalence, for every simplicial set X.* □

The subdivision process must be iterated. Formally, this will be done as follows. Define inductively functors $\mathrm{Sd}^n : \mathit{SiSets} \to \mathit{SiSets}$, and natural transformations $\mathrm{d}^n : \mathrm{Sd}^n \dot{\to} \mathbf{i}_{\mathit{SiSets}}$, for all $n \in \mathbf{N}$, by taking $\mathrm{Sd}^0 = \mathbf{1}_{\mathit{SiSets}}$, $\mathrm{Sd}^{n+1} = \mathrm{Sd} \circ \mathrm{Sd}^n$ and $\mathrm{d}^0 X = \mathbf{1}_X$, $\mathrm{d}^{n+1} X = \mathrm{d}^n X \circ \mathrm{d}' \mathrm{Sd}^n X$, for all simplicial sets X. The functor Sd^n is called n-th *normal subdivision* and gives rise also to a relative subdivision. For this, let X be a simplicial set, let A be a simplicial subset of X and let n be a natural number greater than 0. The simplicial set $\mathrm{Sd}^n A$ can again be viewed as a simplicial subset of the simplicial set $\mathrm{Sd}^n X$. Thus, the natural simplicial map $\mathrm{d}^n A : \mathrm{Sd}^n A \to A$ can be considered as a partial simplicial map $\mathrm{Sd}^n X \mathrel{-/\!\!\to} A$; forming the corresponding simplicial attaching, one obtains a simplicial set $X^{(n)}$ which is called the n-th *normal subdivision of X rel. A*. Again, one has a unique *canonical* simplicial map $\tilde{d} : X^{(n)} \to X$ whose composition with the characteristic map $\bar{d} : \mathrm{Sd}^n X \to X^{(n)}$ of this simplicial attaching is just the natural simplicial map $\mathrm{d}^n X$. Now, the obvious analogue of Theorem 4.6.4 also holds true.

Proposition 4.6.6 *The geometric realization $|X^{(n)}|$ of the n-th normal subdivision $X^{(n)}$ of a simplicial set X rel. to a simplicial subset A of X is homeomorphic to the geometric realization $|X|$ of the simplicial set X itself, via a homeomorphism which is homotopic rel. $|A|$ to the geometric realization $|\tilde{d}|$ of the canonical simplicial map $\tilde{d} : X' \to X$ and which allows one to* *consider the CW-complex $|X'|$ as a subdivision of the CW-complex $|X|$ (in the sense of Section* 2.3).

Proof By induction on n. For $n = 0$, there is nothing to prove; assume the statement is true for some n. Take $\hat{X}$ as the simplicial set which is obtained from $\mathrm{Sd}^n A$ by simplicially attaching $\mathrm{Sd}^{n+1} X$ via $d'(\mathrm{Sd}^n A)$; thus, the geometric realization $|\hat{X}|$ of $\hat{X}$ is homeomorphic to the geometric realization $|\mathrm{Sd}^n X|$ of the simplicial set $\mathrm{Sd}^n X$ via a homeomorphism which is homotopic rel. $|\mathrm{Sd}^n A|$ to the geometric realization $|\hat{d}|$ of the canonical simplicial map $\hat{d} : \hat{X} \to \mathrm{Sd}^n X$ and which allows one to consider the CW-complex $|\hat{X}|$ as a subdivision of the CW-complex $|\mathrm{Sd}^n X|$ (apply Theorem 4.6.4). Let $\hat{H} : |\hat{X}| \times I \to |\mathrm{Sd}^n X|$ denote a homotopy from such a homeomorphism $\hat{h}$ to $\hat{d}$. By the law of horizontal composition (see Section A.4) X^{n+1} may be viewed as obtained from A by simplicially attaching $\hat{X}$ via d^n_A; let $\bar{d}^{n+1} : \hat{X} \to X^{(n+1)}$ denote a corresponding characteristic map. The universal property of the simplicial attaching yields a simplicial map $g : X^{(n+1)} \to X^{(n)}$ such that $g|A$ is the inclusion of A into $X^{(n)}$ and $g \circ \bar{d}^{n+1}$ is the composition of $\hat{d}$ with the characteristic map $\bar{d}^n : \mathrm{Sd}^n X \to X^{(n)}$. Now, $|X^{(n+1)}| \times I$ is obtained from $|A| \times I$ by attaching $|\hat{X}| \times I$ via $|d^n_A| \times \mathbf{1}_I$ (see Proposition 4.3.10 and Proposition A.4.8(i)). The universal property of this attaching yields a homotopy $H : |X^{(n+1)}| \times I \to |X^{(n)}|$ such that $H \| A| \times I$ is the composition of the projection onto $|A|$ with the inclusion of $|A|$ into $|X^{(n)}|$ and $H \circ |\bar{d}^{n+1}| \times \mathbf{1}_I$ is the composition of $\hat{H}$ with the geometric realization $|\bar{d}^n|$ of $\bar{d}^n$. Since $\hat{H}(-,0) = \hat{h}$ is a homeomorphism the map $h = H(-,0)$ is at least bijective. Because $h^{-1} \| A|$ is nothing but the inclusion of $|A|$ into $|X^{(n+1)}|$ and $h^{-1} \circ |\bar{d}^n| = |\bar{d}^{n+1}| \circ \hat{h}^{-1}$ the inverse function h^{-1} is also continuous; thus, h is a homeomorphism, which, moreover, is homotopic to the geometric realization $|g|$ of the simplicial map g and allows one to consider the CW-complex $|X^{(n+1)}|$ as a subdivision of the CW-complex $|X^{(n)}|$. Thus, the result follows from the application of the inductive hypothesis. □

Normal subdivision has some special properties when restricted to presimplicial sets. The normal subdivision of a presimplicial set is again a presimplicial set and the subdivision functor applied to a presimplicial map yields a presimplicial map. Thus, one has an induced normal subdivision functor on the category *PSiSets*, which also will be denoted by Sd, yielding the compatibility relation

$$E\,\mathrm{Sd}\,X = \mathrm{Sd}\,EX,$$

for all $X \in \mathrm{Ob}\,PSiSets$. The main feature of $\mathrm{Sd}\,X$ for a presimplicial set X

is that in order to construct the homeomorphism $h : |\mathrm{Sd}\, X| \to |X|$ (see Theorem 4.6.4), all the maps h_x can be taken as $\theta^{\dim x}$. This implies that one obtains a natural family of homeomorphisms, and, moreover, although the natural simplicial maps d'_{EX} fail to be presimplicial, the homotopies connecting h to $|d'_{EX}|$ are also natural.

The next objective is to study a special natural transformation of the composite functor $|EPS-|$, i.e., the composition of the singular functor with the fat realization into itself. For any space T, let

$$h_T : |E\,\mathrm{Sd}\, PST| \to |EPST|$$

denote the natural homeomorphism described before and take $g_T : \mathrm{Sd}\, PST \to PST$ as the adjoint of $j'_T \circ h_T$, i.e., its composition with the co-unit of the adjointness $|E-| \dashv PS$. Then, define the announced natural map as

$$b_T = |Eg_T| \circ h_T^{-1} : |EPST| \to |EPST|.$$

Before proceeding, one should try to understand what this map is doing. To this end, consider a singular simplex $x : \Delta^n \to T$, which is an element of ST as well as of PST. It corresponds to a cell e_x of $|EPST|$ giving rise to a canonical map $c_x : \Delta^n \to |EPST|$, which, up to homeomorphism, may be viewed as a characteristic map. On the other hand, x induces a presimplicial map $\Delta[n] \to PST$ whose normal subdivision $x' : \Delta'[n] \to \mathrm{Sd}\, PST$ can be composed with the presimplicial map g_T to yield a presimplicial map $b_x : \Delta'[n] \to PST$.

Lemma 4.6.7 *For all $x \in (PST)_n$*

$$|Eb_x| = b_T \circ c_x \circ \theta^n.$$

Proof The map c_x may be viewed as the geometric realization of a presimplicial map, and then the naturality of the homeomorphisms h_T, θ^n yields $c_x \circ \theta^n = h_T \circ x'$. Consequently,

$$|Eb_x| = |Eg_T| \circ |x'| = |Eg_T| \circ (h_T)^{-1} \circ c_x \circ \theta^n = b_T \circ c_x \circ \theta^n. \qquad \square$$

The importance of this result lies in the following fact: the cell e_x is mapped by b_T into the union of the cells corresponding to the singular simplices $x \circ \theta^n \circ c_\mu$ where $c_\mu \circ \Delta^m \to |\Delta'[n]|$ denotes the map associated to the non-degenerate m-simplex $\mu \in \Delta'[n]$. Thus, the image of the cell e_x by the kth iteration of b_X is contained in the union of the cells corresponding to the singular simplices $x \circ \theta^n \circ c_\mu$, where μ runs through the non-degenerate simplices of the k-fold normal subdivision of $\Delta[n]$. This also explains the choice of the letter 'b' for these maps: it refers to 'barycentric' (subdivision).

Another interesting property of the map b_T is the following:

Lemma 4.6.8 *The map b_T is naturally homotopic to the identity of the fat realization of ST.*

Proof Since the co-unit $p_{ST} : EPST \to ST$ is a weak homotopy equivalence (see Theorem 4.4.5) and therefore $|p_{ST}|$ is a homotopy equivalence, it suffices to show that $|p_{ST}| \circ b_T \simeq |p_{ST}|$. This will be done by simple but lengthy computations using units and co-units of the adjoitness relations involved. In order not to overload the formulae, the subscripts indicating the respective spaces or (pre)simplicial sets will be dropped from the notation. Start with

$$p \circ Eg = p \circ EPS(j' \circ h) \circ E\eta' = S(j' \circ h) \circ p \circ EP\eta \circ Eu$$
$$= S(j' \circ h) \circ \eta \circ p \circ Eu = Sj' \circ Sh \circ \eta.$$

Now use the fact that h is naturally homotopic to the geometric realization of a simplicial map d, implying (see Proposition 4.3.18):

$$p \circ Eg \simeq Sj' \circ S|d| \circ \eta = Sj' \circ \eta \circ d = Sj \circ S|p| \circ \eta \circ d = Sj \circ \eta \circ p \circ d = p \circ d.$$

Consequently,

$$|p| \circ b = |p| \circ |Eg| \circ h^{-1} = |p \circ Eg| \circ h^{-1} \simeq |p| \circ |d| \circ h^{-1} \simeq |p|. \qquad \square$$

All these considerations prepare the way for the *Simplicial Excision Theorem*:

Theorem 4.6.9 *Let T be a space and let $\boldsymbol{U} = \{U_\gamma : \gamma \in \Gamma\}$ be a family of subsets of T whose interiors form a covering of T. Then, the geometric realization of*

$$S(T, \boldsymbol{U}) = \bigcup_{\gamma \in \Gamma} SU_\gamma$$

is a strong deformation retract of $|ST|$.

Proof It is enough to show that the inclusion $|S(T, \boldsymbol{U})| \to |ST|$ is a homotopy equivalence (see Proposition A.4.2(v)). Since the co-unit $p : EP \to \mathbf{1}$ is a weak homotopy equivalence (see Theorem 4.4.5), it suffices to prove that the inclusion of the corresponding fat realizations is n-connected, for all $n \in \mathbf{N}$ (see Theorem A.8.9 and Whitehead's realizability theorem, Theorem 2.5.1).

To begin with, consider a singular simplex $x : \Delta^n \to T$. The inverse image of the family $\boldsymbol{U}$ under the map x forms a covering of Δ^n for which there is a k-fold barycentric subdivision of Δ^n whose open simplices form a finer covering (see Proposition 3.2.14). The means that the cell $e_x \subset |EPST|$ corresponding to x is mapped under the kth iterate of b_T into the subspace $|EPS(T, \boldsymbol{U})|$.

Now consider an arbitrary map $b : B^n \to |EPST|$ such that $b(S^{n-1}) \subset |EPS(T, U)|$. By compactness, the image of b is contained in a finite union of cells of $|EPST|$ (see Proposition 1.5.2). Thus there is a number $k \in \mathbf{N}$ such that $(b_T)^k \circ b$ factors through $|EPS(T, U)|$. By naturality, the map b_T transforms each $|EPSU_\gamma|$ (and consequently $|EPS(T, U)|$) into itself; the same holds true for the homotopy deforming b_T to $\mathbf{1}_{|EPST|}$. Thus, there is a homotopy

$$H : (|EPST| \times I, |EPS(T, U)| \times I) \to (|EPST|, |EPS(T, U)|)$$

from $(b_T)^k$ to $\mathbf{1}_{|EPST|}$; now, the composition $H \circ b \times \mathbf{1}_I \circ v^n \circ h^n$, where v^n and h^n denote the standard maps (defined in Section 1.0), shows that b is homotopic rel. S^{n-1} to a map factoring through $|EPS(T, U)|$, proving the desired n-connectivity. □

CW-complexes are not triangulable, in general; a sufficient condition assuring this property is regularity (see Theorem 3.4.1). Here is its simplicial analogue. A non-degenerate n-simplex x in a simplicial set X is *regular* if the simplicial subset $\tilde{X}$ of X which is generated by x may be obtained from the simplicial subset $\tilde{X}_n$ which is generated by $x\delta_n$ by simplicially attaching $\Delta[n]$ via the simplicial map $f_n : \Delta[n-1] \to \tilde{X}_n$, $\alpha \mapsto x\delta_n\alpha$; clearly, $\Delta[n-1]$ is considered as a simplicial subset of $\Delta[n]$ via the simplicial injection $\Delta\delta_n$. A simplicial set is *regular* if all its non-degenerate simplices are regular. There is an easy way to obtain a regular simplicial set out of an arbitrary one.

Proposition 4.6.10 *The normal subdivision of any simplicial set is a regular simplicial set.*

Proof Let X be a simplicial set. A non-degenerate n-simplex of the normal subdivision $\operatorname{Sd} X$ of X may be represented by a pair $(x, \boldsymbol{\mu})$, with x a non-degenerate simplex in X and $\boldsymbol{\mu}$ a non-degenerate interior n-simplex in $\Delta'[\dim x]$ (see Proposition 4.6.3). For $\alpha \in \Delta[n] \backslash \Delta[n-1]$, the simplex $\boldsymbol{\mu}\alpha$ is still an interior simplex of $\Delta'[\dim x]$, which implies the claim (again by Proposition 4.6.3). □

The more crucial property of regularity is that it commutes with geometric realization.

Proposition 4.6.11 *The geometric realization of a regular simplicial set is a regular CW-complex.*

Proof Let X be a regular simplicial set. Take a non-degenerate n-simplex x; without loss of generality, assume X to be generated by x. If no face of x is degenerate, then the corresponding simplicial map $\Delta[n] \to X$ is injective and there is nothing to show. Otherwise, regularity implies the existence of a maximal degenerate face $x_1 = x\mu_1$, with $\mu_1 = \delta_n \delta_{n-1} \cdots \delta_k$, for a $k \in [n]$. Take $y_1 = (x_1)^{\#} = x_1 \nu_1$, with $\nu_1 = (x_1)^{\flat\perp}$; recall that $(x_1)^{\flat\perp}$ denotes the maximal right inverse face operator to the degeneracy operator $(x_1)^{\flat}$ (see Section 4.1). The same construction applied to y_1 instead of $x = y_0$ yields $x_2 = y_1 \mu_2$ and $y_2 = x_2 \nu_2$; the process continues to end up at $x_p = y_{p-1} \mu_p$ and $y_p = x_p \nu_p$ where y_p has only non-degenerate faces. Take $m_j = \dim x_j$, $n_j = \dim y_j$. Define inductively pairs of spaces (Z_j, Δ^{n_j}) in the following fashion. Firstly, take $Z_0 = \Delta^n$. Secondly, given (Z_j, Δ^{n_j}), regard $\Delta^{n_{j+1}}$ as a subspace of Δ^{n_j} via $\Delta^{\mu_{j+1}}$ and so as a subspace of Z_j; then, attach Z_j to $\Delta^{n_{j+1}}$ via $\Delta^{(x_{j+1})\flat}$ to obtain Z_{j+1}. Observe that Z_p can be identified with $|X|$. Now, by induction, it follows that every pair (Z_j, Δ^{n_j}) can be identified with the pair (Δ^n, Δ^{n_j}), where in the latter case the inclusion is induced by the face operator $\mu_1 \nu_1 \mu_2 \cdots \mu_j \nu_j$ (see Lemma 3.1.1). □

Corollary 4.6.12 *The geometric realization of any simplicial set can be triangulated.*

Proof See Theorem 3.4.1, Theorem 4.6.4 and Proposition 4.6.10. □

Remark This shows that simplicial sets do not cover all CW-complexes, because not every CW-complex can be triangulated (see the Example in Section 3.4). □

This section is continued with two technical lemmas needed in the preparatory work for the proof of the relative simplicial approximation theorem (see Lemma 4.6.15 and Proposition 4.6.19).

Lemma 4.6.13 *For any $n \in N$ and any $k \in [n]$, there is a simplicial map*

$$\chi_{n,k} : \mathrm{Sd}^2 \Delta[n] \to \mathrm{Sd}\, \Lambda^k[n]$$

such that $\chi_{n,k} | \mathrm{Sd}^2 \Lambda^k[n] = \mathrm{Sd}(\mathrm{d}' \Lambda^k[n])$.

Proof The objective is to construct a simplicial map

$$\chi : \mathrm{Sd}^2 \Delta[n] \to \mathrm{Sd}\, \Delta[n]$$

which factors through $\mathrm{Sd}\, \Lambda^k[n]$ and such that the induced map $\chi_{n,k}$ satisfies the required property.

In order to obtain a simplicial map with the desired domain and codomain, it is sufficient to define a function

$$\chi_0 : (\mathrm{Sd}^2 \Delta[n])_0 \to (\mathrm{Sd}\, \Delta[n])_0$$

such that the images of two vertices in $\mathrm{Sd}^2\,\Delta[n]$ spanning a 1-simplex either coincide or span a 1-simplex in $\mathrm{Sd}\,\Delta[n]$ (in the same order). The vertices of $\mathrm{Sd}^2\,\Delta[n]$ correspond to the non-degenerate simplices of $\mathrm{Sd}\,\Delta[n]$; thus, they may be represented by strongly increasing (with respect to $\subset$) sequences $\boldsymbol{\mu}=(\mu_0,\ldots,\mu_r)$ of face operators with codomain $[n]$. On the other hand, the vertices of $\mathrm{Sd}\,\Delta[n]$ correspond to the face operators with codomain $[n]$, and so they may be considered as subsets of $[n]$. In this sense, define

$$\chi_0(\boldsymbol{\mu})=\{s(\mu_0),\ldots,s(\mu_r)\}$$

where, for any face operator μ with codomain $[n]$,

$$s(\mu)=\begin{cases}\mu(\dim\mu), & \mu\neq\iota,\delta_k,\\ k, & \text{otherwise.}\end{cases}$$

The resulting simplicial map χ extends the simplicial map $\mathrm{Sd}\,(\mathrm{d}'\Lambda^k[n])$; it remains to show that its image is contained in $\mathrm{Sd}\Lambda^k[n]$, i.e., that $[n]$ and $[n]\backslash\{k\}$ are not in the image of χ_0.

Firstly, assume $[n]=\chi_0(\boldsymbol{\mu})$ for some $\boldsymbol{\mu}\in(\mathrm{Sd}^2\,\Delta[n])_0$. Since $[n]$ contains $n+1$ elements, one must have $r=\dim\boldsymbol{\mu}=n$, $\mu_n=\iota$ and since $s(\iota)=k$,

$$s(\mu_i)=\begin{cases}i, & 0\leqslant i<k,\\ i+1, & k\leqslant i<n.\end{cases}$$

Now, let j denote the smallest index such that $k\in\mathrm{image}\,\mu_j$. The definition of the numbers $s(\mu_i)$ implies $j>k$ and $s(\mu_j)=s(\mu_{j-1})$ if $j<n$ or $s(\mu_{n-1})=k$ otherwise; one has a contradiction in either case.

Secondly, assume $[n]\backslash\{k\}=\chi_0(\boldsymbol{\mu})$ for some $\boldsymbol{\mu}\in(\mathrm{Sd}^2\,\Delta[n])_0$. Since $[n]\backslash\{k\}$ contains n elements and μ_r must be different from ι, one must have $r=n-1$ and $\mu_{n-1}=\delta_j$ for some $j\in[n]\backslash\{k\}$, implying $j\notin\chi_0(\boldsymbol{\mu})$; this is a contradiction. □

Lemma 4.6.14 *The inclusion* $\mathrm{Sd}\,\Lambda^k[n]\subset\mathrm{Sd}\,\Delta[n]=\Delta'[n]$ *is an anodyne extension, for every* $n\in\mathbf{N}$ *and* $k\in[n]$.

Proof Since $\mathrm{Sd}\,\Lambda^k[n]\cong\mathrm{Sd}\,\Lambda^n[n]$, for every $k\in[n]$, one may assume $k=n$. The first step consists in defining simplicial inclusions

$$f:\Delta[1]^n\to\Delta'[n]$$

and

$$g:\Delta'[n-1]\times\Delta[1]\to\Delta'[n].$$

To obtain f, identify the vertex operators ε_i^1 with the numbers i, which may be 0 or 1, and consequently the vertices of $\Delta[1]^n$ with the n-tuples $(i_0,\ldots,i_{n-1})$. Then, require the vertex $(i_0,\ldots,i_{n-1})$ to be mapped into the

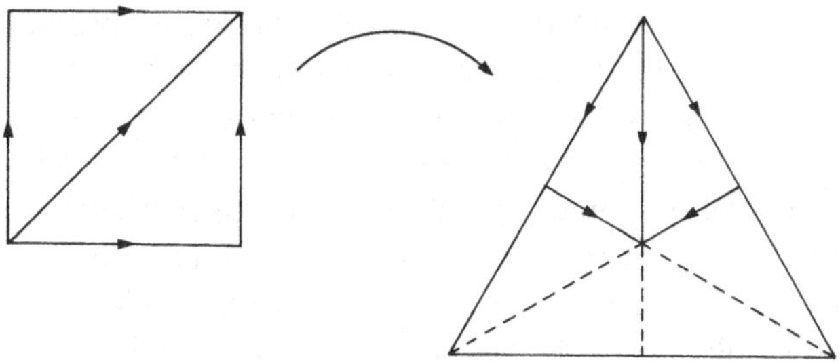

Figure 14

vertex of $\Delta'[n]$ that corresponds to the face operator μ, satisfying

$$\text{image}\, \mu = [n] \backslash \{ j : i_j = 0 \}.$$

This assignment extends in a unique manner to the desired simplicial map (the case $n = 2$ is illustrated by Figure 14). It also suffices to describe the simplicial map g on the vertices. The vertices of $\Delta'[n-1] \times \Delta[1]$ can be considered as pairs (μ, ε_i) consisting of a face operator μ with codomain $[n-1]$ and a vertex operator $\varepsilon_i = \varepsilon_i^1$ with $i \in [1]$. In this sense, take

$$g((\mu, \varepsilon_0)) = (\delta_n \mu)$$
$$g((\mu, \varepsilon_1)) = (\hat{\mu})$$

(see Figure 15). Since the simplicial maps f and g are injective, one may consider $\Delta[1]^n$ and $\Delta'[n-1] \times \Delta[1]$ as simplicial subsets of $\Delta'[n]$. Then, $\Lambda = \mathrm{Sd}\, \Lambda^n[n] \cap \Delta[1]^n$ consists of those simplices of $\Delta[1]^n$ which do not contain the vertex $(1, \ldots, 1)$.

To continue the proof, one needs to know that the inclusion $\Lambda \subset \Delta[1]^n$ is an anodyne extension. For this, observe that the simplices of $\Delta[1]^n$ can be considered as matrices with entries 0 and 1, non-decreasing columns and each row representing a vertex. A non-degenerate r-simplex corresponds to an $(r+1) \times n$-matrix with pairwise distinct rows; it belongs to Λ if its matrix contains at least one column whose entries are all 0 and

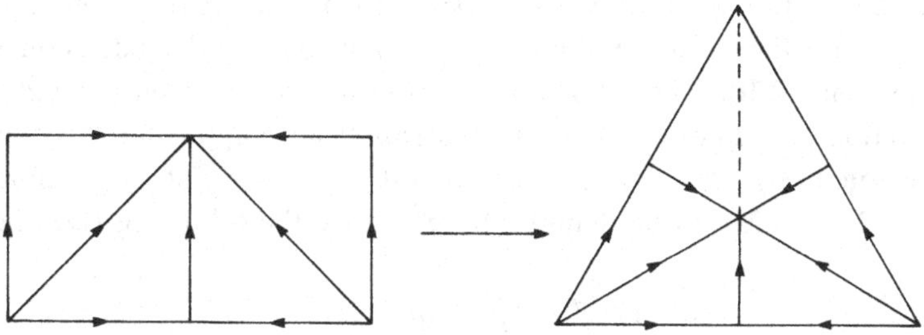

Figure 15

it is an *interior* simplex if the first row of its matrix is constantly 0 and the last row is constantly 1. Now, $\Delta[1]^n$ can be obtained from Λ by successive attachings of the interior (non-degenerate) simplices in order of increasing dimension. The procedure starts with the attaching of the 'diagonal', i.e., the only interior 1-simplex. Notice that before attaching an interior $(r+1)$-simplex x the faces $x\delta_1, \dots, x\delta_r$ are interior simplices which already have been attached and the face $x\delta_{r+1}$ belongs to Λ, while the face $x\delta_0$ is still outside the reached object; thus, one has a horn whose filling gives rise to an anodyne extension (see condition (2) in the definition of anodyne extensions, Section 4.5); but anodyne extensions form a category and so the composition of all these attachings yields again an anodyne extension.

Since $\Gamma = \mathrm{Sd}\,\Lambda^n[n] \cup \Delta[1]^n$ may be considered as obtained from $\mathrm{Sd}\,\Lambda^n[n]$ by attaching $\Delta[1]^n$ via a simplicial map with domain Λ, it follows that the inclusion $\mathrm{Sd}\,\Lambda^n[n] \subset \Gamma$ is an anodyne extension (see again condition (2) in the definition of anodyne extensions, Section 4.5). Consider

$$K = \Gamma \cap \Delta'[n-1] \times \Delta[1] = \Delta'[n-1] \times \Lambda^1[1] \cup \mathrm{Sd}\,\delta\Delta[n-1] \times \Delta[1];$$

then, the inclusion

$$K \subset \Delta'[n-1] \times \Delta[1]$$

is an anodyne extension (see Proposition 4.5.8) and moreover, $\Delta'[n]$ may be thought of as obtained from Γ by attaching $\Delta'[n-1] \times \Delta[1]$ via a simplicial map with domain K. Thus, the inclusion $\Gamma \subset \Delta'[n]$ is also an anodyne extension. □

As in the case of simplicial complexes, subdivision of simplicial sets can be used in order to approximate (continuous) maps by simplicial ones. There is no difficulty to extend the classical simplicial approximation theorem (see Theorem 3.3.17) to presimplicial sets. More refined techniques are necessary for the treatment of simplicial approximation in its most general form, namely the relative approximation of maps between geometric realizations of simplicial sets. To this end, it is necessary to consider the right adjoint functor $\mathrm{Ex} : \mathit{SiSets} \to \mathit{SiSets}$ to normal subdivision, called *extension* (for the existence of the right adjoint to normal subdivision, see Proposition 4.2.10). The following notation will be enforced throughout this section: the adjoint $Y \to \mathrm{Ex}\,Y$ of a simplicial map $f : \mathrm{Sd}\,Y \to X$ will be denoted by f^*, while the adjoint $\mathrm{Sd}\,Y \to X$ of a simplicial map $g : Y \to \mathrm{Ex}\,X$ will be denoted by $g^{\mathscr{C}}$; hence the following rules hold true:

$$(f^*)^{\mathscr{C}} = f, \qquad (g^{\mathscr{C}})^* = g.$$

The natural transformation $\mathrm{d}' : \Delta' \dot{\to} \Delta -$ induces a natural transformation

$\mathrm{e} : \mathbf{1}_{SiSets} \dot{\rightarrow} \mathrm{Ex}$ given by

$$\mathrm{e}_X = (\mathrm{d}'X)^* : X \rightarrow \mathrm{Ex}\, X,$$

for any simplicial set X. The simplicial maps e_X are explicitly given by the formula

$$\mathrm{e}_X(x) = \varphi_x \circ \mathrm{d}'n$$

for any $x \in X_n$; here φ_x denotes the simplicial map $\Delta[n] \rightarrow X$ corresponding to the simplex x via the Yoneda embedding (see Lemma 4.2.1). Since all the simplicial maps $\mathrm{d}'n$ are surjective, the simplicial maps e_X are injective.

First of all, one must list some preservation properties of the functor Ex.

Lemma 4.6.15 *The functor* Ex *preserves*

(i) *the simplicial set* $\Delta[0]$;
(ii) *simplicial homotopies;*
(iii) *simplicial homotopy equivalences, in particular, simplicial contractibility;*
(iv) *Kan fibrations; and*
(v) *Kan sets.*

Proof (i) is trivial. For (ii), let $H : Y \times \Delta[1] \rightarrow X$ be a simplicial homotopy from a simplicial map f to a simplicial map g. Since the functor Ex is right adjoint, it commutes with products, and so $\mathrm{Ex}\, H$ may be considered as a simplicial map $\mathrm{Ex}\, Y \times \mathrm{Ex}\, \Delta[1] \rightarrow \mathrm{Ex}\, X$ whose composition with the simplicial map $\mathbf{1}_{\mathrm{Ex}\, Y} \times \mathrm{e}_{\Delta[1]}$ is the desired homotopy from $\mathrm{Ex}\, f$ to $\mathrm{Ex}\, g$. Statement (iii) is an immediate consequence of (i) and (ii).

(iv): Let $p : Z \rightarrow X$ be a Kan fibration and let $\bar{f} : \Lambda^k[n] \rightarrow \mathrm{Ex}\, Z$ be a simplicial map such that the composition $\mathrm{Ex}\, p \circ \bar{f}$ has an extension $f : \Delta[n] \rightarrow \mathrm{Ex}\, X$; in view of the definition of Kan fibrations, one has to look for a simplicial map $g : \Delta[n] \rightarrow \mathrm{Ex}\, Z$, such that $g | \Lambda^k[n] = \bar{f}$ and $\mathrm{Ex}\, p \circ g = f$. By the naturality of the adjointness, one obtains

$$f^{\mathscr{C}} | \mathrm{Sd}\, \Lambda^k[n] = p \circ \bar{f}^{\mathscr{C}}.$$

Since the inclusion $\mathrm{Sd}\, \Lambda^k[n] \subset \mathrm{Sd}\, \Delta[n]$ is an anodyne extension (see Lemma 4.6.14), there is a simplicial map $k : \mathrm{Sd}\, \Delta[n] \rightarrow Z$ such that $k | \mathrm{Sd}\, \Lambda^k[n] = \bar{f}^{\mathscr{C}}$ and $p \circ k = f^{\mathscr{C}}$ (see Proposition 4.5.3); the adjoint k^* of k has the desired properties. Now (v) is an immediate consequence of (i) and (iii). □

Proposition 4.6.16 *For any simplicial set X, the geometric realization $|\mathrm{e}_X|$ of the natural simplicial map e_X induces an isomorphism between the fundamental groups.*

Proof By construction, the natural simplicial map

$$\mathrm{d}'X = \mathrm{e}_X^{\mathscr{C}} : \mathrm{Sd}\,X \to X$$

factors through

$$\mathrm{Sd}\,(\mathrm{e}_X) : \mathrm{Sd}\,X \to \mathrm{Sd}\,(\mathrm{Ex}\,X);$$

since $|\mathrm{d}'X|$ is a homotopy equivalence (see Corollary 4.6.5), this implies that $|\mathrm{Sd}\,(\mathrm{e}_X)|$ induces monomorphisms between the corresponding homotopy groups (see Corollary A.8.2); in particular, a monomorphism between the fundamental groups. Because $|\mathrm{d}'(\mathrm{Ex}\,X)|$ is also a homotopy equivalence and

$$\mathrm{e}_X \circ \mathrm{d}'X = \mathrm{d}'(\mathrm{Ex}\,X) \circ \mathrm{Sd}\,(\mathrm{e}_X),$$

by the naturality of d′, it follows that

$$|\mathrm{e}_X|_1 : \pi_1(|X|) \to \pi_1(|\mathrm{Ex}\,X|)$$

is a monomorphism, for every selection of a base point of $|X|$ (omitted in the notation).

To prove that $|\mathrm{e}_X|_1$ is an epimorphism, note first that the 1-cells of $|\mathrm{Ex}\,X|$ which correspond bijectively to the non-degenerate 1-simplices of $\mathrm{Ex}\,X$ can be taken as generators for $\pi_1(|\mathrm{Ex}\,X|)$ (see Theorem 2.6.8). Thus, it suffices to check that every 1-cell of $|\mathrm{Ex}\,X|$ considered as a path is homotopic rel. end points to a path in the image of the map $|\mathrm{e}_X|$. Given a 1-simplex in $\mathrm{Ex}\,X$, i.e., a simplicial map $x : \Delta'[1] \to X$ set $x_i = x((\varepsilon_i, \iota))$, for $i \in [1]$ and form the 2-simplex $y : \Delta'[2] \to X$ by taking

$$y((\varepsilon_0, \delta_1, \iota)) = y((\varepsilon_0, \delta_2, \iota)) = x_0 \sigma_1$$

$$y((\varepsilon_1, \delta_0, \iota)) = y((\varepsilon_1, \delta_2, \iota)) = x_1 \sigma_1$$

$$y((\varepsilon_2, \delta_0, \iota)) = y((\varepsilon_2, \delta_1, \iota)) = x_0 \delta_0 \sigma_0 \sigma_1 = x_1 \delta_0 \sigma_0 \sigma_1.$$

Then, the path corresponding to $x = y\delta_2$ is homotopic rel. end points to the path formed by the 1-simplices $y\delta_1 = \mathrm{e}_X(x_0)$ and $y\delta_0 = \mathrm{e}_X(x_1)$ (the latter taken in the inverse direction). □

This result suggests that one should ask how the functor Ex behaves with respect to simplicial universal coverings.

Proposition 4.6.17 *Let X be a simplicial set and let φ be a twisting function for X. Then, there are a twisting function ψ for* $\mathrm{Ex}\,X$ *and an isomorphism* $h : \mathrm{Ex}\,\tilde{X} \to (\mathrm{Ex}\,X)^{\sim}$ *over* $\mathrm{Ex}\,X$, *where* $\tilde{X}$, $(\mathrm{Ex}\,X)^{\sim}$ *denote the simplicial universal coverings of* X, $\mathrm{Ex}\,X$ *with respect to the twisting functions* φ, ψ.

Proof By the previous proposition, the fundamental groups of $|X|$ and $|\mathrm{Ex}\,X|$ can be identified to the same group π. Now it suffices to define

ψ on the 1-simplices of $\operatorname{Ex} X$, i.e., the simplicial maps $\Delta'[1] \to X$. Given such a simplicial map x, set

$$\psi(x) = \varphi \circ x((\varepsilon_0, \iota)) \cdot \varphi \circ x((\varepsilon_1, \iota))^{-1}.$$

Next consider any simplex of $\operatorname{Ex} \tilde{X}$, i.e., a simplicial map $x : \Delta'[n] \to \tilde{X}$. Its 0th vertex $x\varepsilon_0$ corresponds to a pair $(\omega, z) \in \pi \times X_0$. Taking this ω, define $h(x) = (\omega, p \circ x)$ where $p : \tilde{X} \to X$ denotes the simplicial covering projection. □

The process of applying the functor Ex can be iterated, giving rise to an infinite sequence of functors and natural transformations

$$\mathbf{1} \xrightarrow{\mathrm{e}} \operatorname{Ex} \xrightarrow{\mathrm{e}_{\operatorname{Ex}}} \operatorname{Ex}^2 \xrightarrow{\mathrm{e}_{\operatorname{Ex}^2}} \cdots;$$

its colimit is a functor $\operatorname{Ex}^\infty : \mathit{SiSets} \to \mathit{SiSets}$, which is connected to the identity functor by the induced natural transformation $\mathrm{e}^\infty : \mathbf{1} \to \operatorname{Ex}^\infty$. The functor Ex^∞ has similar preservation properties as the functor Ex:

Lemma 4.6.18 *The functor* Ex^∞ *preserves*

(i) *the simplicial set* $\Delta[0]$;
(ii) *simplicial homotopies*;
(iii) *simplicial homotopy equivalences, in particular, simplicial contractibility; and*
(iv) *Kan fibrations.*

Proof (i) is trivial. For (ii), note first that Ex^∞ also preserves finite products; then, use a similar argument as for (ii) of Lemma 4.6.15. Again, (ii) implies immediately (iii). Finally, (iv) is a simple consequence of the corresponding property of the functor Ex. □

The attentive reader will ask for the preservation of Kan sets by the functor Ex^∞; this, clearly, is also trivial but is part of a much stronger result.

Proposition 4.6.19 *The simplicial set* $\operatorname{Ex}^\infty X$ *is a Kan set, for any simplicial set* X.

Proof Let X be a simplicial set and let

$$f : \Lambda^k[n] \to \operatorname{Ex}^\infty X$$

be a given simplicial map, for some $n \in \mathbf{N}$ and $k \in [n]$. Since $\Lambda^k[n]$ contains only finitely many non-degenerate simplices, there is an $r \in \mathbf{N}$ such that f factors through $\operatorname{Ex}^r X$ and also through $\operatorname{Ex}^{r+1} X$. To simplify the notation, set $\Lambda^k[n] = \Lambda$, $\operatorname{Ex}^r X = Y$ and let $g : \Lambda \to \operatorname{Ex} Y$ denote the induced simplicial map; observe that $\mathrm{Ex}^\infty X \cong \mathrm{Ex}^\infty Y$. Define

$$h = g^{\mathscr{C}} \circ \chi : \mathrm{Sd}^2(\Delta[n]) \to Y,$$

where $\chi : \mathrm{Sd}^2(\Delta[n]) \to \mathrm{Sd}\,\Lambda$ is an extension of $\mathrm{Sd}(\mathrm{d}'\Lambda)$ (for the existence of χ, see Lemma 4.6.13). Then, the computation

$$\begin{aligned}(h^{**}|\Lambda)^{\mathscr{C}\mathscr{C}} &= h|\mathrm{Sd}^2\,\Lambda \quad \text{(by the naturality of the adjointness)}\\ &= g^{\mathscr{C}} \circ \mathrm{Sd}(\mathrm{d}'\Lambda) \quad \text{(by the defining property of } \chi)\\ &= (g \circ \mathrm{d}'\Lambda)^{\mathscr{C}} \quad \text{(by the naturality of the adjointness)}\\ &= (\mathrm{d}'(\mathrm{Ex}\,Y) \circ \mathrm{Sd}\,g)^{\mathscr{C}} \quad \text{(by the naturality of d}')\\ &= ((\mathrm{e}_{\mathrm{Ex}\,Y})^{\mathscr{C}} \circ \mathrm{Sd}\,g)^{\mathscr{C}} \quad \text{(by the definition of } \mathrm{e}_{\mathrm{Ex}\,Y})\\ &= (\mathrm{e}_{\mathrm{Ex}\,Y} \circ g)^{\mathscr{C}\mathscr{C}} \quad \text{(by the naturality of the adjointness)}\end{aligned}$$

shows that

$$h^{**}|\Lambda = \mathrm{e}_{\mathrm{Ex}\,Y} \circ g.$$

Therefore, the composition

$$\mathrm{e}^\infty_{\mathrm{Ex}^2 Y} \circ h^{**} : \Delta[n] \to \mathrm{Ex}^\infty Y \cong \mathrm{Ex}^\infty X$$

is the desired extension of the given map

$$f : \Lambda^k[n] \to \mathrm{Ex}^\infty X. \qquad \square$$

The next lemma, needed to establish the simplicial approximation theorem for simplicial sets, relies on homology theory; it is a pity that – up to the writing of this book – there has been no purely combinatorial argument for it.

Lemma 4.6.20 *For any simplicial set X, the natural map $|\mathrm{e}_X| : |X| \to |\mathrm{Ex}\,X|$ is a homotopy equivalence; thus, $|X|$ can be considered as a strong deformation retract of $|\mathrm{Ex}\,X|$ via $|\mathrm{e}_X|$.*

Proof Without loss of generality, one may consider X as a connected simplicial set. The map $|\mathrm{e}_X|$ induces an isomorphism between the fundamental groups (see Proposition 4.6.16), and, since the functors Ex and $|\text{—}|$ commute with the formation of simplicial universal coverings (see Propositions 4.6.17, 4.5.35), $|\mathrm{e}_X|$ also induces an isomorphism on the homology of the universal coverings (see Proposition A.7.3). Thus, $|\mathrm{e}_X|$ is a homotopy equivalence (see Proposition A.8.8).

Moreover, $|\mathrm{e}_X|$ is a closed cofibration (see Corollary 4.3.8 (iii)), and so $|X|$ is embedded in $|\mathrm{Ex}\,X|$ as a strong deformation retract (see Proposition A.4.2 (v)). $\square$

Corollary 4.6.21 *For any Kan set X, the natural simplicial map $\mathrm{e}_X : X \to \mathrm{Ex}\,X$ is a simplicial homotopy equivalence.*

Proof If X is a Kan set, so is $\mathrm{Ex}\,X$ (see Lemma 4.6.15 (v)). But a simplicial map between Kan sets is a simplicial homotopy equivalence iff its geometric realization is a homotopy equivalence (see Corollary 4.5.31 (iii)). □

Corollary 4.6.22 *For any simplicial set X, the natural map $|\mathrm{e}_X^\infty| : |X| \to |\mathrm{Ex}^\infty X|$ is a homotopy equivalence. If X is a Kan set, the natural simplicial map $\mathrm{e}_X^\infty : X \to \mathrm{Ex}^\infty X$ is a simplicial homotopy equivalence.*

Proof Each $|\mathrm{Ex}^n X|$ is embedded in $|\mathrm{Ex}^{n+1} X|$ as a strong deformation retract (see Lemma 4.6.20); moreover, $|\mathrm{Ex}^\infty X|$ is the union space of the resulting expanding sequence (see Proposition 4.3.12). Thus, $|X|$ is a strong deformation retract of $|\mathrm{Ex}^\infty X|$ (see Corollary A.5.8) via the embedding $|\mathrm{e}_X^\infty|$, which therefore is a homotopy equivalence.

The second statement now follows as in the previous corollary. □

Corollary 4.6.23 *For every simplicial set X,*

$$|\mathrm{e}_X \circ \mathbf{1}^{\mathscr{C}}_{\mathrm{Ex}\,X}| \simeq |\mathrm{d}'(\mathrm{Ex}\,X)| \quad \mathrm{rel.}\,|\mathrm{Sd}\,X|$$

via the embedding $|\mathrm{Sd}\,(\mathrm{e}_X)| : |\mathrm{Sd}\,X| \to |\mathrm{Sd}\,(\mathrm{Ex}\,X)|$.

Proof The naturality of the transformation d′– gives the equation

$$\mathrm{d}'(\mathrm{Ex}\,X) \circ \mathrm{Sd}\,(\mathrm{e}_X) = \mathrm{e}_X \circ \mathrm{d}'X,$$

where the maps $\mathrm{d}'(\mathrm{Ex}\,X)$, $\mathrm{d}'X$ and e_X are weak homotopy equivalences (see Proposition 4.6.3 (iii) and Lemma 4.6.20); thus, $|\mathrm{Sd}(\mathrm{e}_X)|$ is a homotopy equivalence. Moreover, $|\mathrm{Sd}\,(\mathrm{e}_X)|$ is a closed cofibration (see Proposition 4.6.3 (ii) and Corollary 4.3.8 (iii)), and therefore $|\mathrm{Sd}\,X|$ is embedded as a strong deformation retract into $|\mathrm{Sd}(\mathrm{Ex}\,X)|$. Then, two maps with domain $|\mathrm{Sd}\,(\mathrm{Ex}\,X)|$ are homotopic rel. $|\mathrm{Sd}\,X|$ if their restrictions to $|\mathrm{Sd}\,X|$ coincide. In order to prove this equality for the maps under consideration, notice that the simplicial map $\mathbf{1}^{\mathscr{C}}_{\mathrm{Ex}\,X}$ is the co-unit of the adjunction, and therefore

$$\mathrm{d}'X = (\mathrm{e}_X)^{\mathscr{C}} = \mathbf{1}^{\mathscr{C}}_{\mathrm{Ex}\,X} \circ \mathrm{Sd}\,(\mathrm{e}_X);$$

substituting $\mathbf{1}^{\mathscr{C}}_{\mathrm{Ex}\,X} \circ \mathrm{Sd}\,(\mathrm{e}_X)$ for $\mathrm{d}'X$ in the equation displayed before gives the desired result. □

The following technical result is more than just a simple consequence of the preceding theory.

Lemma 4.6.24 *Let Y, X be simplicial sets, D a simplicial subset of Y, $f : |Y| \to |X|$ a map and $\bar{f} : Y \to \mathrm{Ex}\,X$ a simplicial map such that*

$$|\bar{f}| \simeq |\mathrm{e}_X| \circ f \quad \mathrm{rel.}\ |D|.$$

Then

$$|\bar{f}^{\mathscr{C}}| \simeq f \circ |\mathrm{d}'Y| \quad \text{rel. } |\mathrm{Sd}\, D|.$$

Proof The naturality of the transformation d′– implies:

$$|\mathrm{d}'(\mathrm{Ex}\, X)| \circ |\mathrm{Sd}\, \bar{f}| = |\bar{f}| \circ |\mathrm{d}'Y| \simeq |\mathrm{e}_X| \circ f \circ |\mathrm{d}'Y| \quad \text{rel. } |\mathrm{Sd}\, D|.$$

From the previous corollary, one now obtains that

$$|\mathrm{e}_X| \circ |\bar{f}^{\mathscr{C}}| = |\mathrm{e}_X \circ \mathbf{1}^{\mathscr{C}}_{\mathrm{Ex}\, X} \circ \mathrm{Sd}\, \bar{f}| \simeq |\mathrm{d}'(\mathrm{Ex}\, X)| \circ |\mathrm{Sd}\, \bar{f}|$$
$$\simeq |\mathrm{e}_X| \circ f \circ |\mathrm{d}'Y| \quad \text{rel. } |\mathrm{Sd}\, D|.$$

But the map $|\mathrm{e}_X|$ has a left inverse r (see Lemma 4.6.20); composing (on the left) the first and last terms of the previous homotopy with r, one obtains the wanted homotopy. □

Now one has all the technology necessary to prove the relative simplicial approximation theorem for simplicial sets.

Theorem 4.6.25 *Let Y be a simplicial set with only finitely many non-degenerate simplices, D a simplicial subset of Y, X an arbitrary simplicial set, $f : |Y| \to |X|$ a map and $g : D \to X$ a simplicial map such that $f \,|\, |D| = |g|$. Then there are a simplicial set Y' containing D as simplicial subset, a homeomorphism $h : |Y'| \to |Y|$ which is the identity on $|D|$ and identifies the CW-complex $|Y'|$ with a subdivision of the CW-complex $|Y|$, and a simplicial map $k : Y' \to X$, such that*

$$k \,|\, D = g$$

and

$$|k| \simeq f \circ h \quad \text{rel. } |D|.$$

Moreover, the homeomorphism h can be chosen homotopic to the geometric realization of a simplicial map inducing the identity on D.

Proof Choose a simplicial map $\tilde{f} : Y \to \mathrm{Ex}^\infty X$, with

$$\tilde{f} \,|\, D = \mathrm{e}^\infty_X \circ g$$

and

$$|\tilde{f}| \simeq |\mathrm{e}^\infty_X| \circ f \quad \text{rel. } |D|$$

(see Theorem 4.5.27).

Since Y has only finitely many non-degenerate simplices $|Y| \times I$ is compact, and, consequently, there are an $r \in \mathbf{N}$ and a simplicial map $\bar{f} : Y \to \mathrm{Ex}^r X$, such that

$$\bar{f} \,|\, D = \mathrm{e}^r \circ g$$

and

$$|\bar{f}| \simeq |\mathrm{e}^r| \circ f \quad \text{rel. } |D|,$$

where $\mathrm{e}^r : X \to \mathrm{Ex}^r X$ denotes the canonical inclusion. Taking r times the adjoints, one obtains a simplicial map $\tilde{k} : \mathrm{Sd}^r Y \to X$ such that

$$|\tilde{k}| \simeq f \circ |\mathrm{d}^r Y| \quad \text{rel. } |\mathrm{Sd}^r D|$$

(see Lemma 4.6.24); let $\tilde{H} : |\mathrm{Sd}^r Y| \times I \to |X|$ denote a suitable homotopy.

Take the simplicial set Y' to be the rth normal subdivision of Y rel. D. Let $\bar{d} : \mathrm{Sd}^r Y \to Y'$ denote the corresponding characteristic simplicial map and $h : |Y'| \to |Y|$ a homeomorphism with the following properties:

(1) h is homotopic rel. $|D|$ to the geometric realization of the induced simplicial map $\tilde{d} : Y' \to Y$; and
(2) h identifies the CW-complex $|Y'|$ with a subdivision of the CW-complex $|Y|$

(see Proposition 4.6.6). The universal property of Y' yields a simplicial map $k : Y' \to X$ such that $k|D = g$ and $k \circ \bar{d} = \tilde{k}$.

Since the functors $|—|$ and $— \times I$ preserve attachings (see Propositions 4.3.10 and A.4.8 (i)), one obtains $|Y'| \times I$ as an adjunction space, and, consequently, a homotopy $H : |Y'| \times I \to |X|$ such that $H||D| \times I$ is the composition of the projection $|D| \times I \to |D|$ with $|g|$ and $H \circ (|\bar{d}| \times \mathbf{1}_I) = \tilde{H}$. This is a homotopy rel. $|D|$ between $|k|$ and $f \circ |\tilde{d}|$, which, in view of (1), implies the desired poperties. □

Exercises

1. Show that the two-fold normal subdivision of a presimplicial set is the associated simplicial set of an ordered simplicial complex.
2. Show that $\mathrm{Sd} \circ \mathrm{Sd}^{\mathrm{op}} = \mathrm{Sd} \circ \mathrm{Sd}$.

Preface to Exercises 3–8: Corollary 4.6.12 has a refinement. The proof of Theorem 3.4.1 was based on an idea that has a purely combinatorial analogue described by the so-called *star functor*. Given a simplicial set X, the binary relation 'contains as a face' on the set $X^\#$ of the non-degenerate simplices of X is reflexive and antisymmetric; thus, it gives rise to an ordered simplicial complex X^* (see Section 3.3, Example 2).

3. Extend the construction of X^* to a functor $* : \mathit{SiSets} \to \mathit{OSiCo}$.
4. In general, the star functor has very bad geometric properties. Verify that, for every $p > 0$, $|S[p]^*| \approx \Delta^1$.
5. However, if X is a regular simplicial set, prove that X^* is isomorphic to the simplicial complex triangulating $|X|$ in the sense of 4.6.12.
6. For a simplicial set X let X^* also denote the simplicial set associated to the ordered simplicial complex X^* (see Section 4.2, Exercise 4). Show that for a regular simplicial set X the natural simplicial map $\mathrm{d}''_X : \mathrm{Sd}^{\mathrm{op}} X \to X$ factors through a natural simplicial map $\mathrm{t}_X : X^* \to X$.

7. Moreover, show that $(\mathrm{Sd}\, X)^*$ is a subdivision of X, for any simplicial set X, in the following general sense: A simplicial set X' is called a *subdivision of the simplicial set X* if there are a simplicial map $d : X' \to X$ and a homeomorphism $h : |X'| \to |X|$ such that the CW-complex $|X'|$ becomes a subdivision of the CW-complex $|X|$ and $|d| \simeq h$.
8. Show that the simplicial map of Exercise 7 may be constructed in such a way that it depends naturally on X; however, naturality is impossible for the homeomorphism.
9. Let G be a group and let BG denote its classifying set. Show that the simplicial map

$$e_{BG} : BG \to \mathrm{Ex}\, BG$$

is a weak homotopy equivalence.

Notes to Chapter 4

The simplicial sets studied in this chapter are precisely the 'complete semi-simplicial complexes' introduced in Eilenberg & Zilber (1950). Since the inception of the theory of complete semi-simplicial complexes (c.s.s. complexes as abbreviated in Kan (1957)), there has been a great deal of confusion about the correct terminology. At Klaus Lamotke's talk given during the Moscow 1976 International Congress of Mathematicians, the audience exhorted the mathematical community to call the objects of our categories *SiSets* and *PSiSets* (see definitions in Sections 4.2 and 4.4) by the names 'semi-simplicial sets' and 'simplicial sets', respectively; the argument was that *PSiSets* could be viewed as a subcategory of *SiSets* (via the embedding functor E), and thus the passage from presimplicial to simplicial sets weakens the defining conditions. This suggestion was not followed up, probably because at a first glance simplicial sets (with their degeneracy operators) have a richer structure than presimplicial sets. The objects of *PSiSets* have also been called Δ-sets in Rourke & Sanderson (1971). Today, the expression 'simplicial set' is almost universally used to indicate the objects of *SiSets*; we follow this trend in the present book.

The theory of simplicial sets was, in large part, developed by Daniel M. Kan (see Kan 1955, 1957, 1958a, 1958b, 1958c, 1970). The first comprehensive textbooks about it were written by John Peter May (May, 1967), Peter Gabriel and Michel Zisman (Gabriel & Zisman, 1967) – who emphasized a strict categorical treatment of the theory – and Klaus Lamotke (Lamotke, 1968); the latter text contains a fairly complete list of references up to the time of its printing. Survey articles broadening the scope of our exposition are Schubert (1958), Gugenheim (1968) and Curtis (1971).

Owing to their strong geometric flavour, face and degeneracy operators have made their presence felt since the start of the theory of simplicial sets. A short but systematic treatment of the category of finite ordinals can be found in MacLane (1971) under the name 'the simplicial category'. Our category of finite ordinals is not exactly the same as MacLane's; however, it is isomorphic to the subcategory obtained from MacLane's category by removing its initial object. The idea of taking the maximal section for a degeneracy operator proved itself fruitful in Fritsch & Puppe (1967) and Fritsch (1969); in an embryonal form the 2-category structure of Δ was first given in Fritsch (1972).

Section 4.2 gives an extract from Lamotke (1968). The development of category theory – in particular, the famous Yoneda Lemma (attributed to Nobuo Yoneda by Peter Freyd, 1964) – strongly influenced the way in which this section was developed. The condition for the Eilenberg–Zilber property for cosimplicial sets (Proposition 4.2.6) can be found in Ruiz Salguero & Ruiz Salguero (1978); it was also obtained by Dieter Puppe (unpublished). The unique representation of elements in a tensor product (Proposition 4.2.7) was proved in full generality in Fritsch (1983), along the lines of the special case dealt with earlier (Fritsch, 1969/I).

At the outset, it was not clear how one should have interpreted the 'geometric realization' of a simplicial set; indeed, this doubt was clearly revealed, for instance, in the title of Kodama's paper (Kodama, 1957). However, after that time, the interpretations given in Milnor (1957) and Puppe (1958) were generally accepted. The comparison theorem (Theorem 4.3.20) was first stated in its full strength in Fritsch & Latch (1981); the compatibility of geometric realization and local triviality (Proposition 4.3.22) is due to Gabriel & Zisman (1967).

The notion of cone functor in the category of presimplicial sets came to our attention via a letter from Dieter Puppe to Tammo tom Dieck; the purely combinatorial proof of Theorem 4.4.5 presented in this book was sketched in the aforementioned letter; the theorem itself was first stated in Kodama (1957).

The defining condition for Kan sets, i.e., the requirement that horns can be filled, is often referred to as the 'Kan condition'; it is also called the 'extension condition', mostly by Daniel M. Kan himself, who originally formulated it for 'cubical sets' (see Kan, 1955) and showed that it leads to a combinatorial homotopy theory. The passage to simplicial sets was done in Kan (1956, 1958c); the latter paper also contains the definition of 'Kan fibration'. In a certain sense, anodyne extensions – invented by Gabriel and Zisman (see Gabriel & Zisman (1967) – are the combinatorial 'strong deformation retracts'. The question of deciding if the geometric realization of a Kan fibration is a fibration was open for a long time; the affirmative answer to this problem (see Theorem 4.5.25) was given in Quillen (1968); in that paper, Daniel Quillen also introduced the notion of 'acyclic fibration'. The concept of minimality and the local triviality of minimal fibrations can be found on the road leading to the proof of Theorem 4.5.25. Minimal strong deformation retracts of singular sets (in the simplicial sense; cf. Proposition 4.5.23) were constructed in Eilenberg & Zilber (1950); in Gabriel & Zisman (1967), one already finds a proof for Proposition 4.5.21. The approximation property of Kan fibrations (see Theorem 4.5.27) was proved in its full generality in (Fritsch, 1976); it leads to the simple proof, given in this book, of the fact that the counits j_T are weak homotopy equivalences (Theorem 4.5.30). The latter result was stated in Milnor (1957), where it is credited to Giever (1950); it was completely proved for the first time in Lamotke (1963). Alternative proofs for the fact that j_T is a homotopy equivalence for CW-complexes may be found in Gabriel & Zisman (1967) and Puppe (1983). Twisting functions appeared in Moore (1958).

The normal subdivision of simplicial sets was introduced in Kan (1956, 1957); it was also discussed in Barratt (1956). The compatibility between geometric realization and normal subdivision (see Corollary 4.6.5) was first proved, within a more general context, in Fritsch (1969/II). The explicit formula in the proof of Theorem 4.6.4 is due to Dieter Puppe (Fritsch & Puppe, 1967); the relative version is contained in Fritsch (1974). The simplicial excision theorem (Theorem 4.6.9) was developed in Puppe (1983) in order to get the mentioned alternative proof of Theorem 4.5.30. The proof of the triangulability of the geometric realizations of simplicial sets given in this book follows the lines of Barratt (1956). The extension functor Ex and its properties were first announced in Kan (1956b); they were studied in detail in Kan (1957), where one also finds the proof of the (absolute) approximation theorem for finite domain. The relative version of that theorem was studied in Fritsch (1974).

5

Spaces of the type of CW-complexes

5.1 Preliminaries

This chapter's work takes place mostly within the category TCW of spaces with the type of CW-complexes and maps.

Proposition 5.1.1 *A space X has the type of a CW-complex iff X is dominated by a CW-complex.*

Proof The necessity of the condition is obvious. To prove the sufficiency, let Y be a CW-complex which dominates X, with maps $j : X \to Y$ and $r : Y \to X$ such that $r \circ j \simeq \mathbf{1}_X$. Form the commutative diagram

$$\begin{array}{ccccc} X & \xrightarrow{j} & Y & \xrightarrow{r} & X \\ \uparrow j_X & & j_Y \uparrow & & j_X \uparrow \\ |SX| & \xrightarrow[|Sj|]{} & |SY| & \xrightarrow[|Sr|]{} & |SX| \end{array}$$

with $|Sr| \circ |Sj| \simeq \mathbf{1}_{|SX|}$. Because Y is a CW-complex, j_Y is a homotopy equivalence (see Corollary 4.5.31 (i)). Let μ_Y be a homotopy inverse for j_Y and define

$$\mu_Y = |Sr| \circ \mu_Y \circ j : X \to |SX|.$$

Then, one can check that $\mu_X \circ j_X \simeq \mathbf{1}_{|SX|}$ and $j_X \circ \mu_X \simeq \mathbf{1}_X$. □

Proposition 5.1.2 *Let X be a space with the type of a CW-complex. Then, X has a covering $\{U_\lambda : \lambda \in \Lambda\}$ which admits a subordinated locally finite partition of unity and such that the inclusion maps $U_\lambda \to X$ are homotopic to constant maps.*

Proof Let $f : X \to Y$ be a homotopy equivalence, with Y a CW-complex; let $g : Y \to X$ denote a homotopy inverse for f. Because Y is locally contractible (see Theorem 1.3.2), it has an open covering $\{V_\lambda : \lambda \in \Lambda\}$, where all the V_λ's are contractible. Define the covering $\{U_\lambda\}$ of X by taking, for every $\lambda \in \Lambda$, $U_\lambda = f^{-1}(V_\lambda)$. Now, if H is a contracting homotopy for V_λ, then the composition of $g | V_\lambda \circ H$ with the induced map $U_\lambda \times I \to V_\lambda \times I$ is a homotopy between $g \circ f | U_\lambda$ and the constant map; because g is a homotopy inverse to f, it follows that $g \circ f | U_\lambda$ is homotopic to the inclusion of U_λ into X. Observe that the open sets U_λ are not necessarily contractible.

Finally, let $\{\mu_\lambda\}$ be a locally finite partition of unity subordinated to the covering $\{V_\lambda\}$ (see Theorem A.3.3). Then, $\{\mu_\lambda \circ f\}$ is a locally finite partition of unity on X subordinated to the covering $\{U_\lambda\}$. □

Proposition 5.1.3 *Let X be a space with the type of a CW-complex. Then, the following implications hold true*:

(i) *X totally disconnected $\Rightarrow X$ discrete*;

(ii) *X connected $\Rightarrow X$ path-connected*;

(iii) *X weakly contractible $\Rightarrow X$ contractible.*

Proof (i): Since X has the type of a CW-complex, its path-components are open (see Proposition 1.4.14). But X being totally disconnected, each point of X is a path-component, and thus open.

(ii): If X has the type of a CW-complex Y, the hypothesis implies that Y also is connected, since connectivity is a homotopy invariant. Then, Y is path-connected (see Corollary 1.4.12); but path-connectivity is also a homotopy invariant, and thus X is path-connected.

(iii): If X has the type of a CW-complex Y, the hypothesis implies that Y also is weakly contractible (see Corollary A.8.2), and therefore contractible (see Theorem 2.5.1). Since contractibility is again a homotopy invariant, X is contractible. □

The remainder of this section is used for some examples.

Example 1 Let X be the subspace of $\mathbf{R}$ consisting of the points 0 and $1/n$, for all integers $n \geqslant 1$. This space is totally disconnected but not discrete, and thus not of the type of a CW-complex (see Proposition 5.1.3 (i)). □

Example 2 The Cantor set (or middle third set) does not have the type of a CW-complex because it is also totally disconnected and non-discrete. □

Example 3 Let S be the graph of the function $f(x) = \sin(1/x), 0 < x \leqslant 1$ in $\mathbf{R}^2$ and $A = \{(0, y) \in \mathbf{R}^2 : -1 \leqslant y \leqslant 1\}$. Take the set $B = A \cup S$ and give it the subspace topology in $\mathbf{R}^2$; the space B is clearly connected, but not path-connected, and thus not of the type of a CW-complex (see Proposition 5.1.3 (ii)). □

The next is an example of a space with the homotopy type of a CW-complex but which is not a CW-complex.

Example 4 For each $n \in \mathbf{N} \setminus \{0\}$, let A_n be the segment of $\mathbf{R}^2$ with vertices $(-1,0)$ and $(0,1/n)$; let D be the segment with vertices $(-1,0)$ and $(0,0)$. Now define X to be the set

$$X = \bigcup_{n \in \mathbf{N} \setminus \{0\}} A_n \cup D$$

with the subspace topology. Since X is contractible, it has the homotopy type of a CW-complex. Suppose there is a CW-structure for X. The points $(0,0)$ and $(0,1/n)$, for all $n \in \mathbf{N} \setminus \{0\}$, cannot be interior points of open cells of dimension >0, in view of the theorem of invariance of domain (see Theorem A.9.6). Thus, they must belong to the 0-skeleton, contradicting its discreteness (cf. Example 1).

Moreover, note that although X is contractible the singleton space $\{(0,0)\}$ is not a strong deformation retract of X. In other words, there is no based homotopy equivalence between $(X,(0,0))$ and the singleton space. □

This example gives rise to another space not having the type of a CW-complex:

Example 5 For each $n \in \mathbf{N} \setminus \{0\}$, consider, besides the segments A_n defined in Example 4, the segments B_n having vertices $(0,-1/n)$ and $(1,0)$. Let C be the segment with vertices $(-1,0)$ and $(1,0)$; define the space X as the set

$$X = \left(\bigcup_{n \in \mathbf{N} \setminus \{0\}} A_n \right) \cup C \cup \left(\bigcup_{n \in \mathbf{N} \setminus \{0\}} B_n \right)$$

with the subspace topology. This space is weakly contractible but not contractible (see Section A.8, Example 3); if it were of the type of a CW-complex, it would be contractible (see Proposition 5.1.3 (iii)). □

Example 6 For any based space (X, x_0), the based path-space (PX, ω_0) (see Section A.4, Example 6) has the type of a CW-complex. In fact, PX is contractible, and thus it has the type of a CW-complex. □

Exercises

1. A space X is *semilocally contractible* if each point of X has a neighbourhood V, such that the inclusion $V \to X$ is homotopic to a constant map. Show that any space in the category TCW is semilocally contractible (indeed, any space having the type of a locally contractible space is semilocally contractible). (Dydak & Geoghegan, 1986).
2. Let $\{U_\lambda : \lambda \in \Lambda\}$ be a numerable covering of a space X such that $U_\lambda \in TCW$, for each $\lambda \in \Lambda$. Show that $X \in TCW$.

5.2 CW-complexes and absolute neighbourhood retracts

As seen in Chapter 3, simplicial complexes with the strong topology are ANRs; this result has an inverse, up to type.

Theorem 5.2.1 *The following conditions are equivalent for a space X:*
(i) *X is dominated by a CW-complex;*
(ii) *X has the type of a CW-complex;*
(iii) *X has the type of a simplicial complex;*
(iv) *X has the type of a simplicial complex with the strong topology;*
(v) *X has the type of an ANR.*

Proof (i)⇒(ii): Proposition 5.1.1.
(ii)⇒(iii): Corollary 4.6.12.
(iii)⇒(iv): Proposition 3.3.7.
(iv)⇒(v): Theorem 3.3.10.
(v)⇒(i): Without loss of generality, assume X to be an ANR; as a metric space, it can be viewed as a subspace of the normed linear space $L = C(X, \mathbf{R})$, closed in its convex hull $Z = H(X)$ (see Proposition A.6.1); then, because X is an ANR, there is a retraction $r : U \to X$, where U is a neighbourhood of X in Z. For each $u \in U$, let $\eta = \eta(u) > 0$ be a real number such that the convex set $B(u, \eta) = \{z \in Z : d(z, u) < \eta\}$ is contained in U. Take $\tilde{U} = \{u \in U : B(u, \eta(u)/2) \cap X \neq \varnothing\}$ and note that $\{V_u : u \in \tilde{U}\}$, with $V_u = B(u, \eta(u)/2) \cap X$ is an open covering of X. Let P be the nerve of this covering; it will be proved that the simplicial complex P dominates X.

Choose a locally finite partition of unity $\{\mu_u : u \in \tilde{U}\}$ subordinated to the covering $\{V_u\}$ (see Theorem A.3.3, noting that, as a metric space, X is paracompact) and let $f : X \to P$ be the canonical map given by $f(x) = \{\mu_u(x) : u \in \tilde{U}\}$ (see Lemma 3.3.4 (ii)). Define a map $\tilde{g} : P \to U$ by $\tilde{g}(V_u) = u$ and the linear extension over all simplices of P. To see that $\tilde{g}$ indeed takes values only in U, proceed as follows. Let $\{V_{u_0}, \ldots, V_{u_n}\}$ be a simplex of P and take $x \in \bigcap_{j=0}^{n} V_{u_j}$. Relabelling the indices, if necessary, assume that $\eta(u_j) \leqslant \eta(u_0), j = 0, \ldots, n$. It follows that $d(x, u_j) < \eta(u_j)/2 \leqslant \eta(u_0)/2$, and thus, $d(u_0, u_j) < \eta(u_0)$; i.e., $u_j \in B(u_0, \eta(u_0))$. This shows that $\tilde{g}(\{V_{u_0}, \ldots, V_{u_n}\}) \subset B(u_0, \eta(u_0)) \subset U$. Now define the map $g : P \to X$ as the composition $g = r \circ \tilde{g}$.

Take the affine connection between $\tilde{g} \circ f$ and $\mathbf{1}_X$, i.e., the homotopy $\tilde{H} : X \times I \to Z$ given by

$$\tilde{H}(x, t) = tx + (1 - t)\tilde{g} \circ f(x)$$

for every $(x, t) \in X \times I$; $\tilde{H}$ also takes its values in U. For a given $x \in X$, let $u_0, \ldots, u_n$ denote the points in $\tilde{U}$ with $\mu_{u_j}(x) \neq 0, j = 0, \ldots, n$; as before, assume $\eta(u_j) \leqslant \eta(u_0)$. Then, the line segment connecting x to $\tilde{g} \circ f(x)$ is totally contained in $B(u_0, \eta(u_0)) \subset U$. The composition of the homotopy $\tilde{H} : X \times I \to U$ with the retraction $r : U \to X$ gives the desired homotopy from $g \circ f$ to $\mathbf{1}_X$. □

Corollary 5.2.2 *A compact ANR is dominated by a finite CW-complex.*

Proof See Corollary 1.5.5. □

Remark In contrast to Theorem 5.2.1 and Corollary 5.2.2, it is not true, in general, that a space dominated by a finite CW-complex has the type of a compact ANR. However, in the presence of countability, one can refine Theorem 5.2.1.

Theorem 5.2.3 *For a space X the following conditions are equivalent:*
(i) *X is dominated by a countable CW-complex;*
(ii) *X has the type of a countable CW-complex;*
(iii) *X has the type of a countable simplicial complex;*
(iv) *X has the type of a countable locally finite simplicial complex;*
(v) *X has the type of an ANR satisfying the second axiom of countability.*

Proof (i)⇒(iii): Let $X \xrightarrow{j} Y \xrightarrow{r} X$ be given with $r \circ j \simeq \mathbf{1}_X$ and Y a countable CW-complex. It is known that $X \simeq |SX| \in CW$ (see Proposition 5.1.1).

Let (K, h) be a triangulation of $|SX|$ (see Corollary 4.6.12). Using the same notation as in 5.1.1, let $\mu_Y : Y \to |SY|$ denote a homotopy inverse for the natural map $j_Y : |SY| \to Y$. Form the map

$$\mu = h^{-1} \circ |Sr| \circ \mu_Y : Y \to K$$

and observe that $\mu \circ j = h^{-1} \circ |Sr| \circ \mu_Y \circ j$ is a homotopy equivalence with inverse $j_X \circ h$. Because Y is countable, its image by μ is contained in a countable subcomplex $L^{(0)}$ of K. In fact, for any cell e of Y, $\mu(\bar{e})$ is compact, and thus is contained in a finite subcomplex (see Corollary 1.5.4); clearly, a countable union of finite subcomplexes is countable.

Let

$$H : K \times I \to K$$

be a homotopy from $\mu \circ j \circ j_X \circ h$ to the identity map of K. Because $L^{(0)} \times I$ is countable, there is a countable subcomplex $L^{(1)} \subset K$ such that $H(L^{(0)} \times I) \subset L^{(1)}$. By iteration, one obtains a sequence of countable subcomplexes of K, say

$$L^{(0)}, L^{(1)}, L^{(2)}, \ldots,$$

such that $H(L^{(n)} \times I) \subset L^{(n+1)}$. The subspace

$$L^* = \bigcup_{n \in \mathbf{N}} L^{(n)}$$

of K is a subcomplex of K (see Corollary 1.4.5); furthermore, L^* is countable,

$$(\mu \circ j)(X) \subset L^{(0)} \subset L^*,$$

and

$$H(L^* \times I) \subset L^*.$$

Set $j^* = j_X \circ h | L^*$ and take μ^* as the map from X to L^* induced by $\mu \circ j$; then the homotopy

$$H : L^* \times I \to L^*$$

shows that $\mu^* \circ j^* \simeq \mathbf{1}$. On the other hand, $j^* \circ \mu^* = j_X \circ h \circ \mu \circ j \simeq \mathbf{1}_X$, and therefore X has the type of the countable simplicial complex L^*.

(iii)$\Rightarrow$(ii) and (ii)$\Rightarrow$(i): Obvious.

(iii)$\Rightarrow$(iv): A countable simplicial complex Y, viewed as a countable regular CW-complex (see Theorem 3.3.2), has the type of a regular, countable and locally finite CW-complex T (see Proposition 2.2.5). Moreover, T is triangulable (see Theorem 3.4.1), and, because of the topological invariance of local finiteness (see Proposition 1.5.10) and countability (see Proposition 1.5.20), the simplicial complex obtained has the desired properties.

(iv)$\Rightarrow$(v): Suppose that X has the type of a countable locally finite simplicial complex Y. This complex is metrizable and satisfies the second axiom of countability (see Theorem 1.5.15); moreover, it is an ANR (see Theorem 3.3.10).

(v)$\Rightarrow$(i): The proof for this is similar to that for the corresponding part of Theorem 5.2.1. One has only to use the fact that the open covering $\{B(u, \eta(u)/2) \cap X : u \in U\}$ of X has a countable star-finite refinement $\{V_n : n \in \mathbf{N}\}$ (see Theorem A.3.2). Its nerve is a countable, locally finite simplicial complex (see Proposition 3.3.12), which, like its counterpart in the previous theorem, dominates X. □

Some consequences of Theorem 5.2.3 will be proved next.

Corollary 5.2.4 *Every n-manifold satisfying the second axiom of countability has the type of a countable CW-complex.*

Proof Let X be an n-manifold, i.e., a Hausdorff space such that each point $x \in X$ has an open neighbourhood V_x homeomorphic to $\mathbf{R}^n$; assuming that X also satisfies the second axiom of countability, then X is a Lindelöf space, and hence the open covering $X = \bigcup_{x \in X} V_x$ has a countable subcovering. The open sets V_x are all ANRs and so is X (see Proposition A.6.9). The implication (v)$\Rightarrow$(ii) of Theorem 5.2.3 completes the proof. □

Corollary 5.2.5 *If X has the type of a countable CW-complex and C is a compact metric space, the function space*

$$X^C = \{f : C \to X : f \text{ continuous}\}$$

with the compact-open topology has the type of a countable CW-complex.

Proof Let K be an ANR satisfying the second axiom of countability which has the type of X (see Theorem 5.2.3). Then, K^C is an ANR (see Proposition A.6.10 with $C_0 = \varnothing$), has the type of X^C and satisfies the second axiom of countability as a subspace of $(I^\infty)^C$ (see Theorem A.9.8 and Proposition A.9.9). □

Remark The condition of compactness on C in Corollary 5.2.5 is essential, as can be seen by the following example: $(S^0)^{\mathbb{N}}$ is homeomorphic to the Cantor set, and thus is not of the type of a CW-complex (see Section 5.1, Example 2).

Corollary 5.2.6 *If a Lindelöf space X has the type of a CW-complex, then X has the type of a countable CW-complex.*

Proof Let $f : X \to Y$ be a homotopy equivalence, where Y is a CW-complex. The subcomplex $L = Y(f(X))$ of Y (see Section 1.4, Example 1) clearly dominates X. Moreover, as the continuous image of a Lindelöf space, $f(X)$ is a Lindelöf subspace of Y, and therefore L is countable (see Proposition 1.5.21). □

5.3 *n*-ads and function spaces

An *n-ad* is a space together with a sequence of n subspaces. If one wishes to be perfectly clear, the notation

$$\{X; X_0, X_1, \ldots, X_{n-1}\}$$

should be used to describe the n-ad consisting of the *global* space X and the corresponding sequence; otherwise, if the sequence is clearly understood, just write X instead of the previous lengthy expression. By abuse of language, a space X is said to be an n-ad if a sequence of n subspaces is implicitly given.

An n-ad $\{X; X_0, X_1, \ldots, X_{n-1}\}$ is a

CW-n-ad if X is a CW-complex and all the X_i's are subcomplexes of X;

simplicial-n-ad if X is a simplicial complex and all the X_i's are subcomplexes of X;

simplicial-n-ad with the strong topology if X is a simplicial complex with the strong topology and all the X_i's are subcomplexes of X (with the strong topology);

ELCX-n-ad if X is an ELCX-space and every X_i is an ELCX-subspace.

The product of an n-ad $\{X; X_0, \ldots, X_{n-1}\}$ and a space Z is the n-ad $\{X \times Z; X_0 \times Z, \ldots, X_{n-1} \times Z\}$. If X and Y are n-ads, then an *n-ad map* $f : Y \to X$ is a map such that, for every $0 \leqslant i \leqslant n-1$, $f(Y_i) \subset X_i$; its *mapping cylinder* $\{M; M_0, \ldots, M_{n-1}\}$ is obtained by taking the mapping cylinder $M = M(f)$ of f, and, for each $0 \leqslant i \leqslant n-1$, the mapping cylinder M_i of the induced map $Y_i \to X_i$. If Y, X are CW-n-ads and f is a cellular map, its mapping cylinder is also a CW-n-ad (see Section 2.3, Example 2). The set of all n-ad maps $Y \to X$ forms the *n-ad function space* $X^{Y,n}$ as a subspace of the function space X^Y. An *n-ad homotopy* is a homotopy which is also an n-ad map. The concepts of retraction, deformation retraction, deformation and homotopy equivalence are appropriately defined and suggested by the usual definitions. In particular, an *n-ad homotopy equivalence* is an n-ad map $f : Y \to X$ for which there is an n-ad map $g : X \to Y$ such that:

(1) g is a homotopy inverse of f in the ordinary sense, and
(2) the homotopies connecting $g \circ f$ and $f \circ g$ to the respective identity maps move all the X_i's and the Y_i's respectively, within themselves.

The category whose objects are n-ads of the *type* of CW-n-ads, i.e., homotopy equivalent to CW-n-ads, and whose morphisms are n-ad maps, will be denoted by TCW^n. Conditions for an n-ad to belong to TCW^n will be examined next; in particular, it will be shown that certain function space constructions, like the construction of loop spaces, do not lead outside the category $TCW^0 = TCW$.

A sufficient criterion for an n-ad to belong to the category TCW^n is given in the following result.

Lemma 5.3.1 *Let $\{X; X_0, \ldots, X_{n-1}\}$ be an n-ad such that:*
(i) $X_{i+1} \subset X_i$, *for every* $i \in [n-2]$,[†]
(ii) *the inclusions $X_i \subset X$ are closed cofibrations, for every $i \in [n-1]$, and*
(iii) $X \in TCW$, $X_i \in TCW$, *for every* $i \in [n-1]$.
Then, $\{X; X_0, \ldots, X_{n-1}\} \in TCW^n$.

[†] As in Section 4.1, the symbol $[n-2]$ denotes the set of integers from 0 to $n-2$.

Proof The proof is done by induction on n. It is trivial for $n = 0$. Assume the statement of the lemma holds true for n. Let $\{X; X_0, \ldots, X_n\}$ be a given $(n+1)$-ad satisfying the hypothesis of the lemma. Then, the n-ad $\{X_0; X_1, \ldots, X_n\}$ also satisfies the hypothesis of the lemma (for (ii) use Proposition A.4.2 (vi)). Thus, by the induction hypothesis, $\{X_0; X_1, \ldots, X_n\} \in TCW^n$. Let $\{Y_0; Y_1, \ldots, Y_n\}$ be a CW-n-ad which is n-ad homotopy equivalent to $\{X_0; X_1, \ldots, X_n\}$ via maps $f : X_0 \to Y_0, g : Y_0 \to X_0$ and homotopies H from $f \circ g$ to $\mathbf{1}_{Y_0}$ and H^* from $g \circ f$ to $\mathbf{1}_{X_0}$. By assumption, the homotopy H moves any subcomplex Y_i within itself; without loss of generality, one may further assume that $H(y, t) = y$, for all $y \in Y_0$ and all $t \in [\frac{1}{2}, 1]$. Similar assumptions can be made with respect to H^*.

Let $\tilde{f} : X \to \tilde{Y}$ be a homotopy equivalence whose codomain $\tilde{Y}$ is a CW-complex. Choose a cellular approximation j to $\tilde{f} | X_0 \circ g$ and construct the mapping cylinder $Y = M(j)$; this is a CW-complex containing Y_0 as a CW-complex (see Section 2.3, Example 2). Let $i, \tilde{\imath}$ denote the inclusions of $Y_0, \tilde{Y}$ into Y, respectively. Since the inclusion $i_0 : X_0 \to X$ is a closed cofibration, there is a map $\bar{f} : X \to Y$ which is homotopic to $\tilde{\imath} \circ \tilde{f}$ and whose restriction to X_0 coincides with $i \circ f$; notice that $\bar{f}$ is also a homotopy equivalence. Moreover, $\bar{f}$ is indeed an $(n+1)$-ad homotopy equivalence of pairs (see Lemma A.5.10). □

Condition (ii) in the statement of this lemma is necessary. In fact, Example 4 in Section 5.1 exhibits a 1-ad $\{X, (0,0)\}$ whose components have the type of a CW-complex but which does not belong to TCW^1.

Consider an n-ad $\{X; X_0, \ldots, X_{n-1}\}$ and for each non-empty set $S \subset [n-1]$, define the space

$$X_S = \bigcap_{i \in S} X_i.$$

If $S = \varnothing$, define $X_\varnothing = X$; an n-ad map $f : Y \to X$ is a *complete weak homotopy equivalence* if the 2^n induced maps $f_S : Y_S \to X_S$ are weak homotopy equivalences.

Lemma 5.3.2 *A complete weak homotopy equivalence in TCW^n is a homotopy equivalence.*

Proof Let $f : Y \to X$ be a complete weak homotopy equivalence in TCW^n. Without loss of generality, assume that X and Y are CW-n-ads; furthermore, it is possible to suppose that f is cellular (see Corollary 2.4.12).

In order to show that f is a honest homotopy equivalence in TCW^n, one constructs, as in the proof of Whitehead's realizability theorem (see Theorem 2.5.1), a deformation retraction $M(f) \to Y$ in TCW^n. The hypothesis of complete weak homotopy equivalence on f assures that the deformation of a cell e belonging to several M_i's can be chosen to take place within the corresponding M_S. Therefore, the procedure yields an n-ad deformation retraction as desired. □

The next result gives a characterization of the category TCW^n. First, recall that the metric topology on a simplicial complex Y coincides with the trace of the product topology on Y, namely, the initial topology with respect to all projections $p_\lambda : Y \to I$, $\lambda \in \Lambda = Y_0$ (see Proposition 3.3.3).

Theorem 5.3.3 *Let X be a given n-ad. Then, the following statements are equivalent*:

(i) *X is dominated by a CW-n-ad*;
(ii) *X is an object of TCW^n*;
(iii) *X has the type of a simplicial n-ad*;
(iv) *X has the type of a simplicial n-ad with the strong topology*;
(v) *X has the type of an ELCX-n-ad.*

Proof (ii)⇒(i): This is an obvious consequence of the definitions.

(i)⇒(ii): Suppose X is dominated by a CW-n-ad Y; let $f : X \to Y$ and $g : Y \to X$ be n-ad maps such that $g \circ f \simeq \mathbf{1}_X$. Observe that the functor $|S-|$ can be viewed as a functor from the category of n-ads into TCW^n which preserves the n-ad homotopy relation (see Corollary 4.3.19). The naturality of the co-unit $j : |S-| \dot{\to} Id$ assures that the maps j_X and j_Y are n-ad maps; moreover, all the maps $j_Y, j_{Y,S}$, for all $S \subset [n-1]$, are homotopy equivalences (see Corollary 4.5.31 (i)). Thus, j_Y is a complete weak homotopy equivalence and the proof can be completed along the lines of the proof for Proposition 5.1.1.

(ii)⇒(iii): From the preceding argument, it follows that X has the type of a CW-n-ad whose global CW-complex is the geometric realization of a simplicial set; this CW-complex has a subdivision which is a simplicial complex (see Corollary 4.6.12) and contains the subspaces involved as subcomplexes (see Proposition 2.3.6).

(iii)⇒(ii): Trivial.

(iii)⇔(iv): see Proposition 3.3.7 and its proof.

(iv)⇒(v): see Corollary 3.3.9.

(v)⇒(i): Let $\{X; X_0, \ldots, X_{n-1}\}$ be an ELCX-n-ad. Let the ELCX-structure on X be given by the convex covering $\{V_\gamma : \gamma \in \Gamma\}$ of X and the equiconnecting homotopy $E : U \times I \to X$. The space X is paracompact as a metric space; hence, it is possible to choose a locally finite refinement for X, which in turn gives rise to a barycentric refinement $\{U_\lambda : \lambda \in \Lambda\}$ of $\{V_\alpha\}$ (see Proposition A.3.1).

One may assume that the covering $\{U_\lambda\}$ consists only of non-empty sets, and, moreover, has the following property: given a pair (λ, S) such that $\lambda \in \Lambda$ and $S \subset [n-1]$, then, $U_\lambda \cap X_i \neq \varnothing$, for all $i \in S$, implies that $U_\lambda \cap X_S \neq \varnothing$. (Otherwise, replace U_λ by the sets $U_{\lambda,S} = U_\lambda \setminus \cup \{X_j : j \notin S\}$ defined for all the subsets $S \subset [n-1]$ which are maximal with respect to the property $U_\lambda \cap X_S \neq \varnothing$.) Notice that these conditions mean that for every λ there is a *unique* maximal set $S_\lambda \subset [n-1]$ such that $U_\lambda \cap X_{S_\lambda} = \varnothing$.

Let Y denote the nerve of the covering $\{U_\lambda : \lambda \in \Lambda\}$. Notice that because $\{U_\lambda\}$ is a barycentric refinement of $\{V_\gamma\}$, for any simplex $\{\lambda_0, \ldots, \lambda_k\}$ of Y, there is a convex set V_γ such that $\bigcup_{j=0}^{k} U_{\lambda_j} \subset V_\gamma$. Now, for each $i \in [n-1]$, let Y_i be the subcomplex of Y consisting of the simplices $\{\lambda_0, \ldots, \lambda_r\}$ satisfying the condition

$$X_i \cap \left(\bigcap_{j=0}^{r} U_j \right) \neq \varnothing.$$

It will be shown that the CW-n-ad $\{Y; Y_0, \ldots, Y_{n-1}\}$ dominates the given n-ad X.

Using the paracompactness of the space X, once more choose a locally finite partition of unity $\{\mu_\lambda : \lambda \in \Lambda\}$ subordinated to the open covering $\{U_\lambda\}$. This partition of unity determines the canonical map $\psi : X \to Y$ (see Lemma 3.3.4 (ii)), which, according to the choice of the subcomplexes Y_i, is indeed an n-ad map. The objective is to construct an n-ad map $q : Y \to X$ such that $q \circ \psi \simeq \mathbf{1}_X$.

Turn the simplicial Y into an ordered simplicial complex by choosing a total order of the set Λ (see Section 3.3, Example 2). Define $q | Y^0 = \Lambda$ by selecting, for each $\lambda \in \Lambda$, a point $q(\lambda) \in U_\lambda \cap X_{S_\lambda}$. Given $s = (s_0, \ldots, s_k) \in \Delta_{\{\lambda_0, \ldots, \lambda_k\}}$, with $\lambda_0 < \lambda_1 < \cdots < \lambda_k$, define

$$q(s) = E(E(\cdots E(E(q(\lambda_0), q(\lambda_1), s_1), q(\lambda_2), s_2) \cdots), q(\lambda_k), s_k).$$

Since $\{\lambda_0, \ldots, \lambda_k\}$ is a simplex of Y, there is a convex set V_γ containing all the points $q(\lambda_0), \ldots, q(\lambda_k)$; the successive evaluations of the map E in the definition of $q(s)$ do not lead outside of V_γ, and therefore are always possible. Thus, $q(s)$ is really defined for all points of Y. If some $s_i = 0$, then the given formula reduces to the corresponding formula for $q((s_0, \ldots, s_{i-1}, s_{i+1}, \ldots, s_k))$, showing thereby that q is well defined. The restriction of q to a fixed simplex is a combination of continuous functions, and so is itself continuous; hence, q is continuous.

To see that, for each $1 \leqslant i \leqslant n-1$, q maps Y_i into X_i, consider a simplex

$\{\lambda_0,\ldots,\lambda_k\}$ of Y_i and a point $s\in\Delta_{\{\lambda_0,\ldots,\lambda_k\}}$; by definition,

$$X_i\cap\left(\bigcap_{j=0}^{k} U_{\lambda_j}\right)\neq\varnothing,$$

and so, $q(\lambda_j)\in X_i$, for every $j=0,\ldots,k$. Since X_i is an ELCX-subspace of X, the successive evaluations of E necessary to obtain $q(s)$ do not lead outside of X_i.

Finally, since $\{U_\lambda\}$ is a barycentric refinement of $\{V_\gamma\}$, the map $X\times I\to X\times X\times I, (x,t)\mapsto(x,q\circ\psi(x),t)$ takes values in $U\times I$, and thus can be composed with $E: U\times I\to X$, giving the desired homotopy. □

Function spaces will be focused next. Given the n-ads $\{X;X_0,\ldots,X_{n-1}\}$ and $\{C;C_0,\ldots,C_{n-1}\}$, the *function n-ad* associated to them is the n-ad

$$\{X^C;(X,X_0)^{(C,C_0)},\ldots,(X,X_{n-1})^{(C,C_{n-1})}\},$$

where $(X,X_i)^{(C,C_i)}$ is the subspace of all elements of X^C which take C_i into X_i.

Proposition 5.3.4 *Let $\{X;X_0,\ldots,X_{n-1}\}$ be an ELCX-n-ad and let $\{C;C_0,\ldots,C_{n-1}\}$ be an n-ad. If C is compact Hausdorff, then the function n-ad X^C is an ELCX-n-ad.*

Proof Recall that the compact-open topology on X^C coincides with the metric topology induced by the metric of X (see Appendix A.1), and thus X^C is a metric space.

Let the ELCX-structure on X be given by the convex covering $\{V_\gamma:\gamma\in\Gamma\}$ of X and the equiconnecting homotopy $E: U\times I\to X$.

Define

$$W=\{(f,g)\in X^C\times X^C:(f(x),g(x))\in U,\quad\text{for every } x\in C\};$$

as usual, U denotes the union of all the products $V_\gamma\times V_\gamma$. Under the canonical homeomorphism $X^C\times X^C\to(X\times X)^C$, the set W corresponds to the sub-basic open set $[C,U]$ in $(X\times X)^C$ (in the compact-open topology), and hence W is open in $X^C\times X^C$. Moreover, note that the diagonal $\Delta(X^C)$ is contained in W.

Now define the homotopy $E': W\times I\to X^C$ by

$$E'(f,g,t)(y)=E(f(y),g(y),t)$$

for every $y\in C$ and every $t\in I$. Its continuity follows from the exponential law. In fact, observe that $W\cong[C,U]\cong U^C$, and so, by evaluation, there is a map $ev: C\times W\times I\to U\times I$; but E' is just the adjoint of the composition $E\circ ev$. Clearly, E' is a homotopy as required in condition (2) of the definition of ELCX-spaces.

For every $x \in X$, choose a convex set $V_{\gamma(x)}$ containing x and an open neighbourhood W_x such that $\bar{W}_x \subset V_{\gamma(x)}$. Now, if $h \in X^C, h(C) \subset X$ is compact and is covered by finitely many open sets W_x, say, $W_{x_0}, \ldots, W_{x_k}$; then, the sets $D_i = h^{-1}(W_{x_i})$, $i = 0, \ldots, k$ cover C, are compact and such that $h(D_i) \subset V_{\gamma(x_i)}$. In this way, one obtains an open neighbourhood

$$Z_h = \bigcap_{i=0}^{k} [D_i, V_{\gamma(x_i)}]$$

for every $f \in X^C$, and therefore an open covering $\{Z_h : h \in X^C\}$ of X^C. Consider a pair $(f, g) \in Z_h \times Z_h$. If y is an arbitrary element of $D_i, i = 0, \ldots, k$, then $(f(y), g(y)) \in V_{\gamma(x_i)} \times V_{\gamma(x_i)} \subset U$; because the sets $D_0, \ldots, D_k$ cover C, it follows that $(f, g) \in W$. This conclusion holds true for all $(f, g) \in Z_h \times Z_h$, and hence $Z_f \times Z_f \subset W$. Also, for every pair $(f, g) \in Z_h \times Z_h$ and every $y \in D_i, i = 0, \ldots, k$,

$$E'(f, g, t)(y) = E(f(y), g(y), t) \in V_{\gamma(x_i)},$$

implying $E'(Z_h \times Z_h \times I) \subset Z_h$.

Hence, an ELCX-structure for X^C is obtained by taking the sets Z_h as convex sets and the evident restriction of E' as the equiconnecting homotopy.

It remains to prove that, for every $j = 0, \ldots, n-1$, the sets $(X, X_j)^{(C, C_j)}$ are ELCX-subspaces of X^C. First, observe that if $f, g \in (X, X_i)^{(C, C_i)}$ with $(f, g) \in Z_h \times Z_h$, then the definitions show that $E'(f, g, t) \in (X, X_i)^{(C, C_i)}$ for all $t \in I$. Second, show that $(X, X_j)^{(C, C_j)}$ is closed in X^C. To this end, for each $x \in X_j$, the set

$$H_x = \{f \in X^C : f(x) \in X_j\}$$

is closed because its complement in X^C is the sub-basic open set $[\{x\}, X \backslash X_j]$. Hence, $(X, X_i)^{(C, C_i)} = \bigcap_{x \in C_i} H_x$ is closed. □

Corollary 5.3.5 *If X is an ELCX-n-ad and C is an n-ad with compact global space, then $X^C \in TCW^n$.* □

Corollary 5.3.6 *If $X \in TCW^n$ and C is an n-ad with compact global space, then $X^C \in TCW^n$ and $X^{C,n} \in TCW$.*

Proof If $X \in TCW^n$, then X has the type of an ELCX-n-ad Y. The application of the functor $-^C$ to n-ad maps and n-ad homotopies describing a homotopy equivalence between X and Y yields n-ad maps and n-ad homotopies between X^C and $Y^C \in TCW^n$; these show that the function n-ads in question have the same type.

Finally, notice that these maps also induce homotopy equivalences between

$$X^{C,n} = \bigcap_{j=0}^{n-1} (X, X_j)^{(C, C_j)}$$

and

$$Y^{C,n} = \bigcap_{j=0}^{n-1} (Y, Y_j)^{(C, C_j)};$$

the latter space is an ELCX-space as the intersection of ELCX-subspaces of Y^C, and therefore it has the type of a CW-complex. □

In particular, this applies to loop spaces.

Corollary 5.3.7 *If (X, x_0) is a well-pointed based space and $X \in TCW$, then the loop space $(\Omega X, \omega_0)$ is well pointed, and $\Omega X \in TCW$.*

Proof Since (X, x_0) is well pointed, its loop space $(\Omega X, \omega_0)$ is well pointed (see Corollary A.4.4). Moreover, if $X \in TCW$, it follows that $\{X; \{x_0\}\} \in TCW^2$. Now take the 2-ad $\{S^1, \{e_0\}\}$ and note that the corresponding function 2-ad has the form

$$\{X^{S^1}; \Omega X\};$$

this shows that $\Omega X \in TCW$. □

Under special conditions on X, the preceding result has a converse, as follows:

Proposition 5.3.8 *Let X be a path-connected space with a covering which admits a subordinated locally finite partition of unity such that the inclusions of the members of the covering into X are homotopic to constant maps. Then, if for some base point x_0, the loop space ΩX has the type of a CW-complex, so does X.*

Proof Let $f : Y \to X$ be a weak homotopy equivalence whose domain Y is a CW-complex (see Corollary 2.7.8) and choose a base point $y_0 \in Y$ such that $f(y_0) = x_0$. The induced map Ωf is also a weak homotopy equivalence. But $\Omega Y \in TCW$ (see Corollary 5.3.7) and $\Omega X \in TCW$, by hypothesis; then, Ωf is a homotopy equivalence (see Theorem 2.5.1). This implies that f itself is a homotopy equivalence (see Proposition A.4.24), and thus $X \in TCW$. □

5.4 Spaces of the type of CW-complexes and fibrations

Let $p : Y \to X$ be a fibration with X path-connected and let F be any fibre; recall that all the fibres of p have the same type (see Corollary A.4.21). In this section, the following question is analysed: given that two of the three spaces X, Y, F have the type of a CW-complex, does the third space have the type of a CW-complex?

Proposition 5.4.1 *Let $p : Y \to X$ be a fibration with X path-connected and $X, Y \in TCW$. Then, any fibre $F \in TCW$.*

Proof Let x be a non-degenerate base point of X. If X has no non-degenerate base points, it can be modified as follows. Select a point $x \in X$ and construct the mapping cylinder $M(i)$ of the inclusion $i : \{x\} \to X$; then, $(x, 1)$ is a non-degenerate base point for $M(i)$ and the retraction $r_i : M(i) \to X$ is a homotopy equivalence, and so $M(i) \in TCW$. The induced fibration $\bar{p} : M(i) \sqcap Y \to M(i)$ has fibre $F = \bar{p}^{-1}(x, 1) = p^{-1}(x)$ and the total space of the same type of Y, i.e., the type of a CW-complex (see Corollary A.4.20). Thus, one can always assume the existence of non-degenerate base points.

Consider F to be the fibre of p over x. Since PX is contractible, the map $h_x : \{x\} \to PX$ which takes x into the constant path ω at x is a homotopy equivalence. Take the evaluation map $v_1 : PX \to X, \sigma \mapsto \sigma(1)$ and form the commutative diagram

$$\begin{array}{ccccc} Y & \xrightarrow{p} & X & \xleftarrow{i} & \{x\} \\ \downarrow{\scriptstyle 1_Y} & & \downarrow{\scriptstyle 1_X} & & \downarrow{\scriptstyle h_x} \\ Y & \xrightarrow[p]{} & X & \xleftarrow[v_1]{} & PX \end{array}$$

which shows that $\{x\} \sqcap_i Y = F$ and $PX \sqcap|_{v_1} Y$ have the same type (see the cogluing theorem, theorem A.4.19).

Decompose the map $p : Y \to X$ via the mapping cylinder, to obtain $p = r_p \circ i_Y$, where $r_p : M(p) \to X$ can be considered as a based homotopy equivalence (see Proposition A.4.10 (iv)). Form the commutative diagram

$$\begin{array}{ccccc} PM(p) & \xrightarrow{v_1} & M(p) & \xleftarrow{i_Y} & Y \\ \downarrow{\scriptstyle r_p^I} & & \downarrow{\scriptstyle r_p} & & \downarrow{\scriptstyle 1_Y} \\ PX & \xrightarrow[v_1]{} & X & \xleftarrow[p]{} & Y \end{array}$$

where $r_p{}^I$ is given by $t_p{}^I(\sigma) = r_p \circ \sigma$ is a homotopy equivalence and v_1 is a fibration (in both cases; see Section A.4, Example 6). It proves that $Y \sqcap_p PX$ and $Y \sqcap_{i_Y} PM(p)$ have the same type (see again the Cogluing Theorem A.4.19). Because $Y \sqcap_p PX = PX \sqcap_{v_1} Y$, it follows that $F \cong Y \sqcap_{i_Y} PM(p)$. But $Y \sqcap_{i_Y} PM(p) \in TCW$: to see this, form the 2-ad

$\{M(p); Y, \{x\}\}$, which belongs to TCW^2; next, form the 2-ad $\{I;\{1\},\{0\}\}$ and identify $Y \sqcap_{i_Y} PM(p)$ with the 2-ad function space $M(p)^{I,2} \in TCW$ (see Corollary 5.3.6).

Theorem 5.4.2 *Let $p : Y \to X$ be a fibration with X path-connected and such that X and $F = p^{-1}(x)$ have the type of CW-complexes, for any $x \in X$. Then, Y has the type of a CW-complex.*

Proof Let $f : Z \to Y$ be a weak homotopy equivalence with Z a CW-complex (see Corollary 2.7.8). Consider a factorization $f = p' \circ u$, where $p' : T(f) \to Y$ is a fibration and $u : Z \to T(f)$ is a homotopy equivalence (see Proposition A.4.18). Notice that $T(f) \in TCW$. Because $p \circ p' : T(f) \to X$ is a fibration with $X, T(f) \in TCW$, the fibre $F' = (p \circ p')^{-1}(x) \in TCW$ (see Proposition 5.4.1). Since u is a homotopy equivalence and f is a weak homotopy equivalence, it follows that p' is a weak homotopy equivalence. Construct the diagram formed by the homotopy sequences of the fibrations $p : Y \to X, p \circ p' : T(f) \to X$ (see Proposition A.8.17) and the morphisms induced by the maps $p'|F', p'$ and $\mathbf{1}_X$; then, it follows from the five lemma that $p'|F'$ is also a weak homotopy equivalence, and therefore is a homotopy equivalence (see Theorem 2.5.1). Finally, p' is a homotopy equivalence (apply Proposition 5.1.2 to X and Theorem A.4.23 to the fibrations p and $p \circ p'$). □

If the total space and the fibres of a fibration have the type of a CW-complex, the base space does not necessarily belong to TCW (see Exercise 2). However:

Proposition 5.4.3 *Let $p : (Y, y_0) \to (X, x_0)$ be a based fibration such that X is path-connected, (X, x_0) and (Y, y_0) are well pointed and both $Y, F = p^{-1}(x_0)$ have the type of CW-complexes. Then, the loop space ΩX has the type of a CW-complex.*

In particular, if X has a covering which admits a subordinated locally finite partition of unity and such that the inclusion maps of the members of the covering into X are homotopic to constant maps, then X itself has the type of a CW-complex.

Proof Let $\bar{v}_X : Z \to PY = (Y; \{y_0\})^{(I;\{0\})}$ be the fibration induced from $v_X : PX \to X$ by $p \circ v_Y$ (see Section A.4 for the definition of these maps). The map $p \circ v_Y$ is itself a fibration; its fibre is the 2-ad function space $Y^{I,2}$ of all 2-ad maps $\{I; \{1\}, \{0\}\} \to \{Y; F, \{y_0\}\}$. Because (X, x_0) is well pointed, the inclusion of F into Y is a closed cofibration (see Proposition A.4.17);

moreover, using the fact that (Y, y_0) is well pointed, it follows that $\{Y; F, \{y_0\}\} \in TCW^2$ (see Lemma 5.3.1). Consequently, $Y^{I,2}$ has the type of a CW-complex (see Corollary 5.3.6). But $Y^{I,2}$ is also the fibre of the induced fibration $Z \to PX$; therefore, $Z \in TCW$ (see Corollary A.4.21 and Theorem 5.4.2). On the other hand, the fibre of $\bar{v}_X$ is just ΩX, from which one concludes that $\Omega X \in TCW$ (see Proposition 5.4.1).

If X satisfies the extra hypothesis described in the statement of the proposition, then $X \in TCW$ (see Proposition 5.3.8). □

Exercises

1. Show that the *Polish circle*

$$P = \{(x, y) : x = 0, -2 \leqslant y \leqslant 1\} \cup \{(x, y) : 0 \leqslant x \leqslant 1, y = -2\}$$
$$\cup \{(x, y) : x = 1, -2 \leqslant y \leqslant 0\} \cup \{(x, y) : 0 < x \leqslant 1, y = \sin \frac{2\pi}{x}\} \subset \mathbf{R}^2$$

   does not have the type of a CW-complex. (*Hint*: Show that all homotopy groups of the polish circle vanish, but a map into a singleton space is not a homotopy equivalence.)
2. Show that there are continuous bijections $[0, 1) \to P$ – where P is the Polish circle – and that all these maps are fibrations (this proves the existence of fibrations whose total space and fibre are CW-complexes and whose base does not even have the type of a CW-complex).
3. Let $X_1 \xleftarrow[f_1]{} X_2 \xleftarrow[f_2]{} \cdots$ be a sequence of fibrations where $X_n \in TCW$ for all $n \in \mathbf{N}$, and for any $m \in \mathbf{N}$, all but finitely many fibrations f_n have m-connected fibres. Let X be the inverse limit of the sequence. Show that $X \in TCW$ iff all but finitely many of the f_n are homotopy equivalences. (Dydak & Geoghegan 1986).

Notes to Chapter 5

Since the introduction of CW-complexes almost forty years ago, various questions have been raised concerning their relation to homotopy theory; in a sense, CW-complexes are easily manipulated and their simplicity is often reflected in topological invariants that can be described algebraically. Thus, the question immediately arose of what kinds of spaces could be represented, up to homotopy type, by CW-complexes; this led to Milnor's well-known paper (Milnor, 1959).

The first three sections of Chapter 5 examine closely most of the results contained in Milnor's paper. However, some of these theorems precede Milnor's work: Proposition 5.1.1 is Theorem 23 of Whitehead (1950); in Theorem 5.2.3, the equivalence of (i), (ii) and (iv) are also due to J.H.C. Whitehead (in particular, the implication (i)⇒(iv) is Theorem 24 of Whitehead (1950)), while the implications (iv)⇒(v)⇒(i) are in Hanner (1951). The notation of *ECLX*-space and its generalization to *ECLX*-*n*-ad are due to John Minor.

The development of Section 5.4 is influenced by the paper of Rolf Schön (Schön, 1977). The first attempt to prove Theorem 5.4.2 was made by James Stasheff (see Stasheff, 1963, Proposition (0)). For the case in which $p : Y \to X$ is a covering projection, the result was already known to J.H.C. Whitehead (see Whitehead, 1949a); in this text, the corresponding result is Proposition 2.3.9.

Appendix

This appendix is intended to give the reader an easy access to all the definitions and results which are needed but are not an intrinsic part of the theory in the main body of the book. The definitions are sometimes presented in a systematic way, within a specific context; as a result, it is possible that some concept is used before it has been fully described. The reader is invited to make full use of the index while perusing the results presented here. Precise references or proofs for the results stated are given.

A.1 Weak Hausdorff *k*-spaces

Because of its simplicity, the concept of topological space is an appropriate basis for a number of mathematical disciplines. Nevertheless, topological spaces have some disadvantages when presented in their full generality; in fact, some important constructions have certain useful properties only for restricted classes of topological spaces. For the sake of exposition, *Top* will denote – exceptionally in this section – the category of *all* topological spaces and maps.

The main problem arises from the fact that for an arbitrary space Z the Cartesian product functor $-\times_c Z : Top \to Top$, which associates to a space X the Cartesian product $X \times_c Z$ (endowed with the product topology), and to a map $f : Y \to X$ the product map $f \times \mathbf{1} : Y \times_c Z \to X \times_c Z$, does not preserve pushouts. Even worse, the product functors $-\times_c Z$ do not preserve identification maps, i.e., if $f : Y \to X$ is an identification map then $f \times \mathbf{1}$ may fail to be an identification map.

Another disadvantage concerns mapping spaces. Given spaces Y and X, in general it is not possible to endow the set $C(Y, X)$ of all maps $Y \to X$ with a topology such that the evaluation map

$$\varepsilon : C(Y, X) \times_c Y \to X, \qquad \varepsilon(f, y) = f(y)$$

is continuous. The best approximation to such a topology is the *compact-open topology*, which has as sub-basis for the set of open sets, all sets

$$[K, U] = \{f \in C(Y, X) : f(K) \subset U\}$$

with K a compact subspace of Y and U an open subset of X. The space of continuous functions $Y \to X$ with the compact-open topology is denoted by $C_o(Y, X)$. If Y is compact Hausdorff and X is metric, the compact-open topology for $C(Y, X)$ coincides with the metric topology determined by the metric

$$d'(f, g) = \max_{y \in Y} d(f(y), g(y)),$$

for every $f, g \in C(Y, X)$ and where d is the metric for X (see Dugundji, 1966, Chapter XII, Section 8, Theorem 8.2 (3)).

Several attempts have been made to single out a 'convenient' class of topological spaces, for which these difficulties do not arise. For the purposes of this book, the so-called 'weak Hausdorff k-spaces' seem to be the most appropriate.

k-spaces

Let X be a space. A subset $A \subset X$ is said to be *compactly closed* if, for every compact Hausdorff space K and every map $f : K \to X$, $f^{-1}(A)$ is closed in K, the space X is said to be a *k-space* whenever all its compactly closed subsets are closed. For Hausdorff spaces, this concept was probably first described in writing in Gale (1950) under the name *compactly generated Hausdorff space* and for the development of the general case see Brown (1988, Notes to Chapter 5).

The property of being a k-space is preserved by closed subspaces and identifcations; more precisely, if X is a k-space and

(1) if $A \subset X$ is a closed subspace of X, then A is also a k-space;
(2) if $p : X \to X'$ is an identification map, then X' is also a k-space.

The locally compact Hausdorff spaces and the spaces satisfying the first axiom of countability are k-spaces (see Brown, 1988, 5.9.2). Thus, in particular, Euclidean spaces $\mathbf{R}^n$, balls B^n, spheres S^n, and, more generally, all metric spaces are k-spaces. As non-locally finite CW-complexes, B^∞, S^∞ and $\mathbf{F}P^\infty$, $\mathbf{F} = \mathbf{R}, \mathbf{C}$ or $\mathbf{H}$ are examples of k-spaces which are not locally compact. Moreover, they do not satisfy the first axiom of countability.

For an arbitrary space X, let kX denote its *k-ification*, that is to say, the k-space having the same underlying set as X, but with the topology given by taking as closed sets the compactly closed sets with respect to the topology of X. Note that $k(kX) = k(X)$ for every space X; moreover, since kX has a finer topology than X, the identity function is a map $kX \to X$.

If Y is a k-space, a function $f : Y \to X$ is continuous iff $f : Y \to kX$ is continuous. Therefore, if $f : Y \to X$ is a map, the same function is a map $kY \to kX$. This permits to extend the k-ification to a functor $k : Top \to Top$ by taking $kf = f$ on maps.

The image $k(Top)$ of the functor k is a full subcategory of Top, satisfying the following properties.

(1) The inclusion functor $k(Top) \to Top$ is left-adjoint to the functor k.

(2) The category $k(Top)$ is both complete and cocomplete. In particular, if X and Y are k-spaces, then their product is given by

$$X \times Y = k(X \times_c Y).$$

At this point, note that if at least one of the spaces X or Y is locally compact Hausdorff, or if both spaces are metric, then $X \times Y = X \times_c Y$.

(3) The category $k(Top)$ has mapping spaces satisfying the exponential law. More precisely, for any k-spaces X and Y, let $X^Y = k(C_o(X, Y))$; then, if X, Y and Z are k-spaces, the following exponential law holds true:

$$(X^Y)^Z = X^{Y \times Z}.$$

The exponential law implies that the functors

$$-^Y, - \times Y : k(Top) \to k(Top)$$

are adjoint to each other. Hence, $-^Y$ preserves all limits and $- \times Y$ preserves all colimits. In particular, the functor $-^Y$ preserves embeddings and the functor $- \times Y$ preserves identification maps; both functors preserve homotopies and homotopy equivalences. Notice that the functor $- \times Y$ does not normally preserve subspaces.

The category of k-spaces is larger than the category of compactly generated Hausdorff spaces. The latter category has the advantage that a subset $A \subset X$ is already closed if $A \cap C$ is closed in C for every compact Hausdorff subspace $C \subset X$. This, on the one hand, avoids the nuisance of taking maps $a : K \to X$ and studying $a^{-1}(A)$ away from X. On the other hand, although the category of compactly generated Hausdorff spaces has quotients, since it is cocomplete, these are far from the usual quotients because of the required separation axiom (see Example 3 below). In this text, a middle course is elected, one which gives a subcategory of $k(Top)$ larger than the category of compactly generated Hausdorff spaces, and shares with it the advantage mentioned before, but which also has the usual quotients.

Weak Hausdorff k-spaces

A k-space X is said to be *weak Hausdorff* if the diagonal Δ_X is closed in $X \times X$. Recall that a topological space X is Hausdorff iff the diagonal Δ_X is closed in the Cartesian product $X \times_c X$; thus, a Hausdorff k-space is weak Hausdorff. The weak Hausdorff k-spaces generate a full subcategory of $k(Top)$, denoted by $wHk(Top)$ in this section (but simply by Top in the remainder of this book).

The category $wHk(Top)$ is closed under the formation of closed subspaces and finite products, these taken in $k(Top)$; furthermore, if Y is a weak Hausdorff k-space and $f : Y \to X$ is an identification map, then the k-space X is weak Hausdorff iff $(f \times f)^{-1} \Delta_X$ is closed in $Y \times Y$. Moreover, the functor $- \times Y$ restricted to $wHk(Top)$ preserves subspaces.

There is another useful characterization of the weak Hausdorff property for k-spaces.

Lemma A.1.1 *If X is a k-space, then X is weak Hausdorff iff for every map $a : K \to X$, with K compact Hausdorff, $a(K)$ is closed and compact Hausdorff.*

Proof See McCord (1969, Lemma 2.1 and Proposition 2.3). □

This implies, in particular, that weak Hausdorff k-spaces have the separation property T_1; it further implies the desired property that in a weak Hausdorff k-space X, a subset A is closed if $A \cap C$ is closed in C for every compact Hausdorff space $C \subset X$.

The concept of relative homeomorphism is helpful in explaining why the category $wHk(Top)$ is closed under the formation of quotients. A map of pairs $\bar{f} : (Y, D) \to (X, A)$ is called a *relative homeomorphism* if D is closed in Y, and the map $\bar{f} : Y \to X$ is an identification and induces a homeomorphism $Y \backslash D \to X \backslash A$. Note that, except for the fact that the homotopy extension property is not required for the pair (Y, D), the space X can be viewed as the adjunction space obtained by attaching Y to A via f.

Example 1 Let D be a closed subspace of a topological space Y. Form the quotient Y/D and let $*$ be the point of Y/D to which D is identified. Finally, let $\bar{f} : Y \to Y/D$ be the quotient map. Then, $\bar{f} : (Y, D) \to (Y/D, \{*\})$ is a relative homeomorphism. □

Example 2 Let $f : Y -/\to A$ be a partial map with domain D and take $X = A \bigsqcup_f Y$; then, the projection $A \sqcup Y \to X$ can be viewed as a relative homeomorphism

$$(A \sqcup Y, A \sqcup D) \longrightarrow (X, A).$$ □

Proposition A.1.2 *Let Y and A be weak Hausdorff k-spaces and let $\bar{f} : (Y, D) \to (X, A)$ be a relative homeomorphism. Then, X is a weak Hausdorff k-space.*

Proof See McCord (1969, Proposition 2.5). □

This and Example 1 prove that the category $wHk(Top)$ has the usual quotients; more precisely:

Corollary A.1.3 *Let X be a weak Hausdorff k-space and let A be a closed subspace of X. Then X/A is a weak Hausdorff k-space.* □

In view of Example 2, Proposition A.1.2 also has the following consequences:

Corollary A.1.4 *Let $f : Y -/\rightarrow A$ be a partial map with Y and A weak Hausdorff k-spaces. Then, the adjunction space $A \bigsqcup_f Y$ is a weak Hausdorff k-space.* □

The next example shows that the quotient of a Hausdorff k-space by a closed subspace is not necessarily Hausdorff. This should convince the reader of the relevance of extending the notion of 'Hausdorff k-space' to that of 'weak Hausdorff k-space'.

Example 3 (*The Tychonoff plank*; see Kelley (1955), Problem F, Chapter 4.) Let Ω' be the set of all ordinal numbers not greater than the first uncountable ordinal number Ω, and let ω' be the set of ordinals not greater than the first infinite ordinal ω. Endow both sets Ω' and ω' with the order topology; with this topology, both Ω' and ω' become compact Hausdorff spaces. Hence, $\Omega' \times \omega'$ is compact Hausdorff, and thus normal. The space

$$X = \Omega' \times \omega' \backslash (\Omega, \omega)$$

is locally compact, regular, but not normal. In particular, X is a k-space, because it is locally compact and Hausdorff. Now, let A and B be disjoint closed subsets of X which do not have disjoint neighbourhoods. The regularity of X implies that the quotient space $X' = X/A$ is Hausdorff; moreover, X' is a k-space, but it is no longer regular. Let B' denote the subspace of X' which corresponds to B. Then, the quotient space X'/B' is a k-space which is not Hausdorff. □

Finally, the category $wHk(Top)$ is closed under the mapping space construction which is essential to the development of Chapter 5.

Proposition A.1.5 *Let X and Y be k-spaces; if X is a weak Hausdorff, then X^Y is a weak Hausdorff k-space.*

Proof Use the characterization of the weak Hausdorff property given in

Lemma A.1.1. Let $a : K \to X^Y$ be a map, with K compact Hausdorff; let $\tilde{a} : K \times Y \to X$ denote the adjoint of a. One must show that $X^Y \backslash a(K)$ is open. Let f be an arbitrary element of $X^Y \backslash a(K)$ and note that, for every $y \in Y$, the set

$$K_y = \{k \in K : \tilde{a}(k, y) \neq f(y)\}$$

is open in K. Since $f \notin a(K)$, for each $k \in K$, choose an element $y(k) \in Y$ such that $k \in K_{y(k)}$. The normality of K implies that, for each $k \in K$, there is an open set L_k which contains k and whose closure $\bar{L}_k$ is contained in $K_{y(k)}$. Clearly, the set L_k form an open covering of K; since the latter space is compact, it can be covered by finitely many sets L_k, say by $L_{k_1}, \ldots, L_{k_n}$. Then, because X is weak Hausdorff, for every $i = 1, \ldots, n$, the set $\tilde{a}(\bar{L}_k, \{y(k_i)\})$ is closed (see Lemma A.1.1) and hence,

$$U = \cap[\{y(k_i)\}, X \backslash \tilde{a}(\bar{L}_k, \{y(k_i)\})]$$

is a neighbourhood of f in X^Y which does not meet $a(K)$. □

A.2 Topologies determined by families of subspaces

The topology of a space X – or in short, a space X – is *determined by a family of subspaces* U_λ if a set U is open (closed) in X iff $U \cap U_\lambda$ is open (closed) in U_λ, for every λ; this is equivalent to require that a function $f : X \to Z$, where Z is any space, is continuous iff $f | U_\lambda$ is continuous for every λ.

The following simple result is very useful in dealing with this concept.

Proposition A.2.1 *Let* $\{U_\lambda : \lambda \in \Lambda\}$ *and* $\{V_\gamma : \gamma \in \Gamma\}$ *be coverings of a space* X *such that* $\{V_\gamma\}$ *is a refinement of* $\{U_\lambda\}$. *Then, if* X *is determined by the family* $\{V_\gamma\}$, *it is also determined by the family* $\{U_\lambda\}$.

Proof Suppose that X is determined by $\{V_\gamma\}$; let $f : X \to Z$ be a function from X to a space Z such that $f | U_\lambda$ is continuous, for every λ. Take arbitrarily an element V_γ of the family $\{V_\gamma\}$; since this family is a refinement of $\{U_\lambda\}$, there is an index λ such that $V_\gamma \subset U_\lambda$. Then, $f | V_\gamma = (f | U_\lambda) | V_\gamma$ is continuous, and therefore f is continuous. □

A space X has the *final topology* with respect to a family of maps $f : Y_\lambda \to X$ if a function $f : X \to Z$, where Z is any space, is continuous iff the compositions $f \circ f_\lambda : Y_\lambda \to Z$ are continuous for all λ; again, an equivalent formulation is to say that U is open (closed) in X iff $f_\lambda^{-1}(U)$ is open (closed) in Y_λ, for every λ.

Proposition A.2.2 *Let the space* X *have the final topology with respect to* *the family* $\{f_\lambda : Y_\lambda \to X : \lambda \in \Lambda\}$ *of maps and let* $p : Z \to X$ *be a map. Then, the space* Z *has the final topology with respect to the family* $\{\bar{f}_\lambda : \bar{Y}_\lambda \to Z : \lambda \in \Lambda\}$, *where* $\bar{f}_\lambda$ *is induced from* f_λ *by* p.

Proof See Gabriel & Zisman (1967, Section III.2). The functor $Z \times -$ is compatible with colimits (see Section A.1); thus, the space $Z \times X$ has the final topology with respect to the family $\{\mathbf{1}_Z \times f_\lambda\}$.

Notice that all spaces $\bar{Y}_\lambda$ can be considered as closed subspaces of $Z \times Y_\lambda$. Consider a subset $C \subset Z$ such that $\bar{f}_\lambda^{-1}(C)$ is closed in $\bar{Y}_\lambda$, for each λ. Since the domain of any map is homeomorphic to its graph, the space Z is embedded as a subspace in $Z \times X$, via the assignment $z \mapsto (z, p(z))$; thus, C can be considered as a subset of $Z \times X$. The sets $\bar{f}_\lambda^{-1}(C) \cong (\mathbf{1}_Z \times f_\lambda)^{-1}(C)$ are closed in the respective spaces $Z \times Y_\lambda$; consequently, C itself is closed in $Z \times X$ and also in Z. □

Corollary A.2.3 *Let* X *be a space determined by the family of closed subspaces* $\{U_\lambda : \lambda \in \Lambda\}$ *and let* $p : Z \to X$ *be a map. Then,* Z *is determined by the family* $\{p^{-1}(U_\lambda) : \lambda \in \Lambda\}$.

Proof Since each U_λ is closed in X, the inclusion $p^{-1}(U_\lambda) \subset Z$ is induced from the inclusion $U_\lambda \subset X$ by p. □

Let a set X, a family of spaces $\{Y_\lambda : \lambda \in \Lambda\}$ and a family of functions $\{f_\lambda : Y_\lambda \to X\}$ be given and such that $(f_\lambda \times f_{\lambda'})^{-1}(\Delta_X)$ is closed in $Y_\lambda \times Y_{\lambda'}$, for each pair λ, λ' of indices. Then, there is a unique topology τ for X such that all functions f_λ become continuous and (X, τ) has the final topology with respect to the family $\{f_\lambda\}$; the condition ensures the weak Hausdorff property for (X, τ) (see McCord (1969, Proposition 2.4)). The process of forming the space (X, τ) from the given data is referred to as *providing* or *endowing X with the final topology with respect to the family* $\{f_\lambda\}$. Under some additional hypothesis, X is determined by a family of subspaces:

Lemma A.2.4 *Let the space* X *be endowed with the final topology with respect to the family* $\{f_\lambda\}$. *Furthermore, assume that*

(i) *each* f_λ *is injective;*

(ii) $f_\lambda^{-1}(f_{\lambda'}(Y_{\lambda'}))$ *is closed in* Y_λ, *for all pairs* λ, λ' *of indices, and*

(iii) *the subspaces* $f_\lambda^{-1}(f_{\lambda'}(Y_{\lambda'}))$ *and* $f_{\lambda'}^{-1}(f_\lambda(Y_\lambda))$ *of* Y_λ *and* $Y_{\lambda'}$, *respectively, are homeomorphic, for each pair* λ, λ' *of indices.*

Then, for all $\lambda \in \Lambda$, f_λ *embeds* Y_λ *as a closed subspace into* X, *and consequently,* X *is determined by the family* $\{Y_\lambda\}$ *of subspaces.*

Proof See Dugundji (1966, Proposition VI.8.2). □

The concept of initial topology is dual to that of final topology; to wit, a space Y has the *initial topology* with respect to a family of maps $f_\lambda : Y \to X_\lambda$ if a function $f : Z \to Y$, where Z is any space, is continuous iff the compositions $f_\lambda \circ f : Z \to X_\lambda$ are continuous for all λ.

The following concept is stronger than that of determination of a space by a family of subspaces. A space X is said to be *topologically dominated by a family* $\{C_\lambda : \lambda \in \Lambda\}$ of closed subspaces if a set $C \subset X$ is closed in X iff there is a subset $\Lambda' \subset \Lambda$ such that:

(1) $\{C_\lambda : \lambda \in \Lambda'\}$ covers C and
(2) $C \cap C_\lambda$ is closed in C_λ, for every $\lambda \in \Lambda'$.

Clearly, if a space X is topologically dominated by the family $\{C_\lambda : \lambda \in \Lambda\}$ of closed subspaces then X is determined by such a family; the converse does not hold true (see Section 1.5, Exercises).

Theorem A.2.5 *A space topologically dominated by a family of paracompact closed subspaces is paracompact.*

Proof See Morita (1954) and Michael (1956). □

A.3 Coverings

A family $\{U_\lambda : \lambda \in \Lambda\}$ of subsets of a space X is said to be

countable if Λ is countable;

locally finite, if each point $x \in X$ has a neighbourhood which meets only finitely many U_λ;

star-finite if, for each $\lambda \in \Lambda$, $U_\lambda \cap U_\mu \neq \varnothing$ for only a finite number of $\mu \in \Lambda$;

a covering of a subset $A \subset X$ if $A \subset \bigcup_{\lambda \in \Lambda} U_\lambda$ (in this case, one also says that $\{U_\lambda\}$ *covers* A);

an open (closed) covering of X if it is a covering of X and all sets U_λ are open (closed);

a refinement of the covering $\{V_\gamma : \gamma \in \Gamma\}$ if, for every $\lambda \in \Lambda$, there is a $\gamma \in \Gamma$ with $U_\lambda \subset V_\gamma$;

a barycentric refinement of the covering $\{V_\gamma : \gamma \in \Gamma\}$ if it is a covering of X and if the family $\{\tilde{U}_x : x \in X\}$ is a refinement of $\{V_\gamma\}$, where

$$\tilde{U}_x = \bigcup_{x \in U_\lambda} U_\lambda.$$

Clearly, a star-finite covering is locally finite.

Proposition A.3.1 *Every locally finite open covering of a normal space has a barycentric refinement.*

Proof See Dugundji (1966, Chapter VIII, Theorem 3.2). □

The space X is *paracompact* if it is Hausdorff and if every open covering of X has an open, locally finite refinement. Every paracompact space is normal; every closed subspace of a paracompact space is paracompact.

Every metric space is paracompact. A sharper statement holds true for metric spaces satisfying the second axiom of countability:

Theorem A.3.2 *Every open covering of a metric space satisfying the second axiom of countability has a countable, star-finite refinement.*

Proof See Kaplan (1947, Theorem 1). □

A family $\{\mu_\lambda : \lambda \in \Lambda\}$ of maps $\mu_\lambda : X \to I$ is a (*point finite*) *partition of unity* on X, if, for each point $x \in X$, $\mu_\lambda(x) = 0$ for all but a finite number of indices λ, and, furthermore, if, for all $x \in X$,

$$\sum_\lambda \mu_\lambda(x) = 1.$$

(The sum exists since it contains only finitely many nonvanishing summands.)

The set of all partitions of unity for X has an interesting binary operation. To wit, the *product* of the partitions of unity $\{\mu_\lambda\}$ and $\{v_\gamma\}$ is the partition of unity $\{\mu_\lambda \cdot v_\gamma\}$ defined by all possible multiplications $\{\mu_\lambda \cdot v_\gamma\}$, and indexed by the set of all pairs (λ, γ).

The partitions of unity have a strong connection to the open coverings of the space X: if $\{\mu_\lambda\}$ is a partition of unity for X, then the family $\{\mu_\lambda^{-1}((0, 1])\}$ is an open covering of X. A partition of unity $\{\mu_\lambda\}$ is *locally finite* if the covering $\{\mu_\lambda^{-1}((0, 1])\}$ is locally finite. A partition of unity $\{\mu_\lambda\}$ is said to be *subordinated to* (or *dominated by*) a given open covering $\{U_\lambda\}$ of X, if, for every λ, the closure of $\{\mu_\lambda^{-1}((0, 1])\}$ is contained in U_λ. A covering $\{U_\lambda : \lambda \in \Lambda\}$ of a space X is said to be *numerable* if it admits a subordinated locally finite partition of unity.

The following result is true (see Dugundji (1966, Chapter VIII, Theorem 4.2)):

Theorem A.3.3 *If X is a paracompact space, and if $\{U_\lambda\}$ is an open covering of X, then there exists a localy finite partition of unity for X which is subordinated to the covering $\{U_\lambda\}$.* □

CW-complexes, as paracompact spaces, satisfy Theorem A.3.3.

There is an interesting property of spaces which implies paracompactness. Let X be a space with a topology τ (i.e., τ is the set of all open sets of X). A *stratification* of X is a function

$$\tau \mapsto \tau^{\mathbf{N}}, U \to \{U_n : n \in \mathbf{N}\}$$

such that:

(1) $\{U_n : n \in \mathbf{N}\}$ covers U; and,
for every $n \in \mathbf{N}$,
(2) the closure $\bar{U}_n$ of U_n is contained in U; and
(3) $U \subset V$, $U, V \in \tau$ implies that $U_n \subset V_n$.

The space X is called *stratifiable* if there is a stratification for X.

Proposition A.3.4 *A stratifiable space is paracompact and normal.*

Proof See Ceder (1961). □

A.4 Cofibrations and fibrations; pushouts and pullbacks; adjunction spaces

This section reviews some important concepts and results of homotopy theory. Recall that in this book the category of weak Hausdorff k-spaces is denoted by *Top* (except in Section A.1, where it is denoted by *wHk* (*Top*)).

Cofibrations

A pair (X, A) of spaces, i.e., a **1**-ad in the terminology of Section 5.3, has the *homotopy extension property* if for any commutative diagram (full arrows)

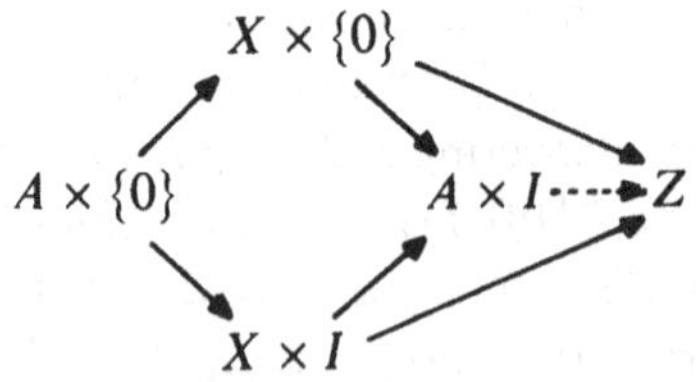

there is a map $X \times I \to Z$ (dotted arrow) which makes the resulting triangles commutative.

Proposition A.4.1 *Let A be a closed subset of a space X. The following conditions are equivalent:*

(i) *the pair (X, A) has the homotopy extension property;*
(ii) *the space $\hat{X} = X \times \{0\} \cup A \times I$ is a retract of $X \times I$;*
(iii) *$\hat{X}$ is a strong deformation retract of $X \times I$;*

(iv) *there are a neighbourhood U of A in X which is deformable rel. A to A in X and a map $\alpha : X \to I$ such that $\alpha^{-1}(0) = A$ and $\alpha|(X \setminus U) = \mathbf{1}$.*

Proof The equivalence of (i), (ii) and (iii) follows from Brown (1968, Propositions 7.2.2 and 7.2.3). The proof of the equivalence of (i) and (iv) can be found in Strøm (1966, Theorem 2). □

Rather than just dealing with pairs (X, A), one expands the scope of Proposition A.4.1 by taking embeddings; more precisely, an embedding $i : Y \to X$ is a *closed cofibration* if $i(Y)$ is closed in X and the pair $(X, i(Y))$ satisfies any of the equivalent conditions of Proposition A.4.1.

Example 1 The map $r^{n+1} : B^{n+1} \times I \to B^{n+1} \times \{0\} \cup S^n \times I$ introduced in Section 1.0 (pages 7, 8) shows that the inclusion $S^n \to B^{n+1}$, $n \in \mathbf{N}$, is a closed cofibration. □

The following is an example of an embedding of a closed subspace which is not a closed cofibration.

Example 2 Take X to be the subspace of $\mathbf{R}$ consisting of the points 0 and $1/n$, for all integers $n \geqslant 1$, and take A to be the closed subspace of X consisting of the single number 0. Clearly, A has no neighbourhood in X which is deformable to A in X; thus, the inclusion $A \subsetneq X$ cannot be a closed cofibration (see Proposition A.4.1 (iv)). □

The following facts about closed cofibrations are known.

Proposition A.4.2 (i) *The closed cofibrations form a subcategory of Top containing all isomorphisms and all initial morphisms; i.e., all identities and all other homeomorphisms are closed cofibrations, any composition of closed cofibrations is a closed cofibration, all maps whose domain is the empty space are closed cofibrations.*

(ii) *If $\{A_\lambda \to X_\lambda : \lambda \in \Lambda\}$ is a set of closed cofibrations, then $\sqcup A_\lambda \to \sqcup X_\lambda$ is a closed cofibration; i.e., a topological sum of closed cofibrations is a closed cofibration.*

(iii) *If $A \to X$ is a closed cofibration and Z is any space, then, $A \times Z \to X \times Z$ is a closed cofibration; i.e., the product functor $- \times Z$ preserves closed cofibrations.*

(iv) Product theorem: *If $A_\lambda \to X_\lambda$, $\lambda = 0, 1$ are both closed cofibrations then,*

$$X_0 \times A_1 \cup A_0 \times X_1 \to X_0 \times X_1$$

is a closed cofibration.

(v) *Let $i : A \to X$ be a closed cofibration. Then i is a homotopy equivalence iff A is a strong deformation retract of X.*

(vi) *If $j : A \to X$ and $i : X \to V$ are maps such that i and ij are closed cofibrations, then j is a closed cofibration.*

(vii) *If $A_0 \to X$, $A_1 \to X$ and $A_0 \cap A_1 \to X$ are closed cofibrations, so is $A_0 \cup A_1 \to X$.*

(viii) *Let $A_i \to X$, $i \in [n]$*[†] *be closed cofibrations. Suppose that for every subset $S \subset [n]$ the inclusion of $A_S = \bigcap_{i\in S} A_i$ into X is a closed cofibration. Then $\cup_{i=0}^{n} A_i \to X$ is a closed cofibration.*

(ix) *Let $A \to X$ be a closed cofibration and let C be a compact Hausdorff space. Then $A^C \to X^C$ is a closed cofibration.*

Proof (i): Brown (1988, 7.3.2); (ii):easy; (iii):Brown (1988, 7.2.4 Corollary 2); (iv):Brown (1988, 7.3.8); (v):Brown (1988, 7.2.9 Corollary 1); (vi):Strøm (1972, Lemma 5); (vii):Lillig (1973, Corollary 2); (viii):Lillig (1973, Corollary 3); (ix);Strøm (1972, Lemma 4).

□

A based space (X, x_0) is said to be *well-pointed* whenever the inclusion $\{x_0\} \subsetneq X$ is a closed cofibration.

Example 3 The based spaces (B^{n+1}, e_0), $n \in \mathbf{N}$ are well-pointed, since the inclusions, $\{e_0\} \to B^{n+1}$ are the compositions of the closed cofibrations $\{e_0\} \to S^n$ and $S^n \to B^{n+1}$. □

Indeed, any based CW-complex is well-pointed (see Corollary 1.3.7 and Lemma 2.3.7).

The space X described in Example 2 together with the point $0 \in X$, gives an example of a based space, namely $(X, 0)$, which is not well pointed. There are spaces for which no choice of a base point yields a well-pointed based space:

Example 4 The Cantor set is totally disconnected and every point is a limit point (see Bourbaki, 1966, Chapter IV, Section 2.5, Example); thus, no one of its points has a neighbourhood which can be contracted to it (cf. Proposition A.4.1 (iv)). □

There is a relative version of Proposition A.4.2 (ix), namely:

[†] As in Section 4.1, the symbol $[n]$ denotes the set of integers from 0 to n.

Proposition A.4.3 *Let $A \to X$ be a closed cofibration and let (C, D) be a pair of spaces with C compact Hausdorff. Then, the inclusion $A^C \to (X, A)^{(C,D)}$ is a closed cofibration.*

Proof Recall that $(X, A)^{(C,D)}$ is the subspace of X^C formed by all functions $C \to X$ whose restrictions to D take values in A. Let U be a neighbourhood of A in X which is deformable to A in X, rel. A and let $\alpha : X \to I$ be a map such that $\alpha^{-1}(0) = A$ and $\alpha|(X \setminus U) = 1$. Then, $(U, A)^{(C,D)}$ is a neighbourhood of A^C in $(X, A)^{(C,D)}$, which is deformable to A^C rel. A^C. Furthermore, the composition

$$\beta = \max \circ (\alpha^C | (X, A)^{(C,D)}),$$

where $\max : I^C \to I$ denotes the map which assigns to every map $f : C \to I$ its maximal value, is such that $\beta^{-1}(0) = A^C$ and its restriction to the complement of $(U, A)^{(C,D)}$ takes the constant value 1. □

A particular case of the previous proposition deserves to be mentioned and applies to spaces of based maps into well-pointed spaces.

Corollary A.4.4 *If (X, x_0) is a well-pointed based space and (C, c_0) is a compact based space, then the based space $((X, x_0)^{(C,c_0)}, \omega_0)$, where ω_0 is the constant map with value x_0, is well pointed.* □

For any space X, the diagonal map $\Delta : X \to X \times X$ is an embedding of a closed subspace. If, moreover, this embedding is a closed cofibration, the space X is called *locally equiconnected* (LEC, in short). 'Roughly speaking, X is LEC if there are paths between sufficiently nearby points such that the paths depend continuously on the end points' (see Fox (1943)). An important feature of LEC spaces is the following:

Proposition A.4.5 *LEC spaces are locally contractible.*

Proof See Dugundji (1965, Lemma 2.3). □

Discrete spaces are clearly LEC. The deformation retraction

$$d^n : (B^n \times B^n) \times I \to B^n \times B^n$$

of $B^n \times B^n$ onto ΔB^n introduced in pages 2, 3 together with the map

$$\alpha : B^n \times B^n \to I, \qquad (\mathbf{s}, \mathbf{s}') \mapsto |\mathbf{s} - \mathbf{s}'|$$

shows that the ball B^n, $n \geqslant 1$ is an LEC space (see Proposition A.4.1 (iv)). Moreover, arbitrary coproducts of balls are LEC: this follows from the equation

$$\Delta(\sqcup B^{n_\lambda}) = \sqcup(\Delta B^{n_\lambda})$$

(see Proposition A.4.2 (ii)).

Proposition A.4.6 *If* $i : A \to X$ *is a closed cofibration and* X *is an LEC space, then* A *is LEC.*

Proof The diagonal map $\Delta : A \to A \times A$ is a closed cofibration (see Proposition A.4.2 (i), (iii) and (vi)). □

As a consequence of the previous result, the sphere $S^{n-1} = \delta B^n$ is LEC, for every $n \geqslant 1$.

Fibrations

The concept of fibration is dual to that of cofibration.[†] A map $p : Y \to X$ is a *fibration* if, for any commutative diagram (full arrows), in *Top*, where v_0

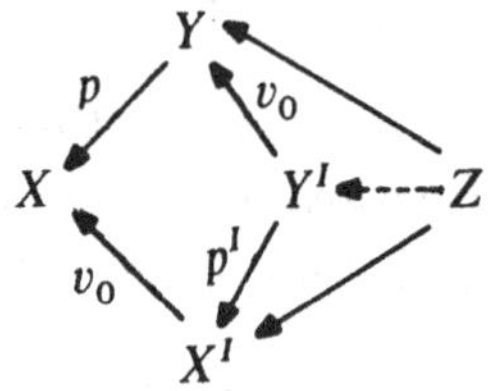

denotes the evaluation at 0, there is a map $Z \to Y^I$ (dotted arrow) which makes the resulting triangles commutative. Using the adjointness between the functors $- \times I$ and $-^I$ one obtains the description of fibrations by means of the *homotopy lifting property*: the map $p : Y \to X$ is a fibration if, for

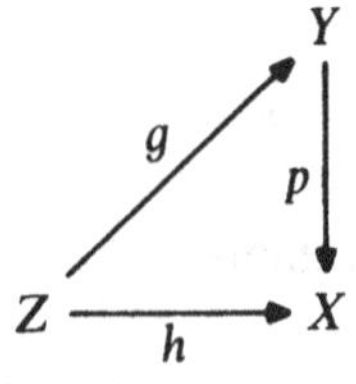

every commutative diagram in *Top*, every homotopy H starting at h *lifts* to a homotopy G starting at g, i.e., gives rise to a homotopy G starting at

[†] Closed cofibrations are essential for the attaching process which plays an important role throughout the book from the very beginning; fibrations show up in Chapters 1, 4 and 5.

g and such that $p \cdot G = H$. Under these conditions, if X is path-connected and $Y \neq \emptyset$, then p is a surjection. The space X is the *base space* or *base* of the fibration; the space Y is the *total space* of the fibration. For any point $x \in X$, the inverse image $p^{-1}(x)$ is the *fibre over* x; a subspace of Y is a *fibre* of the fibration if it is the fibre over some point $x \in X$.

If, in the definition of fibration, the lifting property is required only for $Z = B^n(\Delta^n)$, $n \in N$, one obtains the weaker notion of *Serre fibration* (see Serre, 1951); on the other hand, if Z is any topological space (not necessarily a weak Hausdorff k-space), one obtains the notion of *Hurewicz fibration.*

Given the fibrations $p : Y \to X$, $p' : Y' \to X$, a map $f : Y \to Y'$ is a *fibre map* if $p' \circ f = p$; a homotopy $H : Y \times I \to Y'$ is a *fibre homotopy* if $p' \circ H$ is nothing but the composition of the projection $Y \times I \to Y$ with p (this is the dual concept to that of homotopy rel. a subspace). Two maps $f, g : Y \to Y'$ are *fibre homotopic* if there is a bifre homotopy connecting them; note that in this case the maps f, g are themselves fibre maps. A fibre map $f : Y \to Y'$ is a *fibre homotopy equivalence* if there is a fibre map $g : Y' \to Y$ such that the two compositions are fibre homotopic to the respective identities.

Proposition A.4.7 (i) *The fibrations form a subcategory of Top containing all isomorphisms and all terminal morphisms; i.e., all identities and all other homeomorphisms are fibrations, any composition of fibrations is a fibration, all maps whose codomain is a singleton space are fibrations.*

(ii) *If $Y \to X$ is a fibration and D is a retract of Y over X, the restriction $D \to X$ is a fibration.*

(iii) *If $Y \to X$ and $Z \to W$ are fibrations $Y \times Z \to X \times W$ is a fibration.*

(iv) *Let $A \to X$ be a closed cofibration and let Z be any space. Then, $Z^X \to Z^A$ is a fibration.*

Proof (i) and (iii) are trivial.

(ii): Recall that for a given map $p : Y \to X$, a subspace $D \subset Y$ is a *retract of Y over X* if there is a map $r : Y \to D$ such that $r|D = \mathbf{1}_D$ and $p|D \circ r = p$. The claim follows immediately from the definitions.

(iv): See Spanier (1966, Theorem 2.8.2). □

Example 5 If Y and Z are any spaces, the projections from $Y \times Z$ to Y and Z respectively are fibrations. This follows from statements (i) and (iii) of Proposition A.4.7. □

Example 6 Given any space X, the map $v_0 : X^I \to X$ is a fibration (see Proposition A.4.7 (iv)); for a fixed base-point $x_0 \in X$, the fibre over x_0 is denoted by PX. The map $v_X : PX \to X$ obtained by evaluation at 1 is again a fibration whose fibre over x_0 is the *loop space* ΩX of the based space (X, x_0) (see Spanier, 1966, Corollary 2.8.8); the elements of ΩX are *loops* of X (based at x_0). The loop space ΩX has a canonical base point, namely, the constant loop ω_0 at the point x_0. The based space $(\Omega X, \omega_0)$ is well pointed if (X, x_0) is well pointed (see Corollary A.4.4).

The construction of the based pair of spaces $(PX, \Omega X, \omega_0)$ extends to a functor. It associates to a based map $f : (Y, y_0) \to (X, x_0)$ the maps $Pf : PY \to PX$ and $\Omega f : \Omega Y \to \Omega X$, defined just by composition. □

Example 7 A map $p : \tilde{X} \to X$ is a *covering projection* if every point $x \in X$ has an open neighbourhood U such that $p^{-1}(U) = \sqcup U_\lambda$, where each U_λ is open and homeomorphic to U via p; in this situation, the space $\tilde{X}$ is called a *covering space* of X. Covering projections are fibrations (see Spanier, 1966, Theorem 2.2.3). □

Pushouts and pullbacks

A commutative diagram in *Top*

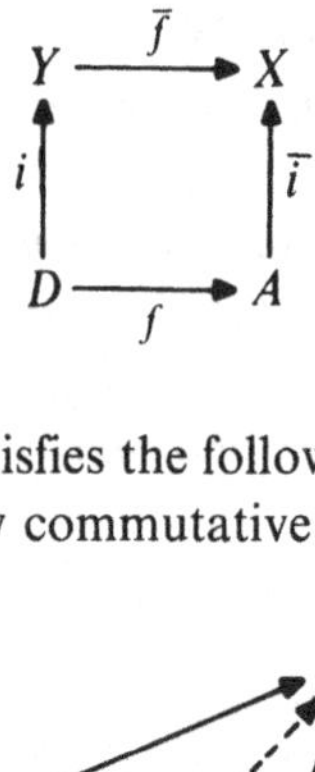

is said to be a *pushout* if it satisfies the following *universal property* (called the *pushout property*): for any commutative diagram (full arrows)

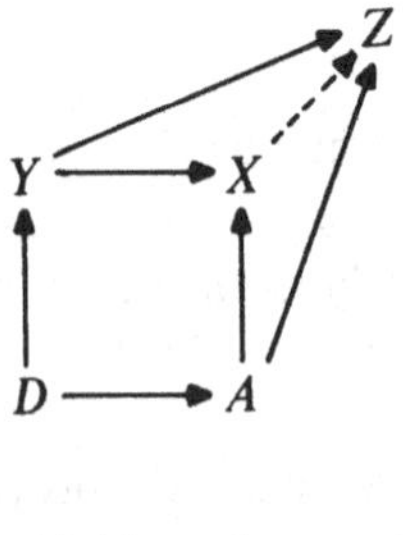

there is a unique map $X \to Z$ (dotted arrow) making commutative the triangles which arise; in other words, given maps $g : A \to Z, h : Y \to Z$ such that $gf = hi$, there is a unique map $k : X \to Z$ such that $k\bar{i} = g, k\bar{f} = h$.

This concept is also useful in other categories, like the categories of groups and groupoids. In such a square, the space (group, groupoid) X is uniquely determined up to homeomorphism (isomorphism) by the diagram

(*)

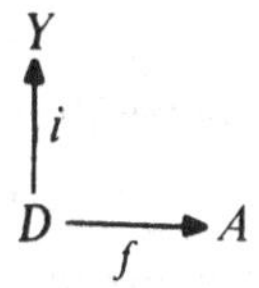

more precisely, if

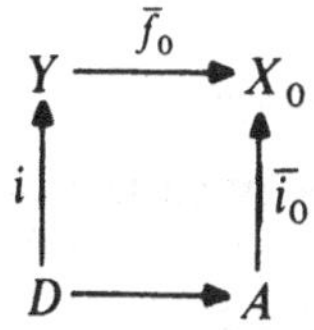

and

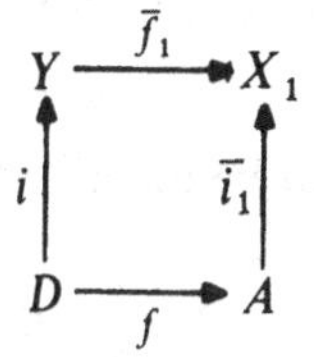

are pushouts then there are unique homeomorphisms (isomorphisms) $h_\lambda : X_\lambda \to X_{1-\lambda}$, $\lambda = 0, \mathbf{1}$, inverse to each other, such that

$$h_\lambda \cdot \bar{f}_\lambda = \bar{f}_{1-\lambda}, \qquad h_\lambda \cdot \bar{i}_\lambda = \bar{i}_{1-\lambda}.$$

Conversely, any diagram (*) of spaces and maps (respectively, groups, groupoids and homomorphisms) can be completed to a pushout.

Pushouts are dualized by *pullbacks*: a commutative diagram in *Top*,

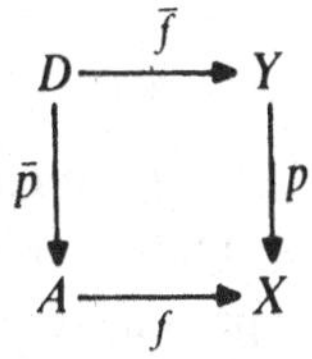

is said to be a *pullback* if it satisfies the following *universal property* (called the *pullback property*): for any commutative diagram (full arrows),

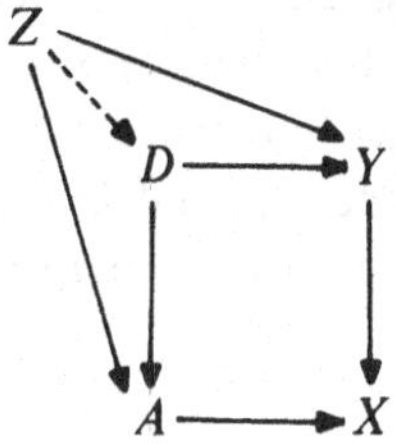

there is a unique map $Z \to D$ (dotted arrow), making commutative the triangles which arise. In this situation, the space D is uniquely determined (up to homeomorphism) and denoted by $A \sqcap_f Y$. The map $\bar{p}$ is said to be *induced from p by f* and the map $\bar{f}$ is *induced from f by p*. The space $A \sqcap_f Y$ can be defined as the inverse image of the diagonal with respect to the map

$$A \times Y \to X \times X, \qquad (a, y) \mapsto (f(a), p(y));$$

thus, because X is weak Hausdorff, the space $A \sqcap_f Y$ can be considered as a closed subspace of $A \times Y$. If f is the imbedding of a closed subspace of X, then $A \sqcap_f Y$ is homeomorphic to the closed subspace $p^{-1}(A)$ of Y; thus, the map $\bar{p}$ can be identified with the map

$$p^{-1}(A) \to A, a \mapsto p(a),$$

and the map $\bar{f}$ can be identified with the inclusion $p^{-1}(A) \subset Y$.

Adjunction spaces

A diagram of spaces and maps,

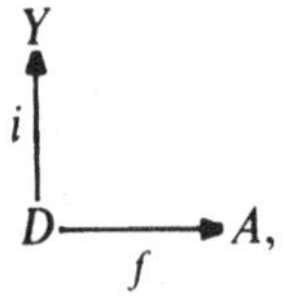

in which i is a closed cofibration, is called a *partial map* from Y to A with *domain D* and is denoted by

$$f : Y -/\to A.$$

When no confusion arises, the letter 'f' represents either the map $f : D \to A$ or the partial map $f : Y -/\to A$; moreover, whenever the letter 'D' is not specified, the notation *dom* f represents the domain of the partial map f. Notice that because the product functor $- \times Z$ preserves closed cofibrations (see Proposition A.4.2 (iii)), for any given partial map $f : Y -/\to A$ and any space Z, there is an induced partial map $f \times \mathbf{1} : Y \times Z \to A \times Z$.

A partial map $f : Y -/\to A$ can be completed to a pushout as follows. For the underlying set X, take a disjoint union $A \sqcup (Y \backslash dom\ f)$.† Next, endow X with the final topology with respect to the canonical functions $\bar{f} : Y \to X$ and $\bar{i} : A \to X$. Thus, by construction, the space X contains A as a closed subspace. Therefore, one says that X is obtained from A by *attaching* or *adjoining Y via f*, or, also, that X is an *attaching space* or an *adjunction*

†Alternatively, one can take the set $A \sqcup Y/d \sim f(d)$.

space. Such an attaching space will be denoted systematically by

$$X = A \bigsqcup\nolimits_f Y;$$

moreover, the map f is called *attaching map* and the map $\bar{f} : Y \to X$ is called *characteristic map*.

Adjunction spaces have been around for a long time; a reasonably complete account on the general properties for adjunction spaces can be found in Brown (1988). In particular, the gluing theorem (see Theorem A.4.12) is presented there; for CW-complexes, that theorem had been developed earlier in Spanier & Whitehead (1957).

Although the previous construction of the adjunction space X seems to be unique, it is not. The ambiguity in the construction – only up to homeomorphism, as already noted – is due first to the fact that A and $Y \setminus D$ may not be disjoint. In order to form the disjoint union, one has to make the sets disjoint, a process which is not unique. Secondly, if X, $\bar{f}$ and $\bar{i}$ are given, replacing $\bar{f}$ and $\bar{i}$ with the compositions $\bar{f} \cdot \tilde{f}$ and $\bar{i} \cdot \tilde{i}$ respectively, where $\tilde{f} : Y \to Y$ and $\tilde{i} : A \to A$ are homeomorphisms such that $\tilde{f} \cdot i = i$ and $\tilde{i} \cdot f = f$, one obtains a new pushout square which is equivalent to the original one. In other words, if the square

$$\begin{array}{ccc} Y & \xrightarrow{\ \bar{f}\ } & X \\ {\scriptstyle i}\uparrow & & \uparrow{\scriptstyle \bar{i}} \\ D & \xrightarrow[\ f\ ]{} & A \end{array}$$

is a pushout, so is the square

$$\begin{array}{ccc} Y & \xrightarrow{\ \bar{f}\circ\tilde{f}\ } & X \\ {\scriptstyle i}\uparrow & & \uparrow{\scriptstyle \bar{i}\circ\tilde{i}} \\ D & \xrightarrow[\ f\ ]{} & A \end{array}$$

A slightly different terminology and notation is used if, in addition, $f : D \to A$ happens to be an inclusion which is also a closed cofibration. Then one may consider the set X to be

$$X = A \setminus D \sqcup D \sqcup Y \setminus D = A \cup Y.$$

In this case, X is called a *union* (*space*) *of A and Y over D* and it is denoted by

$$X = A \bigsqcup\nolimits_D Y.$$

This situation arises whenever one is given a space X together with closed subspaces A, Y such that $A \cup Y = X$ and the inclusions of $A \cap Y$ in A and

Y, respectively, are closed cofibrations. Then the equation

$$X = A \bigsqcup_D Y$$

holds true, in the sense that both sides agree up to a canonical homeomorphism.

Example 8 Because the inclusion $S^{n-1} \to B^n$ is a closed cofibration, the sphere S^n can be viewed as the union space

$$S^n = B^n \bigsqcup_{S^{n-1}} B^n$$

with the maps $i_+, i_- : B^n \to S^n$ considered as characteristic maps. □

Proposition A.4.8 *Let $f : Y -/\to A$ be a partial map and let X denote the adjunction space $A \bigsqcup_f Y$. Then, the following properties hold true:*

(i) *for every space Z, $(A \times Z) \bigsqcup_{f \times 1} (Y \times Z) = (A \bigsqcup_f Y) \times Z$;*

(ii) *the inclusion $A \subsetneq X$ is a closed cofibration;*

(iii) *any characteristic map $\bar{f} : Y \to X$ induces a homeomorphism $Y \backslash D \to X \backslash A$;*

(iv) *if Y and A are (perfectly) normal spaces, then X is (perfectly) normal;*

(v) *if Y and A are normal spaces of dimension $\leqslant n$ then, so is X;*

(vi) *if D is a strong deformation retract of Y then, A is a strong deformation retract of X; if $r : Y \to D$ is a deformation retraction, the unique map $r' : X \to A$ satisfying the conditions $r'|A = \mathbf{1}_A$ and $r' \circ \bar{f} = f \circ r$ is a deformation retraction.*

Proof (i): Follows from the exponential law described in Section A.1.

(ii): It is enough to construct a retraction

$$r_X : X \times I \to \hat{X} = X \times \{0\} \cup A \times I$$

(see Proposition A.4.1 (ii)). But

$$X \times I = (A \times I) \bigsqcup_{f \times 1} (Y \times I)$$

(see property (i)); now take for r_X the unique map such that $r_X | A \times I = \mathbf{1}$ and $r_X \circ (\bar{f} \times \mathbf{1})$ is the composition of the retraction $r_Y : U \times I \to \hat{Y} = Y \times \{0\} \cup D \times I$ with the canonical map from $\hat{Y}$ to $\hat{X}$.

(iii): Is clear from the construction of adjunction spaces.

(iv): In view of Tietze's extension theorem, it is enough to prove that any map $k : C \to I$, where C is a closed subset of X, can be extended to a map over all of X. Because A is normal, there is an extension $g' : A \to I$ of $k | A \cap C$; then, let $h : \bar{f}^{-1}(C) \cup D \to I$ be the map given by $h | \bar{f}^{-1}(C) = k \circ \bar{f}_C$, where $\bar{f}_C : \bar{f}^{-1}(C) \to C$ is the map induced by $\bar{f}$, and

$h|D = g' \circ f$. This map h can be extended to a map $h : Y \to I$ because Y is normal. The pushout property now induces a map $k' : X \to I$ extending k.

A space is *perfectly normal* if every closed subset is the zero set of a map with non-negative real values. Thus, let $C \subset X$ be a closed subset. If A is perfectly normal, one can find a map $g : A \to [0, \infty)$ with zero set $A \cap C$. Then extend the composition $g \circ f$, firstly to a map over $D \cup \bar{f}^{-1}(C)$, taking the value 0 outside D, and, secondly, by means of Tietze's extension theorem, to a map $h' : Y \to [0, \infty)$; for this, Y has only to be normal. Moreover, if Y is perfectly normal, there is a map $h'' : Y \to [0, \infty)$ with zero set $D \cup \bar{f}^{-1}(C)$. Finally, the maps g and $h' + h''$ induce a map $k : X \to [0, \infty)$ with zero set C.

(v): Let $C \subset X$ be a closed subset and let $k' : C \to S^n$ be a map. Since A has dimension $\leqslant n$, the map $k'|A \cap C$ has a continuous extension $g : A \to S^n$ (see Theorem A.9.1). Next, define $h' : D \cup \bar{f}^{-1}(C) \to S^n$ by taking $h'|D = g \circ f$ and $h'|\bar{f}^{-1}(C) = k' \circ \tilde{f}$, where $\tilde{f} : \bar{f}^{-1}(C) \to C$ denotes the map induced by $\bar{f}$. Since Y has dimension $\leqslant n$, the map h' extends to a map $h : Y \to S^n$ (again by Theorem A.9.1). Finally, the maps g and h induce an extension of k' to a map $k : X \to S^n$, proving the desired result (once more using A.9.1).

(vi): Let $H : Y \times I \to Y$ be a retracting homotopy of Y onto D. The maps $\bar{f} \circ H$ and $\bar{i} \circ pr_1$ coincide over $D \times I$; thus, by (i) above, one obtains a retracting homotopy $X \times I \to X$ with the desired properties. □

In the special situation of a union space, statement (ii) of the previous proposition yields the following.

Corollary A.4.9 *If $X = A \sqcup_D Y$, then the inclusions of A, Y and $D = A \cap Y$ into $X = A \cup Y$ are closed cofibrations.* □

Example 8′ From Corollary A.4.9 and Example 8, it follows that the embeddings $i_+, i_- : B^n \to S^n$ introduced on page 3, as well as the inclusions $S^{n-1} \to S^n$, are closed cofibrations. More generally, it follows that the inclusions $S^m \to S^n$ are closed cofibrations, for all pairs of natural numbers (m, n) with $m \leqslant n$. A similar result holds for projective spaces: for $\mathbf{F} = \mathbf{R}$, $\mathbf{C}$ and $\mathbf{H}$, the inclusions $\mathbf{F}P^m \to \mathbf{F}P^n$ are closed cofibrations, for all pairs of natural numbers (m, n) with $m \leqslant n$. □

Remark For every $X_0, X_1 \in Top$, the inclusion $X_0 \to X_0 \sqcup X_1$ is a closed cofibration (see Proposition A.4.2 (i) and (ii)). Moreover, if $A_\lambda \to X_\lambda$, $\lambda = 0, 1$ are both closed cofibrations, all the canonical inclusions in the following diagram are closed cofibrations:

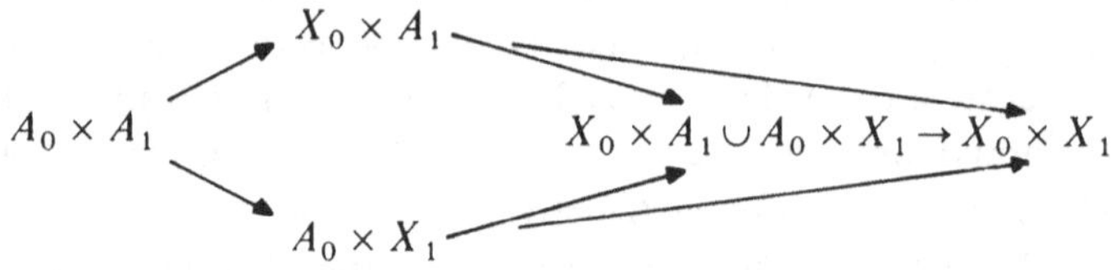

(see Proposition A.4.2 (iii), (iv) and Corollary A.4.9). □

Example 9 The based spaces (S^n, e_0), $n \in \mathbf{N}$ are well pointed. This follows from the fact that $\{e_0\} \to S^0$ is a closed cofibration (see Remark above), and also because the inclusion $S^0 \to S^n$ is a closed cofibration (see Example 8′). □

The following 'algebraic laws' are useful in dealing with adjunction spaces.

(L1) *Horizontal composition* Let $f : Y \text{-/}\!\!\to A$ be a partial map and let $g : A \to A'$ be a map considered as a partial map $A \sqcup_f Y \text{-/}\!\!\to A'$. Then,

$$A' \sqcup_g (A \sqcup_f Y) \cong A' \sqcup_{gf} Y.$$

The following diagram is useful for a better understanding of this law:

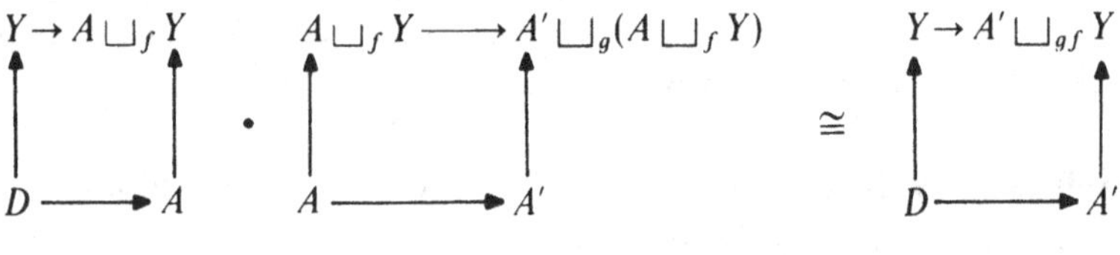

There is a noteworthy relationship between the possible characteristic maps for the adjunctions in this law. Assume $A' \sqcup_g (A \sqcup_f Y)$ and $A' \sqcup_{gf} Y$ to be really the same space X', and assume that one is given fixed inclusions $\bar{\imath} : A \to X = A \sqcup_f Y$ and $\bar{\imath}' : A' \to X'$, as well as a fixed characteristic map $\bar{f} : Y \to X$. Then, on the one hand, for any characteristic map $\bar{g} : X \to X'$, the composition $\bar{g} \circ \bar{f}$ is a characteristic map for the attaching of Y to A' via gf; on the other hand, if a characteristic map $\overline{gf} : Y \to X'$ is given, then the unique map $\bar{g} : X \to X'$ with $\bar{g} \circ \bar{f} = \overline{gf}$ and $\bar{g} \circ \bar{\imath} = \bar{\imath}' \circ g$ is a characteristic map for the attaching of X to A' via g.

(L2) *Vertical composition* Let $f : Y \text{-/}\!\!\to A$ be a partial map, $j : Y \rightarrowtail Y'$ be a closed cofibration, and $\bar{f} : Y \to A \sqcup_f Y$ be a characteristic map. If f is also viewed as a partial map $Y' \text{-/}\!\!\to A$ and $\bar{f}$ as a partial map

$Y' -/\rightarrow A \sqcup_f Y$, then

$$(A \sqcup_f Y) \sqcup_{\bar{f}} Y' \cong A \sqcup_f Y'.$$

This situation may be depicted by the following diagrams:

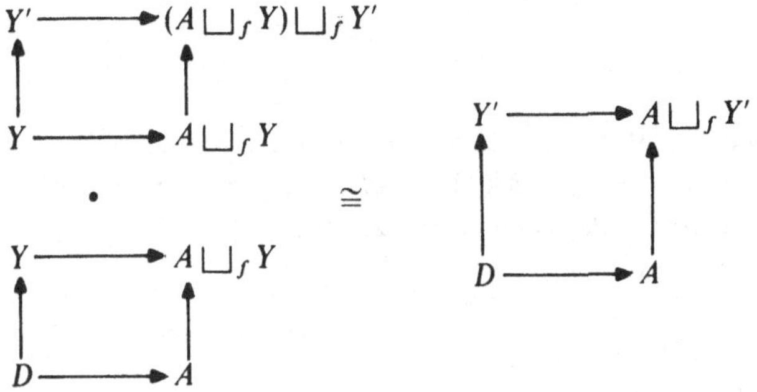

(L3) *Addition* Let $f_\lambda : Y_\lambda -/\rightarrow A$ be partial maps, $\lambda = 0, 1$. Denote by $(f_0, f_1) : Y_0 \sqcup Y_1 -/\rightarrow A$ the induced partial map from the topological sum of Y_0 and Y_1 to A; its domain is the topological sum $dom\ f_0 \sqcup dom\ f_1$, on which the underlying map is defined in the obvious manner. Use the inclusions $A \subsetneq A \sqcup_{f_\lambda} Y_\lambda$ to consider f_λ also as a partial map

$$Y_\lambda -/\rightarrow A \sqcup_{f_{1-\lambda}} Y_{1-\lambda};$$

then,

$$A \sqcup_{(f_0, f_1)} (Y_0 \sqcup Y_1) \cong (A \sqcup_{f_0} Y_0) \sqcup_A (A \sqcup_{f_1} Y_1)$$
$$\cong (A \sqcup_{f_0} Y_0) \sqcup_{f_1} Y_1 \cong (A \sqcup_{f_1} Y_1) \sqcup_{f_0} Y_0.$$

(L4) *Restriction* Let $f : Y -/\rightarrow A$ be a partial map and let U be an open or closed subset of $X = A \sqcup_f Y$; then

$$U \cong (U \cap A) \sqcup_{f'} \bar{f}^{-1}(U),$$

where $\bar{f}$ is a characteristic map for the adjunction space X and f' is the induced partial map $\bar{f}^{-1}(U) -/\rightarrow U \cap A$.

(L5) *Multiplication* Let $f_\lambda : Y_\lambda -/\rightarrow A_\lambda$ be partial maps, $\lambda = 0, 1$. Take $X_\lambda = A_\lambda \sqcup_{f_\lambda} Y_\lambda$, $\lambda = 0, 1$, and define canonically a partial map

$$f = (\bar{f}_0 \times f_1) \cup (f_0 \times \bar{f}_1) : Y_0 \times Y_1 -/\rightarrow X_0 \times A_1 \cup A_0 \times X_1$$

with domain $Y_0 \times dom\ f_1 \cup dom\ f_0 \times Y_1$; then,

$$X_0 \times X_1 = (X_0 \times A_1 \cup A_0 \times X_1) \sqcup_f (Y_0 \times Y_1).$$

Mapping cylinders

Mapping cylinders constitute an important special case of adjunction spaces. Let $f : D \to A$ be any map and let $i : D \to D \times I$ be the embedding $x \mapsto (x, 0)$; consider f as a partial map $D \times I -/\to A$. Then, the *mapping cylinder* of f is defined by

$$M(f) = A \bigsqcup_f (D \times I).$$

Proposition A.4.10 *Let $f : D \to A$ be any map. Then,*
(i) *the inclusion $\bar{i} : A \to M(f)$ is a closed cofibration;*
(ii) *A is a strong deformation retract of $M(f)$;*
(iii) *the composite map*

$$i_D : D \cong D \times \{1\} \subsetneq D \times I \to M(f)$$

is a closed cofibration;
(iv) *the map f factors through i_D; more precisely, $f = r_f \circ i_D$, where $r_f : M(f) \to A$ denotes the deformation retraction determined by $r_f \circ \bar{f} = f \circ pr_1$;*
(v) *$f : D \to A$ is a homotopy equivalence iff the embedding i_D is a homotopy equivalence iff D is a strong deformation retract of $M(f)$ via the embedding i_D;*
(vi) *$f : D \to A$ is a weak homotopy equivalence iff the pair $(M(f), D)$ is n-connected, for every $n \in \mathbf{N}$;*
(vii) *if $j : D' \subsetneq D$ is a closed cofibration, the mapping cyclinder $M(f \circ j)$ is a strong deformation retract of $M(f)$.*

Proof (i): Follows immediately from Proposition A.4.8 (ii).
(ii): Follows from Proposition A.4.8 (vi), because $D \times \{0\}$ is a strong deformation retract of $D \times I$.
(iii): Observe that $A \bigsqcup_f (D \times \dot{I}) = A \sqcup D$ and by the law of vertical composition, $M(f) = (A \sqcup D) \bigsqcup_{\tilde{f}} (D \times I)$, where $\tilde{f} : D \times \dot{I} \to A \sqcup D$ is a characteristic map. Then, i_D is the composition of the closed cofibrations $D \subsetneq A \sqcup D \subsetneq M(f)$.
(iv): Given arbitrarily $x \in D$,

$$r_f i_D(x) = r_f(\bar{f}(x, 1)) = f(pr_1(x, 1)) = f(x).$$

(v): From the previous property, and the fact that r_f is a homotopy equivalence, one concludes that f is a homotopy equivalence iff i_D is a homotopy equivalence. But $i_D : D \to M(f)$ is a closed cofibration; then, i_D is a homotopy equivalence iff D is a strong deformation retract of $M(f)$ (see Proposition A.4.2 (v)).

(vi): The map r_f is a homotopy equivalence, and therefore induces isomorphisms of the homotopy groups; then, the equation $f = r_f \circ i_D$ implies that f induces isomorphisms of the homotopy groups iff i_D induces isomorphisms of the homotopy groups. But the latter condition is equivalent to the n-connectivity of the pair $(M(f), D)$, for every $n \in \mathbf{N}$ (see Corollary A.8.12).

(vii): First note that the mapping cylinder $M(i) = D \bigsqcup_i (D' \times I) \cong D \times \{0\} \cup D' \times I$ is considered to be a subspace of $D \times I$ containing $D \times \{0\}$, with all inclusions involved being closed cofibrations. Then

$$\begin{aligned}
M(f) &= A \textstyle\bigsqcup_f (D \times I) && \text{by definition of } M(f)\\
&= (A \textstyle\bigsqcup_f M(i)) \textstyle\bigsqcup_{\bar f} (D \times I) && \text{by vertical composition}\\
&= (A \textstyle\bigsqcup_f (D \textstyle\bigsqcup_i D' \times I)) \textstyle\bigsqcup_{\bar f} (D \times I) && \text{by definition of } M(i)\\
&= (A \textstyle\bigsqcup_{f'} D' \times I) \textstyle\bigsqcup_{\bar f} (D \times I) && \text{by horizontal composition}\\
&= M(f') \textstyle\bigsqcup_{\bar f} (D \times I) && \text{by definition of } M(f')
\end{aligned}$$

where $dom\ \bar f = D \times \{0\} \cup D' \times I$. Since $i : D' \subsetneq D$ is a closed cofibration $dom\ \bar f$ is a strong deformation retract of $D \times I$. Now Proposition A.4.8 (vi) implies that $M(f')$ is a strong deformation retract of $M(f)$. □

Some important results for the development of this book will be discussed next; their proofs are obtained using mapping cylinders.

Proposition A.4.11 *Let $f : Y -/\rightarrow A$ be a partial map. If $f : D \rightarrow A$ is a homotopy equivalence, so is any characteristic map $\bar f : Y \rightarrow A \bigsqcup_f Y$.*

Proof First notice that D is a strong deformation retract of $M(f)$ via i_D whenever f is a homotopy equivalence (see Proposition A.4.10 (v)); then, observe that Y is a strong deformation retract of $Y \bigsqcup_i M(f) = M(f) \sqcup i_D Y$ (see Proposition A.4.8 (vi)).

Next, compute $A \bigsqcup_f (D \times I \cup Y \times \{1\})$:

$$\begin{aligned}
A \textstyle\bigsqcup_f (D \times I \cup Y \times \{1\}) &= M(f) \textstyle\bigsqcup_{\bar f} (D \times I \cup Y \times \{1\}\} \text{ by vertical composition}\\
&= M(f) \textstyle\bigsqcup_{i_D} Y \text{ by horizontal composition.}
\end{aligned}$$

Let $\bar f : D \times I \cup Y \times \{1\} \rightarrow M(f) \bigsqcup_{i_D} Y$ be a characteristic map. Then, by vertical composition, it follows that

$$\begin{aligned}
\overline{M(f)} &= A \textstyle\bigsqcup_f Y \times I\\
&= (M(f) \textstyle\bigsqcup_{i_D} Y) \textstyle\bigsqcup_{\bar f} Y \times I.
\end{aligned}$$

Since $i : D \to Y$ is a closed cofibration, $D \times I \cup Y \times \{1\}$ is a strong deformation retract of $Y \times I$ (see Proposition A.4.1(iii)); then, $M(f) \sqcup_{i_D} Y$ is a strong deformation retract of $\overline{M(f)}$. Therefore, Y is a strong deformation retract of $M(\tilde{f})$. Finally, $\tilde{f}$ is a homotopy equivalence (see Proposition A.4.10 (v)). □

The next result shows that for every partial map $f : Y \text{-}/\!\to A$ with domain D, the type[†] of $A \sqcup_f Y$ depends on the types of A and (Y, D).

Theorem A.4.12 (*The gluing theorem*) *Let*

$$\begin{array}{ccccc} Y & \xleftarrow{\;i\;} & D & \xrightarrow{\;f\;} & A \\ \downarrow{\scriptstyle h_Y} & & \downarrow{\scriptstyle h_D} & & \downarrow{\scriptstyle h_A} \\ Y' & \xleftarrow{\;i'\;} & D' & \xrightarrow{\;f'\;} & A' \end{array}$$

be a commutative diagram in which i, i' are closed cofibrations, and h_Y, h_D, h_A are homotopy equivalences. Then $A \sqcup_f Y$ and $A' \sqcup_{f'} Y'$ have the same type.

Proof See Brown (1988, 7.5.7). □

Take up again the diagram of the gluing theorem; changing some of its assumptions, one obtains another interesting result.

Proposition A.4.13 *Let*

$$\begin{array}{ccccc} Y & \xleftarrow{\;i\;} & D & \xrightarrow{\;f\;} & A \\ \downarrow{\scriptstyle h_Y} & & \downarrow{\scriptstyle h_D} & & \downarrow{\scriptstyle h_A} \\ Y' & \xleftarrow{\;i'\;} & D' & \xrightarrow{\;f'\;} & A' \end{array}$$

be a commutative diagram in which i, i', h_Y, h_D and h_A are closed cofibrations and $D = D' \cap Y$, where Y, D and D' are considered as subspaces of Y'. Then, the induced map $h : A \sqcup_f Y \to A' \sqcup_{f'} Y'$ is a closed cofibration.

† In this book, 'type' means 'homotopy type'; more precisely, two spaces are said to have the same *type* whenever they have the same homotopy type.

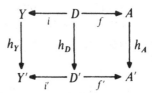

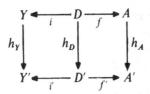

Proof See Lewis (1982, Proposition 2.5). □

The LEC-property for adjunction spaces given in the next corollary is due to Dyer & Eilenberg (1972); the simplified proof presented here is inspired by that of Lewis (1982).

Corollary A.4.14 *Let $f : Y -/\to A$ be a partial map such that the inclusion $i : D = \mathrm{dom}\, f \to Y$ is a closed cofibration. If A and Y are LEC spaces, so is the adjunction space $A \bigsqcup_f Y$.*

Proof Because of the multiplication law (L5), the product space $(A \bigsqcup_f Y) \times (A \bigsqcup_f Y)$ is homeomorphic to the adjunction space of the partial map

$$(\bar{f} \times f) \cup (f \times \bar{f}) : Y \times Y -/\to (A \bigsqcup_f Y) \times A \cup A \times (A \bigsqcup_f Y)$$

with domain $Y \times D \cup D \times Y$ (the inclusion of the latter space in $Y \times Y$ is a closed cofibration; see Proposition A.4.2 (iv)). Consider the partial maps

$$\nabla_D : Y \times D \bigsqcup D \times Y -/\to D \times D$$

and

$$\nabla_A : (A \bigsqcup_f Y) \times A \bigsqcup A \times (A \bigsqcup_f Y) -/\to A \times A$$

whose respective domains are $D \times D \sqcup D \times D$ and $A \times A \sqcup A \times A$ and whose restrictions to these domains are the appropriate folding maps. By constructing their adjunction spaces, one obtains the closed cofibrations

$$j_1 : D \times D \to Y \times D \cup D \times Y$$

and

$$j_2 : A \times A \to (A \bigsqcup_f Y) \times A \cup A \times (A \bigsqcup_f Y).$$

Let $\Delta_A : A \to A \times A$, $\Delta_Y : Y \to Y \times Y$ and $\Delta_D : D \to D \times D$ be the diagonal maps; the hypotheses of the Proposition imply that these three maps are closed cofibrations (see Proposition A.4.6 for Δ_D). The proof is concluded by taking the closed cofibrations

$$\Delta_Y : Y \to Y \times Y,$$
$$j_1 \circ \Delta_D : D \to Y \times D \bigsqcup D \times Y,$$
$$j_2 \circ \Delta_A : A \to (A \bigsqcup_f Y) \times A \bigsqcup A \times (A \bigsqcup_f Y),$$

the diagonal map of $A \bigsqcup_f Y$ into $(A \bigsqcup_f Y) \times (A \bigsqcup_f Y)$ and applying Proposition A.4.13. □

The next result makes precise the effect of changing the attaching map within its homotopy class.

Proposition A.4.15 *Let $i : D \rightarrowtail Y$ be a closed cofibration and $f_\lambda : D \to A$, $\lambda = 0, 1$, be homotopic maps. Then the adjunction spaces $X_0 = A \bigsqcup_{f_0} Y$ and $X_1 = A \bigsqcup_{f_1} Y$ are homotopically equivalent, via homotopies rel. A.*

Proof Fix inclusions $\bar{i}_\lambda : A \subsetneq X_\lambda$ and characteristic maps $\bar{f}_\lambda : Y \to X_\lambda$. Form the union space $X = X_0 \bigsqcup_A X_1$ and fix inclusions $j_\lambda : X_\lambda \subsetneq X$. Note that $j_0 \circ \bar{i}_0 = j_1 \circ \bar{i}_1 = j$ is an inclusion of A into X. Now take $\tilde{f}_\lambda : Y \times \{\lambda\} \to X$ to be the map obtained by composition of the canonical homeomorphism $Y \times \{\lambda\} \to Y$ with the maps $\bar{f}_\lambda$ and j_λ. Next, let $h : D \times I \to A$ be a homotopy connecting f_0 and f_1. Define a map

$$g : \tilde{Y} = Y \times \{0\} \cup D \times I \cup Y \times \{1\} \to X$$

by

$$g | Y \times \{\lambda\} = \tilde{f}_\lambda, \quad \lambda = 0, 1,$$
$$g | D \times I = j \circ h.$$

(Note that g can also be interpreted as a partial map $Y \times I -/\to X$.) For the moment, note that g takes the subspace $\hat{Y} = Y \times \{0\} \cup B \times I$ of $\tilde{Y}$ into the space X_0; thus, g induces a map $g_0 : \hat{Y} \to X_0$ which will be considered both as a partial map $\tilde{Y} -/\to X_0$ and as a partial map $Y \times I -/\to X_0$. Let $k : D \subsetneq \hat{Y}$ denote the inclusion which takes D into $D \times \{1\}$; this implies that $\tilde{Y} \cong \hat{Y} \bigsqcup_k Y$. The commutative diamond

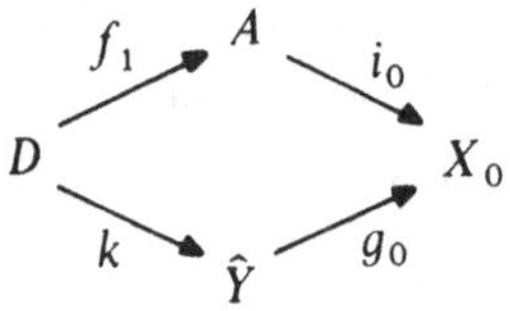

where i_0 denotes the inclusion, gives rise to the following sequence of equations:

$$
\begin{aligned}
X = X_0 \textstyle\bigsqcup_A X_1 = X_0 \bigsqcup_{i_0} (A \bigsqcup_{f_1} Y) \quad & \text{by definition} \\
\cong X_0 \textstyle\bigsqcup_{i_0 f_1} Y \quad & \text{by horizontal composition} \\
= X_0 \textstyle\bigsqcup_{g_0 k} Y \quad & \text{by commutativity of the diamond} \\
\cong X_0 \textstyle\bigsqcup_{g_0} (\hat{Y} \bigsqcup_k Y) = X_0 \bigsqcup_{g_0} \tilde{Y} \quad & \text{by horizontal composition.}
\end{aligned}
$$

Assuming that $g : \tilde{Y} \to X$ is a characteristic map for the attaching of $\tilde{Y}$ to X_0 via g_0, the law of vertical composition guarantees that

$$\tilde{X} = X \textstyle\bigsqcup_g (Y \times I) = (X_0 \bigsqcup_{g_0} \tilde{Y}) \bigsqcup_g Y \times = X_0 \bigsqcup_{g_0} Y \times I.$$

Because $\tilde{Y} = dom\, g_0$ is a strong deformation retract of $Y \times I$, the adjunction space X_0 is a strong deformation retract of the large space $\tilde{X}$. By the symmetric argument, the same is true for the adjunction space X_1. Gluing together the appearing maps and homotopies, both X_0 and X_1 are homotopy equivalent, via homotopies rel. A.

Thus, it remains to show that g is a characteristic map for the attaching of $\tilde{Y}$ to X_0 via g_0. The map $\bar{k} : Y \to \tilde{Y}$ given by $\bar{k}(y) = (y, 1)$ is a characteristic map for the attaching of Y to $\hat{Y}$ via k. Since $j_1 \circ \bar{f}_1 : Y \to X$ is a characteristic map for the attaching of Y to X_0 via $i_0 \circ f_1 = g_0 \circ k$, the equations $g | \hat{Y} = j_0 \circ g_0$ and $g \circ \bar{k} = j_1 \circ \bar{f}_1$ imply that g has the desired property. □

The previous proposition is particularly interesting in the case of the mapping cylinders.

Corollary A.4.16 *Let $f_\lambda : D \to A$, $\lambda = 0, 1$ be homotopic maps; then, the mapping cylinders $M(f_0)$ and $M(f_1)$ are homotopy equivalent via a homotopies rel. $A \square D$.*

Proof The mapping cylinder $M(f)$ of any map $f : D \to A$ can also be viewed as the adjunction space $(A \sqcup D) \bigsqcup_{\bar{f}} (D \times I)$, where $\bar{f} : D \times \{0, 1\} \to A \sqcup D$ is induced by f at the level 0 and by $\mathbf{1}_D$ at the level 1. □

Now for some variations of the construction of mapping cylinders. If $f : D \to A$ is any map, then the quotient space $C(f) = M(f)/D$, where D is thought as embedded into $M(f)$ via the map i_D (see Proposition A.4.10), is called the *mapping cone* of f. The unique point of $C(f)$ corresponding to the shrunken space D is the *peak* of the mapping cone. The composition of the inclusion $A \to M(f)$ with the projection $M(f) \to C(f)$ is again a closed cofibration. In the special case where $f = \mathbf{1}_D$, one has the *cone* $CD = C(\mathbf{1}_D)$ over D which contains D as a subspace in the obvious way. The quotient space $\Sigma D = CD/D$ is the *suspension* of D.

If (D, d_0), (A, a_0) are well-pointed based spaces and $f : (D, d_0) \to (A, a_0)$ is a based map, then one might view the interval $\{d_0\} \times I$ as a subspace of $M(f)$, embedded via the restriction of a characteristic map $\bar{f}$ and form the quotient space $M.(f) = M(f)/(\{d_0\} \times I)$ which has a distinguished point, the class z_0 corresponding to the set $\{d_0\} \times I$; the well-pointed based space $(M.(f), z_0)$ is the *reduced mapping cylinder* of f. At last, analogously to the unreduced case, one obtains the *reduced mapping cone* $(C.(f), z_0)$, the *reduced cone* $(C.D, z_0)$ and the *reduced suspension* $(\Sigma.D, z_0)$. The reduced cone $(C.D, z_0)$ for the case $(D, d_0) = (S^n, e_0)$ was already discussed in Section 1.0; in that context, the reduced suspension was also defined.

Finally, observe that the reduced cone construction can also be considered as a special case of the *smash product* $(D \wedge A, z_0)$ of two based spaces (D, d_0), (A, a_0) given by

$$D \wedge A = (D \times A)/(D \times \{a_0\} \cup \{d_0\} \times A)$$

and the evident base point.

Induced fibrations, mapping tracks and some further results in the theory of fibrations

Recall the definition of pullback and the related notion of 'map induced from a map by another map'.

Proposition A.4.17 *If $p : Y \to X$ is a fibration (resp. a covering projection) and $f : A \to X$ is any map, then the map induced from p by f is also a fibration (resp. a covering projection). If, moreover, f is a closed cofibration, then the map induced from f by p is also a closed cofibration.*

Proof See Spanier (1966, Proposition 2.8.6) and Strøm (1968, Theorem 12). □

The dual of the mapping cylinder is the mapping track: given a map $f : Y \to X$ form the pullback of the fibration $v_0 : X^I \to X$ and the map f to obtain a fibration $\overline{v_0} : T(f) = Y \sqcap_f X^I \to Y$; the space $T(f)$ is a *mapping track* of f; there are two other maps connected to this situation: the cross-section $u : Y \to T(f)$ defined by $u(y) = (y, \omega_{f(y)})$ (where $\omega_{f(y)}$ is the constant path at $f(y)$) for every $y \in Y$, and $p' : T(f) \to X$ defined by $p'(y', \lambda) = \lambda(1)$, for every $(y', \lambda) \in T(f)$.

The following result holds true (compare with Proposition A.4.10).

Proposition A.4.18 *Let $f : Y \to X$ be any map. Then,*

(i) *the map $\overline{v_0} : T(f) \to Y$ is a fibration;*

(ii) *the composition $u \circ \overline{v_0}$ is fibre homotopic to $1_{T(f)}$; in particular, $u : Y \to T(f)$ is a homotopy equivalence;*

(iii) *the map $p' : T(f) \to X$ is a fibration;*

(iv) *$f = p' \circ u$;*

(v) *$f : Y \to X$ is a homotopy equivalence iff p' is a homotopy equivalence.*

Proof For the non trivial parts, see Spanier (1966, Theorem 2.8.9). □

The gluing theorem (Theorem A.4.12) has a dual for fibrations:

Theorem A.4.19 (*The cogluing theorem*) *Let*

$$\begin{array}{ccccc} Y & \xrightarrow{p} & X & \xleftarrow{f} & D \\ \downarrow{\scriptstyle h_Y} & & \downarrow{\scriptstyle h_X} & & \downarrow{\scriptstyle h_D} \\ Y' & \xrightarrow[p']{} & X' & \xleftarrow[f']{} & D' \end{array}$$

be a commutative diagram, in which p, p' are fibrations and h_Y, h_X, h_D are homotopy equivalences. Then, the spaces $D \sqcap_f Y$ and $D' \sqcap_{f'} Y'$ have the same type.

Proof See Brown & Heath (1970, Theorem 1.2). □

This theorem has several interesting consequences.

Corollary A.4.20 *Let $p : Y \to X$ be a fibration and let $f : A \to X$ be a homotopy equivalence. Then, $\bar{f} : A \sqcap Y \to Y$ is a homotopy equivalence.*

Proof See Brown & Heath (1970, Corollary 1.4). For an alternative proof see tom Dieck, Kamps & Puppe (1970, Satz 7.30). □

Corollary A.4.21 *Let $p : Y \to X$ be a fibration with path-connected base space X; then, the fibres of p have the same homotopy type.*

Proof Let $\omega : b_0 \simeq b_1$ be a path in X connecting the points b_0 and b_1. Then, the fibres over b_0 and b_1 are both homotopy equivalent to the total space of the fibration induced by $\omega : I \to X$. For an alternative proof, see Spanier (1966, Corollary 2.8.13). □

A map $p : Y \to X$ is *locally trivial* if every point $x \in X$ has a neighbourhood U such that the map induced from p by the inclusion $U \subset X$ can be chosen as the projection of the product $U \times p^{-1}(x)$ onto U; then, the following statement holds true.

Theorem A.4.22 *A locally trivial map with paracompact codomain is a fibration.*

Proof See tom Dieck, Kamps & Puppe (1970, Satz 5.14). □

Remark Here a word of warning is necessary. The products which appear in the definition of local triviality are taken in the category of weak

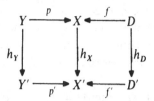

Hausdorff k-spaces. Thus, one can only apply the lifting property to homotopies whose domain also is a weak Hausdorff k-space. That means that one does not obtain Hurewicz fibrations in their most general sense, but still in a somewhat stronger sense than that of Serre fibration. □

The following theorem is an important result in the theory of fibrations; it is an inverse to the fact that a fibre homotopy equivalence induces homotopy equivalences for all fibres and also, in a certain sense, is an inverse of Corollary A.4.21.

Theorem A.4.23 *Let X be a path-connected space with an open covering $\{U_\lambda : \in \Lambda\}$ which admits a subordinated locally finite partition of unity and such that the inclusion maps $U_\lambda \to X$ are homotopic to constant maps. Moreover, let $p : Y \to X$ and $p' : Y' \to X$ be fibrations and let $f : Y \to Y'$ be a map such that $p' \circ f = p$. Then, f is a fibre homotopy equivalence if the restriction of f to some fibre is a homotopy equivalence.*

Proof See Dold (1963, Theorem 6.3). □

The assumptions on X in the previous theorem are satisfied, e.g., by all spaces having the type of a CW-complex (see Theorem 5.2.1). The theorem also allows 'delooping homotopy equivalences'.

Proposition A.4.24 *Let Y, X be path-connected spaces with locally finite open coverings which admit subordinated locally finite partitions of unity and such that the inclusions of the members of the coverings into the respective spaces are homotopic to constant maps. Then, a based map $f : (Y, y_0) \to (X, x_0)$ is a homotopy equivalence iff the induced map $\Omega f : \Omega Y \to \Omega X$ is a homotopy equivalence.*

Proof '$\Rightarrow$': By direct computation.

'$\Leftarrow$': See Allaud (1972, Theorem 1). Let $\overline{v_X} : Z \to Y$ denote the fibration induced from $v_X : PX \to X$ by f whose fibre over y_0 is ΩX. The unique map $g : PY \to Z$ satisfying $\bar{v}_X \circ g = v_y$ and $\bar{f} \circ g = Pf$ induces the homotopy equivalence Ωf when restricted to the fibres over y_0; hence, g is a fibre homotopy equivalence (see Theorem A.4.23). Because PY is contractible, this shows that Z is contractible.

Next, note that the fibre $\tilde{P}X$ over x_0 of the fibration $v_1 : X^I \to X, \omega \mapsto \omega(1)$, is contractible. The fibre over x_0 of the canonical fibration $p' : T(f) \to X$ is also contractible because it is homeomorphic to Z; moreover, p' factors through v_1 and the induced map $T(f) \to X^I$ is a homotopy equivalence (see again Theorem A.4.23). But v_1 itself is a homotopy equivalence and so is p'. Since f is the composition of a homotopy equivalence and p', it is also a homotopy equivalence. □

A.5 Union spaces of expanding sequences

An *expanding sequence* of spaces is a sequence $\{X_n : n \in \mathbf{N}\}$ of spaces such that, for every $n \in \mathbf{N}$, X_n is a subspace of X_{n+1}, and every inclusion $X_n \to X_{n+1}$ is a closed cofibration; the *union space* of the expanding sequence is the space $X = \bigcup_{n=0}^{\infty} X_n$ endowed with the final topology with respect to the family of inclusions $X_n \subset X$. Then, all X_n are closed subspaces of X (see Lemma A.2.4) and X is determined by the family $\{X_n : n \in \mathbf{N}\}$.

Given an expanding sequence $\{X_n : n \in \mathbf{N}\}$ of spaces with union space X, a sequence $\{f_n : n \in \mathbf{N}\}$ of maps $f_n : X_n \to Z$, where Z is any space, is said to be *compatible*, if, for every $n \in \mathbf{N}$, $f_{n+1}|X_n = f_n$; every such sequence induces a unique map $f : X \to Z$ such that $f|X_n = f_n$.

Proposition A.5.1 *Let $\{X_n : n \in \mathbf{N}\}$ be an expanding sequence and let X be its union space. Then,*

(i) *for every space Z, the sequence $\{X_n \times Z : n \in \mathbf{N}\}$ is an expanding sequence with union space $X \times Z$;*

(ii) *$\{X_n \times X_n\}$ is an expanding sequence with union space $X \times X$;*

(iii) *the inclusions $X_n \to X$ are closed cofibrations, for every $n \in \mathbf{N}$;*

(iv) *X is (perfectly) normal iff all spaces X_n are (perfectly) normal;*

(v) *X is a paracompact iff all spaces X_n are paracompact.*

Proof (i): Since all the inclusions $X_n \to X_{n+1}$ are closed cofibrations, so are the inclusions $X_n \times Z \to X_{n+1} \times Z$ (see Proposition A.4.2(iii). In order to show that $X \times Z$ is determined by the spaces $X_n \times Z$, take a function $f : X \times Z \to Y$ such that $f|X_n \times Z$ is continuous, for every $n \in \mathbf{N}$. Then, according to the exponential law, the adjoint functions $f_n : X_n \to Y^Z$ are all continuous and so is the adjoint of f, since X is determined by the family $\{X_n\}$. Again, using the exponential law, it follows that f itself is continuous thus, proving the assertion.

(ii): The inclusions $X_n \times X_n \to X_n \times X_{n+1}$ and $X_n \times X_{n+1} \to X_{n+1} \times X_{n+1}$ are closed cofibrations (see Proposition A.4.2(iii)) and so is the inclusion $X_n \times X_n \to X_{n+1} \times X_{n+1}$ (see Proposition A.4.2(i)); thus, $\{X_n \times X_n\}$ is an expanding sequence. The space $X \times X$ is determined by the subspaces $X_m \times X$ (see (i)) and each $X_m \times X$ is determined by the family of subspaces $\{X_m \times X_k : k \in \mathbf{N}\}$ (again, use (i)); therefore $X \times X$ is determined by the family $\{X_n \times X_k : m, k \in \mathbf{N}\}$. Since every $X_m \times X_k$ is contained in some $X_n \times X_n$, the space $X \times X$ is determined by the family of subspaces $\{X_n \times X_n\}$ (see Proposition A.2.1).

(iii): For every $k \in \mathbf{N}$, take a retraction $r_k : X_{k+1} \times I \to X_{k+1} \times \{0\} \cup X_k \times I$ (see Proposition A.4.1(ii)). According to (i), $X \times I$ is the union space of the expanding sequence $\{X_n \times I\}$; thus, in order to construct a retraction $r : X \times I \to X \times \{0\} \cup X_n \times I$, it is sufficient to exhibit a compatible family of suitable maps $f_k : X_k \times I \to X \times \{0\} \cup X_n \times I$, for all $k \in \mathbf{N}$. For $k \leqslant n$, take f_k to be the canonical inclusion; for larger k's, define inductively

$$f_{k+1}(x,t) = \begin{cases} (x,0), & t=0, \\ r_k(x,t), & r_k(x,t) \in X_{k+1} \times \{0\}, \\ f_k(r_k(x,t)), & \text{otherwise.} \end{cases}$$

(iv)$\Rightarrow$: Closed subspaces of (perfectly) normal spaces are (perfectly) normal.

(iv)$\Leftarrow$: Let C be a closed subset of X and let $f : C \to I$ be an arbitrary map. Using the normality of the spaces X_n, define inductively extensions $f_n : X_n \to I$ of the maps $f'_n : X_{n-1} \cup (C \cap X_n) \to I$, defined by $f'_n | X_{n-1} = f_{n-1}$, $f'_n | C \cap X_n = f | C \cap X_n$. The set of maps $\{f_n : n \in \mathbf{N}\}$ now defines a map $f_\infty : X \to I$ which extends f, thereby proving the normality of X.

Now assume that all X_m are perfectly normal and take a closed subset $C \subset X$. The sets $C_n = C \cap X_n$ are closed in the respective spaces X_n. Construct inductively a compatible sequence $\{f_n : n \in \mathbf{N}\}$ of maps $f_n : X_n \to [0, \infty)$ with zero sets C_n, in the following manner. To begin with, take any suitable map f_0 whose existence is guaranteed by the perfect normality of X_0. If the map f_{n-1} is constructed, first extend it over $C_n \cup X_{n-1}$ by assigning the value 0 to the points outside X_{n-1}, and then extend the latter to a map $f' : X_n \to [0, \infty)$ via Tietze's extension theorem. Now use the perfect normality of X_n to obtain a map $f'' : X_n \to [0, \infty)$ with zero set $C_n \cup X_{n-1}$; the map $h_n = f' + f''$ has the required property.

(v)$\Rightarrow$: Trivial.

(v)$\Leftarrow$: The union space of an expanding sequence is topologically dominated by the sequence, and so it inherits paracompactness (see Proposition A.2.5). □

The reader might question the fact that no mention has been made to the restriction agreed upon at the beginning of this book, namely, that all work be done in the category of weak-Hausdorff k-spaces; the previous proposition is true in the category of topological spaces, but is it true in the more restricted category used here? The next result proves that it is!

Proposition A.5.2 *The union space of an expanding sequence of weak Hausdorff k-spaces is a weak Hausdorff k-space.*

Proof Let $\{X_n : n \in \mathbf{N}\}$ be an expanding sequence of weak Hausdorff k-spaces and let X denote its union space in the category of topological spaces. The space X can be viewed as a colimit in the category of topological spaces of a diagram in the subcategory $k(Top)$. Since $k(Top)$ has colimits, and the inclusion functor of $k(Top)$ into the category of topological spaces preserves colimits, the space X is a k-space.

To prove that X is weak Hausdorff, first note that the product $X \times X$ is the union space of the expanding sequence $\{X_n \times X_n : n \in \mathbf{N}\}$ (see Proposition A.5.1(ii)). Then, observe that, because the spaces X_n are weak Hausdorff, $\Delta X \cap (X_n \times X_n) = \Delta X_n$ is closed in $X_n \times X_n$; hence, the diagonal ΔX is closed in $X \times X$. □

A space may be the union – as a set – of an expanding sequence of subspaces, and yet may fail to be determined by these subspaces; the following result describes a case in which the topology of a space coincides with the topology determined by a family of subspaces.

Proposition A.5.3 *Let $\{X_n : n \in \mathbf{N}\}$ be an expanding sequence of subspaces of a space X such that $X = \bigcup_{n \in \mathbf{N}} X_n$ (as sets) and for every $n \in \mathbf{N}$, $X_n \subset \mathring{X}_{n+1}$ (with respect to X). Then, X is the union space of the expanding sequence $\{X_n : n \in \mathbf{N}\}$.*

Proof The assumption $X_n \subset \mathring{X}_{n+1}$ implies that X is already the union of the open sets $\mathring{X}_n$. Now, take a set $W \subset X$ such that $W \cap X_n$ is open in X_n, for every $n \in \mathbf{N}$. Then, $W \cap \mathring{X}_n$ is open in $\mathring{X}_n$, and thus in X. Therefore,

$$W = W \cap X = W \cap \bigcup_{n \in \mathbf{N}} \mathring{X}_n = \bigcup_{n \in \mathbf{N}} W \cap \mathring{X}_n$$

is open in X. □

In the presence of normality there are still stronger connections between the topologies of the spaces forming the sequence and the topology of the union space.

Proposition A.5.4 *Let $\{X_n : n \in \mathbf{N}\}$ be an expanding sequence of normal spaces and let V be a subspace of its union space X. Then,*

(i) *the closure of V is the union of the closures of all intersections $V \cap X_n$; i.e.,*

$$\bar{V} = \bigcup_{n=0}^{\infty} \overline{V \cap X_n};$$

(ii) *V is determined by the family $\{V \cap X_n : n \in \mathbf{N}\}$.*

Proof (i): Let $x \in X$ be a point such that $x \notin \overline{V \cap X_n}$, for all $n \in \mathbf{N}$, and let $m \in \mathbf{N}$ be such that $x \in X_m$. The normality of X_m implies the existence of a neighbourhood U_m of x in X_m with the property: $\overline{U_m} \cap \overline{V \cap X_m} = \emptyset$. Using the normality condition over and over again, one finds inductively open sets U_n in X_n, for every $n > m$, such that $U_n \supset \overline{U_{n-1}}$, and $\overline{U_n} \cap \overline{V \cap X_n} = \emptyset$. It follows that the set $U = \bigcup_{n=m}^{\infty} U_n$ is a neighbourhood of x in X such that $U \cap V = \emptyset$; hence, $x \notin \bar{V}$. This implies statement (i).

(ii): Because of its universal property, the topology determined by the family $\{V \cap X_n : n \in \mathbf{N}\}$ is finer than the subspace topology. Conversely, let $U \subset V$ be closed in the topology determined by the family $\{V \cap X_n\}$, that is to say, such that $\overline{U \cap X_n} \cap V = U \cap X_n$, for all $n \in \mathbf{N}$. Hence,

$$\bar{U} \cap V \underset{\text{(i)}}{=} \bigcup_{n=0}^{\infty} \overline{U \cap X_n} \cap V = \bigcup_{n=0}^{\infty} U \cap X_n = U \cap \bigcup_{n=0}^{\infty} X_n = U. \qquad \square$$

Proposition A.5.5 *Let $\{A_n : n \in \mathbf{N}\}$ and $\{X_n : n \in \mathbf{N}\}$ be given expanding sequences of normal spaces; suppose that for every $n \in \mathbf{N}$, A_n is a subspace of X_n and the pair (X_n, A_n) has the homotopy extension property. Then, the union space $A = \bigcup_{n \in \mathbf{N}} A_n$ is a subspace of $X = \bigcup_{n \in \mathbf{N}} X_n$ and the pair (X, A) has the homotopy extension property.*

Proof The space A is a subspace of X (see Proposition A.5.4(ii)). Now, suppose one is given an arbitrary space Z and maps $g : X \times \{0\} \to Z$ and $H : A \times I \to Z$ which agree on $A \times \{0\}$. Because (X_0, A_0) has the homotopy extension property, there is a homotopy $K_0 : X_0 \times I \to Z$ such that $K_0 | A_0 \times I = H | A_0 \times I$ and $K_0 | X_0 \times \{0\} = g | X_0 \times \{0\}$. Assume by induction that there is a homotopy $K_n : X_n \times I \to Z$ such that $K_n | A_n \times I = H | A_n \times I$, $K_n | X_n \times \{0\} = g | X_n \times \{0\}$ and $K_n | X_{n-1} \times I = K_{n-1}$. The homotopies $H | A_{n+1} \times I$ and K_n define a homotopy $\tilde{K}_{n+1} : (A_{n+1} \cup X_n) \times I \to Z$. But the pair $(X_{n+1}, A_{n+1} \cup X_n)$ has the homotopy extension property (see Proposition A.4.2(vii)) and therefore $\tilde{K}_{n+1}$ and $g | X_{n+1} \times \{0\}$ induce a homotopy $K_{n+1} : X_{n+1} \times I \to Z$. The homotopies K_n obtained in this way form a compatible sequence, and thus give rise to the desired extension $K : X \times I \to Z$ of the given homotopy H. $\square$

Corollary A.5.6 *Let $\{Y_n : n \in \mathbf{N}\}$ be an expanding sequence of normal LEC spaces; then, the union space $Y = \bigcup_{n \in \mathbf{N}} Y_n$ is an LEC space.*

Proof For every $n \in \mathbf{N}$, take $X_n = Y_n \times Y_n$ and $A_n = \Delta Y_n$. Because all Y_n's are LEC spaces, the inclusions $A_n \to X_n, A_{n+1} \to X_{n+1}$ are closed cofibrations. Moreover, the inclusion $X_n \to X_{n+1}$ is a closed cofibration (see Proposition A.5.1 (ii)); thus, the inclusion $A_n \to X_{n+1}$ is a closed cofibration (see again Proposition A.4.2 (i)) and factors through A_{n+1}. This implies also that the inclusion $A_n \to A_{n+1}$ is a closed cofibration (see Proposition A.4.2 (vi)). Therefore, $\{A_n : n \in \mathbf{N}\}$ and $\{X_n : n \in \mathbf{N}\}$ are expanding sequences satisfying Proposition A.5.5. □

A very useful technical tool in dealing with union spaces of expanding sequences consists in the possibility of 'gluing' homotopies; we borrowed this technique from Schubert (1968, page 202).

Proposition A.5.7 *Let $\{X_n : n \in \mathbf{N}\}$ be an expanding sequence with union space X. Let Z be a space and $\{g_n : n \in \mathbf{N}\}$ be a sequence of maps $g_n : X \to Z$ such that $g_{n+1} \simeq g_n$ rel. X_n. Then, the map $g : X \to Z$ defined by $g|X_n = g_n|X_n$, for every $n \in \mathbf{N}$, is homotopic to g_0 rel. X_0.*

Proof Observe first that, for every $m \geqslant n$, $g_m|X_n = g_n|X_n$, and so g is well-defined and continuous.

For every $n \in \mathbf{N}$, take a homotopy $H_n : X \times I \to Z$ rel. X_n, from g_n to g_{n+1}, and define the function $H : X \times I \to Z$ by

$$H(x,t) = \begin{cases} H_n(x,(n+1)(n+2)t - n(n+2)), & \dfrac{n}{n+1} \leqslant t \leqslant \dfrac{n+1}{n+2} \\ g(x), & t = 1. \end{cases}$$

To prove the continuity of H, one must show that, for every $n \in \mathbf{N}$, $H|X_n \times I$ is continuous (see Proposition A.5.1 (i)). The continuity of $H|X_n \times I$ follows from the fact that, for every $(x,t) \in X_n \times (n/(n+1), 1]$, $H(x,t) = g(x)$. The map H is a homotopy rel. X_0 between g and g_0. □

Corollary A.5.8 *Let $\{X_n : n \in \mathbf{N}\}$ be an expanding sequence with union space X and such that X_n is a strong deformation retract of X_{n+1}, for every $n \in \mathbf{N}$. Then, X_0 is a strong deformation retract of X.*

Proof Let $i_n : X_n \to X_{n+1}, j_n : X_n \to X$ denote the respective inclusions and choose retractions $r_n : X_{n+1} \to X_n$ so that

$$i_n \circ r_n \simeq \mathbf{1}_{X_{n+1}}, \quad \text{rel. } X_n,$$

for every $n \in \mathbf{N}$. Now, for all $n, k \in \mathbf{N}$, define inductively the retractions $r_{n,k} : X_{n+k} \to X_n$ by taking $r_{n,0} = \mathbf{1}_{X_n}$ and $r_{n,k+1} = r_{n,k} \circ r_{n+k}$. For a fixed n, these retractions together yield retractions $s_n : X \to X_n$ such that

$s_n = r_n \circ s_{n+1}$; take $g_n = j_n \circ s_n : X \to X$. Notice that

$$g_n = j_n \circ r_n = j_{n+1} \circ i_n \circ r_n \circ s_{n+1}$$
$$\simeq j_{n+1} \circ s_{n+1} = g_{n+1}, \text{ rel. } X_n.$$

Since $g_n | X_n = j_n = \mathbf{1}_X | X_n$, for all $n \in \mathbf{N}$, it follows that $\mathbf{1}_X \simeq g_0$ rel. X_0. □

Given an expanding sequence of spaces, it is possible to derive from it another one whose union space has the same type but nicer properties than the original one; this is done by means of the telescope construction (see Milnor, 1962). More precisely, let $\{X_n : n \in \mathbf{N}\}$ be an expanding sequence of spaces and let X be its union space. For every $n \in \mathbf{N}$, define $\tilde{T}_n = X_n \times [n, n+1]$ and $T_n = \bigcup_{k=0}^{n} \tilde{T}_k$ as subspaces of $X \times [0, \infty)$. Because the inclusion $X_n \to X$ is a closed cofibration (see Proposition A.5.1 (iii)) and the same holds true for the inclusion $[n, n+1] \to [0, \infty)$, the inclusion $\tilde{T}_n \to X \times [0, \infty)$ is a closed cofibration (see Proposition A.4.2 (iv)). By induction, this shows that every inclusion $T_{n+1} = T_n \cup \tilde{T}_{n+1} \to X \times [0, \infty)$ is a closed cofibration (see Proposition A.4.2 (vii)), and, consequently, every inclusion $T_n \to T_{n+1}$ is a closed cofibration (see Proposition A.4.2 (vi)). Thus, the sequence $\{T_n\}$ is again an expanding sequence whose union space T is called the *telescope* of the expanding sequence $\{X_n\}$. From this definition, it is immediately clear that the telescope T can be considered as a subspace to the product $X \times [0, \infty)$.

The notion of telescope of an expanding sequence was originally introduced in (Milnor, 1962); we use the gluing of homotopies to derive a simple proof of the fundamental property of telescopes:

Corollary A.5.9 *The union space of an expanding sequence and its telescope have the same type.*

Proof Take an expanding sequence $\{Y_n : n \in \mathbf{N}\}$ defined by $Y_0 = T$ and $Y_n = X_n \times [0, \infty) \cup T$, for every $n \in \mathbf{N} \setminus \{0\}$; observe that each term is a strong deformation retract of its successor. Then, T is a strong deformation retract of $X \times [0, \infty)$ (see Corollary A.5.8). □

The final results of this section need some preparatory considerations. A map of pairs $\bar{f} : (Y, D) \to (X, A)$ is a *homotopy equivalence of pairs* if there is a map of pairs $\bar{g} : (X, A) \to (Y, D)$ such that:

(1) $\bar{g}$ is a homotopy inverse of $\bar{f}$ in the ordinary sense; and
(2) the homotopies connecting $\bar{g} \circ \bar{f}$ and $\bar{f} \circ \bar{g}$ to the respective identity maps move A and D respectively, within themselves.

The second condition is neither trivial nor automatic; however, it holds true if the inclusion maps $D \to Y$ and $A \to X$ are closed cofibrations. More precisely:

Lemma A.5.10 *Let $\bar{f} : (Y, D) \to (X, A)$ be a map of pairs which is an ordinary homotopy equivalence and assume that the induced map $f : D \to A$ is also a homotopy equivalence. Furthermore, suppose that the inclusions $i : D \to Y$ and $\bar{i} : A \to X$ are closed cofibrations. Then, $\bar{f}$ is a homotopy equivalence of pairs. Moreover, given a homotopy inverse g for f and a homotopy $H : f \circ g \simeq \mathbf{1}_A$, the needed homotopy inverse $\bar{g}$ of $\bar{f}$ and the homotopy $\bar{H} : \bar{f} \circ \bar{g} \simeq \mathbf{1}_X$ can be chosen to extend g and H, respectively.*

Remark The proof of this result is contained in Brown, 1988, Section 7.4; the proof given here is more direct.

Proof of Lemma A.5.10 Let g be a homotopy inverse of f and let H be a homotopy from $f \circ g$ to $\mathbf{1}_A$. First, take an extension $\bar{H} : X \times I \to X$ of $\bar{i} \circ H$ such that $\bar{H}(-, 1) = \mathbf{1}_X$; then, $\bar{h} = \bar{H}(-, 0)$ extends $f \circ g$. Next, choose a homotopy inverse of $\bar{f}$, say $\bar{k}$, a homotopy $\bar{K} : \bar{k} \circ \bar{f} \simeq \mathbf{1}_Y$ and an extension $\tilde{K}$ of $\bar{K} \circ (i \circ g \times \mathbf{1}_I)$ such that $\tilde{K}(-, 0) = \bar{k} \circ \bar{h}$; then $\bar{g} = \tilde{K}(-, 1)$ is again a homotopy inverse of $\bar{f}$, extends g, and therefore is a map of pairs $(X, A) \to (Y, D)$. The bulk of the work consists in constructing a homotopy from $\bar{f} \circ k$ to $\mathbf{1}_Y$ whose restriction to $D \times I$ is $\bar{i} \circ H$. To this end, one uses several times the homotopy extension property of the pair $(X \times I, X \times \{0, 1\} \cup A \times I)$ given by the product theorem (see Proposition A.4.2 (iv)).

Take $L_0 : X \times I \times I \to Y$ as a homotopy satisfying the following properties:

$(x, t, 1) \mapsto \bar{g}(x)$,	for $x \in X$ and $t \in I$,
$(x, 0, s) \mapsto \bar{K}(\bar{g}(x), s$,	for $x \in X$ and $s \in I$,
$(x, 1, s) \mapsto \tilde{K}(x, s)$,	for $x \in X$ and $s \in I$,
$(x, t, s) \mapsto \bar{K}(i \circ g(x), s)$,	for $x \in A$ and $t, s \in I$.

Thus, $L_0(-, -, 0)$ is a homotopy from $\bar{k} \circ \bar{f} \circ \bar{g}$ to $\bar{k} \circ \bar{h}$, rel. A. Next, let $\hat{H}$ be a homotopy from $\bar{f} \circ \bar{k}$ to $\mathbf{1}_X$ and take $L_1 : X \times I \times I \to X$ as a homotopy satisfying the following properties:

$(x, t, 0) \mapsto \bar{f}(L_0(x, t, 0))$,	for $x \in X$ and $t \in I$,
$(x, 0, s) \mapsto \hat{H}(\bar{f} \circ \bar{g}(x), s)$,	for $x \in X$ and $s \in I$,
$(x, 1, s) \mapsto \hat{H}(\bar{h}(x), s)$,	for $x \in X$ and $s \in I$,
$(x, t, s) \mapsto \bar{H}(\bar{h} \circ \bar{i}(x), s)$,	for $x \in A$ A and $t, s \in I$.

Now, $L_1(-, -, 1)$ is a homotopy from $\bar{f} \circ \bar{g}$ to $\bar{h}$, rel. D.

Let $\bar{G} : X \times I \to X$ be an extension of $i \circ H$ to a homotopy from $\bar{f} \circ \bar{g}$ to a map $\tilde{h}$; then, take $L_2 : X \times I \times I \to X$ as a homotopy such that:

$$\begin{array}{ll} (x,t,0) \mapsto L_1(x,t,1), & \text{for } x \in X \text{ and } t \in I, \\ (x,0,s) \mapsto \bar{G}(x,s), & \text{for } x \in X \text{ and } s \in I, \\ (x,1,s) \mapsto \tilde{H}(x,s), & \text{for } x \in X \text{ and } s \in I, \\ (x,t,s) \mapsto \bar{i} \circ H(x,s), & \text{for } x \in A \text{ and } t, s \in I. \end{array}$$

Without loss of generality, one may assume that the homotopy H has the following property: $H(x,t) = x$, for all $x \in A$ and all $t \in [\frac{1}{2}, 1]$. This allows to find a homotopy $L_3 : X \times I \times I \to X$ satisfying the conditions:

$$\begin{array}{ll} (x,t,1) \mapsto L_2(x,t,1), & \text{for } x \in X \text{ and } t \in I, \\ (x,0,s) \mapsto \bar{G}(x,s), & \text{for } x \in X \text{ and } s \in I, \\ (x,1,s) \mapsto x, & \text{for } x \in X \text{ and } s \in I, \\ (x,t,s) \mapsto \bar{i} \circ H(x,t+s), & \text{for } x \in A, s \in I \text{ and } t+s \leqslant \frac{1}{2}, \\ (x,t,s) \mapsto \bar{i}(x), & \text{for } x \in A, t, s \in I \text{ and and } t+s \geqslant \frac{1}{2}. \end{array}$$

Then, $L_3(-,-,0)$ is a homotopy connecting $\bar{f} \circ \bar{g}$ to $\mathbf{1}_X$ and extending the given homotopy H on A.

Thus, $\bar{g}$ is a right homotopy inverse to $\bar{f}$ and is of the desired kind. Similarly, one finds a right homotopy inverse $\hat{f}$ for $\bar{g}$, such that the homotopy connecting $\bar{g} \circ \hat{f}$ to $\mathbf{1}_Y$ induces a homotopy on D. In the string

$$\bar{g} \circ \bar{f} \simeq \bar{g} \circ \bar{f} \circ \bar{g} \circ \hat{f} \simeq \bar{g} \circ \hat{f} \simeq \mathbf{1}_Y,$$

all homotopies deform A (respectively, D) into itself, showing finally that $\bar{g}$ can be considered as a homotopy inverse for $\bar{f}$ in the category of pairs of spaces. □

The compatibility of homotopy equivalences of expanding sequences and union spaces (see proposition below) has been made explicit in tom Dieck (1971).

Proposition A.5.11 *Let*

$$\begin{array}{ccccccc} Y_0 & \longrightarrow & Y_1 & \longrightarrow & Y_2 & \longrightarrow & \cdots \\ \downarrow{\scriptstyle f_0} & & \downarrow{\scriptstyle f_1} & & \downarrow{\scriptstyle f_2} & & \\ X_0 & \longrightarrow & X_1 & \longrightarrow & X_2 & \longrightarrow & \cdots \end{array}$$

be a commutative ladder between two expanding sequences where f_n is a homotopy equivalence, for every $n \in \mathbf{N}$. Then, the map $f : Y \to X$, where Y, X are the union spaces and f is the induced map, is a homotopy equivalence.

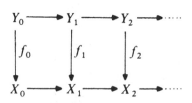

Proof See tom Dieck (1971, Lemma 6). Construct inductively right homotopy inverses g_n for the maps f_n, $n\in\mathbf{N}$ such that g_{n+1} extends g_n and homotopies $H_n : X_n \times I \to X_n$ such that H_{n+1} extends H_n (see Lemma A.5.10). This gives rise to a map $g : X \to Y$ and a homotopy $H : f\circ g \simeq \mathbf{1}_X$. Thus, g is a right homotopy inverse of f. As in the proof of the lemma, one shows that this is also a left homotopy inverse for f. □

A.6 Absolute neighbourhood retracts in the category of metric spaces

Proposition A.6.1 (*Kuratowski–Wojdysławski embedding theorem*) *A metric space X can be embedded in the normed vector space $Z = C(X, \mathbf{R})$ of all bounded maps $X \to \mathbf{R}$ by an isometry $i : X \to Z$ such that $i(X)$ is closed in its convex hull $H(i(X))$.*

Proof See Mardešić & Segal (1982, Chapter I, Theorem 3.2). □

The definition of absolute neighbourhood retract is based on the next result.

Proposition A.6.2 *The following two conditions on a metric space X are equivalent:*

(i) *for every metric space Z, every closed subspace $C\subset Z$ and every map $f : C\to X$, there are a neighbourhood U of C in Z and a map $\bar{f} : U \to X$ such that $\bar{f}|C = f$;*

(ii) *if X is a closed subspace of a metric space Z then there are a neighbourhood U of X and a map $r : U \to X$ such that $r|X = \mathbf{1}_X$.*

Proof See Mardešić & Segal (1982, Chapter I, Theorem 3.1 (ii)). □

A metric space X satisfying the equivalent conditions of Theorem A.6.2 is called an *absolute neighbourhood retract* (abbreviated to ANR). Note specifically that the empty space is an ANR. A wide class of models of ANRs is given by the following:

Proposition A.6.3 (*Dugundji extension theorem*) *A convex subset of a normed linear space is an ANR.*

Proof See Mardešić & Segal (1982, Chapter I, Theorem 3.3). □

As a consequence of this result, Euclidean spaces, balls, geometric simplices and cubes are ANRs.

Proposition A.6.4 *A retract of an ANR is an ANR; an open subset of an ANR is an ANR.*

Proof Let X be an ANR, $Y \subset X$ and $r : X \to Y$ be a retraction. Given any closed subset C of a metric space Z and any map $f : C \to Y$, the composition of f with the inclusion map of Y into X can be extended to a map $\tilde{f} : U \to X$, where U is a neighbourhood of C in Z. Then the map $\bar{f} = r \circ \tilde{f}$ extends f over U.

As for the second part, let W be an open subset of X and let $f : C \to W$ be a map defined on a closed subspace C of a metric space Z. Extend the composition of f with the inclusion $W \subset X$ to a map $\tilde{f} : V \to X$, where V is a neighbourhood of C in Z. Then, $\tilde{f}^{-1}(W)$ is a neighbourhood of C in Z and $\bar{f} = \tilde{f} | \tilde{f}^{-1}(W)$ extends f over $\tilde{f}^{-1}(W)$. □

The class of ANRs has finite products.

Proposition A.6.5 *If X_1 and X_2 are ANRs, so is $X_1 \times X_2$.*

Proof Let $C \subset Z$ be closed and let $f = (f_1, f_2) : C \to X_1 \times X_2$ be a given map. The maps f_i give rise to neighbourhoods U_i of C in Z and maps $\bar{f_i} : U_i \to X_i$ extending f_i, $i = 1,2$. Define $U = U_1 \cap U_2$ and $\bar{f} = (\bar{f_1} | U, \bar{f_2} | U) : U \to X_1 \times X_2$. □

Proposition A.6.6 *Let X be a metric space which is the union of two closed subspaces X_1 and X_2. If $X_0 = X_1 \cap X_2$, X_1 and X_2 are ANRs, so is X.*

Proof See Borsuk (1967, Chapter IV, Theorem 6.1). □

In contrast to open subspaces, closed subspaces of ANRs are not ANRs in general. A condition for this to happen will be given next.

Proposition A.6.7 *Let X be an ANR and let A be a closed subset of X. The following are equivalent:*

(i) *the inclusion $i : A \to X$ is a closed cofibration;*
(ii) *A is an ANR.*

Proof (i)$\Rightarrow$(ii): Take a neighbourhood U of A in X which is deformable to A in X, rel A (see Proposition A.4.1 (iv)). Let $H : U \times I \to X$ be a corresponding deformation. Notice that $H(-,1)| \mathring{U}$ is a retraction of $\mathring{U}$ to A. But $\mathring{U}$ is an ANR, as an open subspace of the ANR X; thus, A is an ANR as a retract of $\mathring{U}$ (see Proposition A.6.4).

(ii)⇒(i): If A is an ANR, then $\hat{X} = X \times 0 \cup A \times I$ is an ANR (see Propositions A.6.5 and A.6.6). Since $\hat{X}$ is a closed subset of $X \times I$, there are a neighbourhood V of $\hat{X}$ in $X \times I$ and a retraction $r : V \to \tilde{X}$.

Observe that, for every $a \in A$ there is an open neighbourhood W_a of a in X such that $W_a \times I \subset V$. It follows that $W = \bigcup_{a \in A} W_a$ is an open neighbourhood of A in X and that $W \times I \subset V$. Using Urysohn's lemma, construct a map $\alpha : X \to I$ with $\alpha(X \backslash W) = \{0\}, \alpha(A) = \{1\}$ and define a map $\varphi : X \times I \to V$ by

$$\varphi(x, t) = (x, \alpha(x)t).$$

The composition $r \circ \varphi : X \times I \to \hat{X}$ is a retraction, and therefore the inclusion of A into X is a closed cofibration (see Proposition A.4.1 (ii)).

□

Corollary A.6.8 *Every ANR is an LEC space.*

Proof If X is an ANR, ΔX is also an ANR, being homeomorphic to X and closed in $X \times X$. □

Theorem A.6.9 *A metric space which is the union of countably many open ANRs is an ANR.*

Proof See Borsuk (1967, Chapter 4, Theorem 10.2). □

The last proposition in this section shows that spaces of functions from compact spaces to ANRs are themselves ANRs; more precisely:

Proposition A.6.10 *Let C_0 be a subset of a compact space C and let x_0 be a point of an ANR space X; then, the function space $(X, x_0)^{(C, C_0)}$ is an ANR.*

Proof See Borsuk (1967, Chapter IV, Theorem 5.1). □

A.7 Simplicial homology

The reader is assumed to be familiar with the basic concepts of homological algebra as developed in Northcott (1960), MacLane (1963) or Hilton & Stammbach (1971).

For any given simplicial set $X = \bigsqcup_{n=0}^{\infty} X_n$ and any $n \in \mathbf{N}$, let $C_n X = FA(X_n)$ be the free abelian group generated by all n-simplices $x \in X_n$; define homomorphisms

$$d_n : C_n X \to C_{n-1} X$$

by setting

$$d_n(x) = \sum_{i=0}^{n} (-1)^i \cdot x\delta_i$$

where $x \in X_n$ and δ_i, $i = 0, 1, \ldots, n$ are the elementary face operators (see Section 4.1). A simplicial map $f: Y \to X$ induces a chain homomorphism

$$Cf = \{C_n f = (Cf)_n : n \in \mathbf{N}\} : CY \to CX;$$

in this way, we obtain the *chain complex functor* $C(-)$ from *SiSets* to the category of chain complexes.[†]

The (integral) *homology* of a simplicial set X is defined as the homology of the chain complex CX (see Northcott, 1960, Section 4.6) and is denoted by

$$H(X) = \{H_n(X) = H_n(CX) : n \in \mathbf{N}\}.$$

This definition can be extended, giving rise to the homology functor from the category *SiSets* to the category of graded abelian groups *GAG*. This latter functor is obviously defined as the composition of the chain complex functor $C(-)$ with the homology functor $H(-)$; the graded homomorphism induced by a simplicial map $f : Y \to X$ will be denoted by

$$f_* = \{f_{*n} : n \in \mathbf{N}\}.$$

Example 1 The simplicial set $\Delta[0]$ has one simplex in every dimension; thus, for every $n \in \mathbf{N}$, $C_n \Delta[0] \cong \mathbf{Z}$; moreover, $d_n = 0$ if n is odd, and $d_n = 1$ if n is even. Consequently, $H_0(\Delta[0]) \cong \mathbf{Z}$, and, for every $n \in \mathbf{N} \backslash \{0\}$, $H_n(\Delta[0]) = 0$. □

Proposition A.7.1 *If* $f_\lambda : Y \to X, \lambda = 0, 1$ *are homotopic simplicial maps, then* $(f_0)_* = (f_1)_*$. *Consequently, simplicial homotopy equivalences induce isomorphisms on homology.*

Proof See Gabriel & Zisman (1967, Appendix II, Lemma 1.4), Northcott (1960, Theorem 4.7) or Lamotke (1968, V.2.3). □

Example 2 The standard simplex $\Delta[p]$ contains the simplex $\Delta[0]$ as a strong deformation retract (see Example 3, Section 4.2). Consequently, $H(\Delta[p]) \cong H(\Delta[0])$, computed in the example above. Similar observations can be made with respect to the normal subdivision $\Delta'[p]$ of $\Delta[p]$ (see Lemma 4.6.1) and the simplicial sets Ex $\Delta'[p]$ (see Lemma 4.6.15 (iii)).

□

[†] Do not confuse the chain complex functor $C(-)$ with the cone functors $C: Top \to Top$ (see Section A.4) or $C : PSiSets \to PSiSets$ (see Section 4.4).

We now turn to geometry. The *singular homology functor* is the functor $Top \to GAG$ obtained by composition of the singular set functor with the homology functor from *SiSets* to *GAG*. The following notation will be used in this context:

$$H(X) = \{H_n(X) = H_n(SX) : n \in \mathbf{N}\}$$

for any space X and

$$f_* = \{f_{*n} = (Sf)_{*n} : n \in \mathbf{N}\}$$

for a map f.

The next result is attributed to S. Eilenberg.

Theorem A.7.2 *There is a natural isomorphism between the homologies of a simplicial set and its geometric realization.*

Proof See Gabriel & Zisman (1967, Appendix II.1). □

In this book only the following fact is needed.

Proposition A.7.3 *The geometric realization* $|\mathrm{e}_X|$ *of the natural map* $\mathrm{e}_X : X \to \mathrm{Ex}\, X$ *induces an isomorphism on homology.*

Proof As in the proof of the fact that $|\mathrm{e}_X|$ induces an isomorphism between the fundamental groups (see Proposition 4.6.16), one concludes, from the existence of a left homotopy inverse for the map $|\mathrm{e}_X|$, that

$$|\mathrm{e}_X|_* : H(|X|) \to H(|\mathrm{Ex}\, X|)$$

is a monomorphism. The remainder of the proof can be done at the simplicial level (see Theorem A.7.2) and is given in the next lemma.

Lemma A.7.4 *For every simplicial set* X, *the natural map* $\mathrm{e}_X : X \to \mathrm{Ex}\, X$ *induces an epimorphism on homology.*

Proof The essence of this proof is to construct a natural right chain homotopy inverse for the chain homomorphism $C(\mathrm{e}_X)$. The chain homomorphism $f : C(\mathrm{Ex}\, X) \to CX$ needed to achieve that goal will be defined inductively. First of all, note that

$$(\mathrm{e}_X)_0 : X_0 \to (\mathrm{Ex}\, X)_0$$

is a natural bijection, whose inverse is used to define f_0 on the generators. Next, given a generator $x : \Delta'[1] \to X$ of $C_1(\mathrm{Ex}\, X)$, define

$$f_1(x) = x((\varepsilon_0, \iota)) - x((\varepsilon_1, \iota)).$$

Assume $f_n:C_n\,(\mathrm{Ex}\,X)\to C_nX$ constructed for $n\geqslant 1$ and all simplicial sets X. Consider the generator $\mathbf{1}:\Delta'[n+1]\to\Delta'[n+1]$ of $C_{n+1}(\mathrm{Ex}\,\Delta'[n+1])$ and choose, once and for all, a chain $d\in C_{n+1}(\Delta'[n+1])$ which is mapped onto $f_n\circ d_{n+1}(\mathbf{1})$ by the boundary homomorphism; the choice of d is possible since the simplicial sets $\Delta'[n]$ have the same homology as $\Delta[0]$ (see Example 2), and therefore the chain complexes $C(\Delta[n])$ are exact at all places with $k>0$. Now, if X is an arbitrary simplicial set and $X:\Delta'[n+1]\to X$ is a generator of $C_{n+1}(\mathrm{Ex}\,X)$, define

$$f_{n+1}(x)=C_{n+1}x(d).$$

This completes the definition of the chain homomorphisms f.

It remains to define a chain homotopy $s=\{s_n : n\in\mathbf{N}\}$ between the composite chain homomorphism $C(e_X)\circ f$ and the identity. By construction, $C_0(\mathrm{e}_X)\circ f_0$ is equal to the identity, and therefore one can take $s_0=0$. Again, assume that

$$s_n : C_n(\mathrm{Ex}\,X)\to C_{n+1}(\mathrm{Ex}\,X)$$

has been suitably constructed. Observe that the simplicial sets $\mathrm{Ex}\,\Delta'[n]$ have the same homology as $\Delta[0]$ (see Example 2). As before, take the generator $\mathbf{1}:\Delta'[n+1]\to\Delta'[n+1]$ of $C_{n+1}(\mathrm{Ex}\,\Delta'[n+1])$ and obtain a chain $c\in C_{n+2}(\mathrm{Ex}\,\Delta'[n+1])$ which is mapped onto

$$C_{n+1}(\mathrm{e}_X)\circ f_{n+1}(\mathbf{1})-\mathbf{1}-s_n\circ d_{n+1}(\mathbf{1})$$

by the boundary homomorphism. For an arbitrary simplicial set X, take a generator $x\in\Delta'[n+1]\to X$ of $C_{n+1}(\mathrm{Ex}\,X)$ and define

$$s_{n+1}(x)=C_{n+2}(c).$$

This completes the definition of the chain homotopy s and the proof of the lemma. □

Remark The method of proof just given can be used under various circumstances. The common features of these proofs are subsumed by referring to a proof 'using acyclic models'. This technique is due to S. Eilenberg and S. Mac Lane (see Eilenberg & Mac Lane, 1953); the reader can find a sophisticated treatment of the theory in Dold (1972).

A.8 Homotopy groups, *n*-connectivity, fundamental groupoid

Homotopy groups: absolute case

A map between based spaces is called a *based map* if it preserves the base point; a homotopy between two based maps is called a *based homotopy* whenever it is constant on the base point. Based spaces and based homotopy classes of based maps form the category $hTop_*$. If (Y,y_0) and

(X, x_0) are based spaces, the set of all based homotopy classes of based maps $(Y, y_0) \to (X, x_0)$ will be denoted by $[Y, X]_*$.

The case in which $(Y, y_0) = (S^n, e_0)$, $n \in \mathbf{N}$ is of special interest; in this situation,

$$\pi_n(X, x_0) = [S^n, X]_*$$

is the *n-th homotopy group* of the based space (X, x_0).

Every homotopy group has a distinguished element, namely, the class of the constant map. Thus, every $\pi_n(X, x_0)$ is a pointed set. Observe that $\pi_0(X, x_0)$ can be considered as the set of all path-components of X; its distinguished element is just the path-component containing the base point. Write $\pi_0(X)$ or simply $\pi(X)$ for the set of all path-components of the space X. The higher homotopy groups (for $n > 0$) live in the distinguished path-component; they have a group structure – induced by the pinching of the sphere (see page 7) – which is abelian for $n > 1$ (see Spanier, 1966, Theorem 1.6.8).

Example 1 $\pi_n(B^n, e_0)$ is the trivial group, for all $n > 0$.

Example 2 The assignment $1 \mapsto [\mathbf{1}_{S^n}]$ induces an isomorphism $\mathbf{Z} \to \pi_n(S^n, e_0)$, for all $n > 0$ (see tom Dieck, Kamps & Puppe, 1970, Section 16).

The group $\pi_1(X, x_0)$ is called the *fundamental* (*Poincaré*) *group of* (X, x_0). The fundamental group of a space does not have to be abelian: the fundamental group of the space $X = S^1 \vee S^1$ – *figure eight* – is the free group in two generators.

A space X is said to be *simply connected* if X is path-connected and its fundamental groups (with respect to any base point) are trivial.

For each $n > 0$, every path $\sigma : I \to X$ gives a natural isomorphism

$$\sigma_n : \pi_n(X, x_1) \to \pi_n(X, x_0), \quad x_0 = \sigma(0), \quad x_1 = \sigma(1),$$

which depends only on the homotopy class of the path σ, rel. end-points (see Spanier, 1966, Theorem 7.3.8).

As an immediate consequence of this fact, one can deduce that the fundamental group $\pi_1(X, x_0)$ acts on $\pi_n(X, x_0)$, $n \geqslant 1$. In particular, if $n = 1$, this is an action by inner automorphisms; if $n > 1$, it extends to an action of the integral group ring $\Lambda = \mathbf{Z}\pi_1(X, x_0)$, thus making $\pi_n(X, x_0)$ a left Λ-module (see Spanier, 1966, Theorem 7.3).

If (Y, y_0) is a second based space and $f : (X, x_0) \to (Y, y_0)$ is a based map, composition with f induces a collection of functions

$$f_\# = \{f_n : \pi_n(X, x_0) \to \pi_n(Y, y_0) : n \geqslant 0\}.$$

The functions f_n are homomorphisms if $n \geqslant 1$; f_0 preserves the distinguished element.

Based spaces and based maps form a category, denoted by Top_*. The constructions described before yield functors

$$\pi_0 : Top_* \to pointed\ sets$$
$$\pi_1 : Top_* \to groups$$
$$\pi_n : Top_* \to Abelian\ groups, \quad n > 1.$$

Homotopic maps induce the same homomorphisms, in the following sense.

Proposition A.8.1 *Let $H : Y \times I \to X$ be a homotopy from f to g. Then, for every $y_0 \in Y$,*

$$g_n = \sigma_n \circ f_n$$

where σ denotes the path $H | \{y_0\} \times I$ from $f(y_0)$ to $g(y_0)$. In particular, if H is a based homotopy, $g_n = f_n$.

Proof See Spanier 1966, Theorem 7.3.14. □

Thus, any map $f : X \to X$ homotopic to the identity map induces isomorphisms of the homotopy groups. More generally, given maps $f : Y \to X$, $g : X \to Y$ with $g \circ f \simeq \mathbf{1}_Y$, it follows that f induces monomorphisms and g induces epimorphisms of the homotopy groups. If the spaces X and Y are related by maps f and g as above, one says that X *dominates* Y (or that Y *is dominated by* X).

The following is a consequence of these observations:

Corollary A.8.2 *If $f : Y \to X$ is a homotopy equivalence, then, for every $y_0 \in Y$ and every $n \in \mathbf{N}$, the homomorphisms*

$$f_n : \pi_n(Y, y_0) \to \pi_n(X, f(y_0))$$

are isomorphisms. □

This suggests a lessening of the notion of homotopy equivalence. A map $f : Y \to X$ is said to be a *weak homotopy equivalence* if $Y \neq \varnothing$ and if for every point $y_0 \in Y$ the induced homomorphisms

$$f_n : \pi_n(Y, y_0) \to \pi_n(X, f(y_0))$$

are isomorphisms; notice that it would suffice to require the latter condition for just one point in every path-component of Y. Clearly, any homotopy equivalence with non-empty domain has this property. The following is an example (often quoted but seldom explained) of a weak homotopy equivalence which is not a homotopy equivalence.

Example 3 In the plane $\mathbf{R}^2$, for every $n\in\mathbf{N}\backslash\{0\}$, consider the following segments:

A_n, the segment with vertices $(-1,0)$ and $(0,1/n)$;
B_n, the segment with vertices $(0,-1/n)$ and $(1,0)$.

Moreover, let C be the segment with vertices $(-1,0)$ and $(1,0)$ and take the subspace of $\mathbf{R}^2$

$$Z=\left(\bigcup_{n\in\mathbf{N}\backslash\{0\}} A_n\right)\cup C\cup\left(\bigcup_{n\in\mathbf{N}\backslash\{0\}} B_n\right),$$

based at the origin $p=(0,0)$ (compare with Example 4, Section 5.1).

Consider the constant map $k:Z\to\{p\}$. Then, the following two statements hold true:

(1) k is a weak homotopy equivalence (because all homotopy groups $\pi_n(Z,p)$ vanish);
(2) k is not a homotopy equivalence (because k has no homotopy inverse).

Proof of (1): Consider a based map $a:(S^n,e_0)\to(Z,p)$ and let

$$\tilde{A}=\left\{s\in S^n : a(s)\in\bigcup_{n=1}^{\infty} A_n\backslash\{(-1,0)\}\right\};$$

notice that $\tilde{A}$ is an open subset of S^n. Let $H:S^n\times I\to Z$ be the function defined by

$$H(s,t)=\begin{cases} a(s), & s\notin\tilde{A},\\ (1-t)\,a(s)-(t,0), & s\in\tilde{A}\end{cases}.$$

The function H is continuous. This follows from the fact that, for all sequences $\{(s_i,t_i)\in\tilde{A}\times I : i\in\mathbf{N}\}$ that converge to a point, $(s,t)\in(S^n\backslash\tilde{A})\times I$, $a(s)=(-1,0)$. Assume that $a(s)\neq(-1,0)$; then, $U=Z\backslash\{(-1,0)\}$ is a neighbourhood of $a(s)\in C$. Since S^n is locally path-connected, there exists a path $\omega:I\to a^{-1}(U)$ connecting s to some s_i; hence, $a\circ\omega$ is a path in U connecting $a(s)$ and $a(s_i)$, which is impossible.

Hence, H is a based homotopy from a to a map a' whose image is contained in $C\cup(\bigcup_{n=1}^{\infty}B_n)$. Similarly, deform the map a' to a map a'' whose image is contained in C; but a'' is clearly homotopic to the constant map from S^n to p. Gluing all these homotopies together, one obtains that $[a]=0\in\pi_n(Z,p)$.

Proof of (2): The existence of a homotopy inverse for k would require the existence of a homotopy $H:Z\times I\to Z$ such that $H_0=\mathbf{1}_Z$ and

$H_1(z) = p = (0,0)$, for every $z \in Z$. Consider the sequences

$$\{z_n = (0, 1/n) : n \in \mathbf{N} \backslash \{0\}\}$$

and

$$\{-z_n = (0, -1/n) : n \in \mathbf{N} \backslash \{0\}\}.$$

Notice that both converge to the point p.

Since $H_0(z_n) = z_n$ and $H_1(z_n) = p$, there must be a $t_n \in I$ such that $H(z_n, t_n) = (-1, 0)$, for every $n \in \mathbf{N} \backslash \{0\}$. The sequence $\{t_n\}$ has a cluster point and hence, without loss of generality, one may assume that it converges, say

$$\lim_{n \to \infty} t_n = t_0.$$

From the continuity of H it follows that

$$\begin{aligned} H(p, t_0) &= H\left(\lim_{n \to \infty} z_n, \quad \lim_{n \to \infty} t_n \right) \\ &= \lim_{n \to \infty} H(z_n, t_n) = (-1, 0). \end{aligned}$$

Thus, the set $\{t \in I : H(p, t) = (-1, 0)\}$ is non-empty and compact. Assume t_0 to be its minimum.

Observe that

$$\lim_{n \to \infty} H(-z_n, t_0) = (-1, 0)$$

implies that, at least for a subsequence of the sequence $\{-z_n\}$, the values $H(-z_n, t_0)$ must be contained in $C \cup A_n$. Again, assume that this is true for all $n \in \mathbf{N} \backslash \{0\}$. Therefore one can find $t'_n \leqslant t_0$ such that $H(-z_n, t'_n) = (1, 0)$, for every $n \in \mathbf{N} \backslash \{0\}$. As above, one obtains a $t'_0 < t_0$ with $H(p, t'_0) = (1, 0)$. Finally, repeating this argument, one obtains a $t''_0 < t'_0$ such that $H(p, t''_0) = (-1, 0)$, contradicting the minimality of t_0. □

Covering projections

Based covering projections form another class of based maps which induce interesting homomorphisms between homotopy groups.

Proposition A.8.3 *If* $p : (\tilde{X}, \tilde{x}_0) \to (X, x_0)$ *is a based covering projection then, for* $n = 1$, *the induced homomorphism* p_1 *is a monomorphism and, for* $n > 1$, p_n *is an isomorphism.*

Proof See Hilton & Wylie (1960, Theorem 6.5.10). □

A partial converse to the previous result says that an inclusion of a subgroup into the fundamental group of a space can be realized, under mild conditions on the space, as the induced homomorphism of a covering projection:

Proposition A.8.4 *Given a connected and locally contractible based space* (X, x_0) *and a subgroup* $\pi \subset \pi_1(X, x_0)$, *there is a based covering projection* $p : (\tilde{X}, \tilde{x}_0) \to (X, x_0)$ *such that* $p_1(\pi_1(\tilde{X}, \tilde{x}_0)) = \pi$.

Proof See Hilton & Wylie (1960, Theorem 6.6.11). □

An important property of covering projections is the following:

Theorem A.8.5 (*Lifting theorem*) *Let* $p : (\tilde{X}, \tilde{x}_0) \to (X, x_0)$ *be a based covering projection and let* $f : (Y, y_0) \to (X, x_0)$ *be a based map such that*

$$f_1(\pi_1(Y, y_0)) \subset p_1(\pi_1(\tilde{X}, \tilde{x}_0)).$$

Then, if Y *is connected and locally path-connected, there is a unique based map* $\tilde{f} : (Y, y_0) \to (\tilde{X}, \tilde{x}_0)$ *which lifts* f, *that is to say, such that* $p \circ \tilde{f} = f$.

Proof See Hilton & Wylie (1960, Lemma 6.6.12). □

If $p : \tilde{X} \to X$ is a covering projection, a homeomorphism $\tilde{\alpha} : \tilde{X} \to \tilde{X}$ such that $p \circ \tilde{\alpha} = p$ is called a *covering transformation.* The covering transformations of a covering projection form a group $G(p)$ under composition.

Theorem A.8.6 *Let* $p : \tilde{X} \to X$ *be a covering projection, with* X *locally path-connected. If* $\tilde{X}$ *is simply connected, then* $G(p) \cong \pi_1(X, x_0)$, *for any choice of base point. Moreover, given any two points* $\tilde{x}_0, \tilde{x}_1 \in \tilde{X}$ *such that* $p(\tilde{x}_0) = p(\tilde{x}_1)$, *there is a unique covering transformation* $\tilde{\alpha} : \tilde{X} \to \tilde{X}$ *such that* $\tilde{\alpha}(\tilde{x}_0) = \tilde{x}_1$.

Proof See Hilton & Wylie (1960, Corollary 6.7.4 and Proposition 6.6.17). □

A simply connected covering space of a space X is called a *universal covering of* X; the corresponding covering projection is called *universal covering projection.*

Propositon A.8.7 *If* $p: \tilde{X} \to X$ *is a universal covering projection and* $f : Y \to X$ *induces an isomorphism of fundamental groups, then* $Y \sqcap_f \tilde{X} \to Y$ *is a universal covering projection.*

Proof Follow from Propositions A.4.17 and A.8.3. □

Proposition A.8.8 *Let $p : \tilde{X} \to X$, $q : \tilde{Y} \to Y$ be universal covering projections. A based map $f : Y \to X$ is a weak homotopy equivalence iff it induces an isomorphism of the fundamental groups and the lifting $\tilde{f} : \tilde{Y} \to \tilde{X}$ induces isomorphisms of the integral homology groups.*

Proof The map $\tilde{f} : \tilde{Y} \to \tilde{X}$ is a lifting of $f \circ q : \tilde{Y} \to X$ (see Theorem A.8.5); now use Proposition A.8.3 and Spanier (1966, Theorem 7.6.25). □

Homotopy groups of maps; relative homotopy groups

Relative homotopy groups as developed by Eckmann & Hilton (1958) play an important role in homotopy theory. The abstract framework for their development is the category Top^2_* *of based maps in* Top_*. The objects of Top^2_* are the based maps $f : (Y, y_0) \to (X, x_0)$; its morphisms, say from f to f', are the *admissible pairs* $(b, a) : f \to f'$, namely, pairs of based maps $a : dom\, f \to dom\, f'$, $b : cod\, f \to cod\, f'$ with $f' \circ a = b \circ f$. A *homotopy* between two admissible pairs, say from (b, a) to (b', a'), is an admissible pair $(H, G) : f \times 1_I \to f'$ such that $H_0 = b$, $H_1 = b'$, $G_0 = a$ and $G_1 = a'$. Clearly, this homotopy is an equivalence relation. The homotopy class of an admissible pair (b, a) is represented by $[b, a]$. Given a based map f, i.e., an object of Top^2_*, the set of all homotopy classes of admissible pairs from the based map $i^{n-1} : (S^{n-1}, e_0) \to (B^n, e_0)$ into f is denoted by $\pi_n(f, y_0)$.† In the special case where f is the inclusion of a subspace A into a space X, one also writes $\pi_n(f, y_0) = \pi_n(X, A; x_0)$; moreover, in this situation the admissible pair $(b, a) : i^{n-1} \to f$ is nothing but a map $b : (B^n, S^{n-1}) \to (X, A)$, and thus the homotopy class of (b, a) is denoted simply by $[b]$. On the other hand, if f is a constant map $Y \to \{y_0\}$, then, $\pi_n(f, y_0) = \pi_{n-1}(Y, y_0)$.

Then set $\pi_n(f, y_0)$ is defined for $n > 0$. It has a distinguished element, namely, the class consisting of the admissible pairs (b, a) for which there exists an extension $b' : B^n \to dom\, f$ of a such that $f \circ b'$ is homotopic to b, rel. S^{n-1}(the proof is similar to that given in Theorem 7.2.1 of Spanier, 1966, for inclusions). Thus, every $\pi_n(f, y_0)$ is at least a pointed set.

As in the absolute case, if $n > 1$, the units, inversions and pinchings introduced on page 6 provide the set $\pi_n(f, y_0)$ with the structure of a group which is abelian if $n \geqslant 3$. The group $\pi_n(f, y_0)$ is the *nth-homotopy group* of (f, y_0).

An admissible pair $(b, a) : f \to f'$ induces, by composition, a homomorphism (of pointed sets or groups)

$$(b, a)_n : \pi_n(f, y_0) \to \pi_n(f', y'_0);$$

thus, π_n becomes a functor on Top^2_*.

†Warning: do not confuse $\pi_n(f, y_0)$ with the induced homomorphism f_n.

Now consider a commutative triangle of based maps

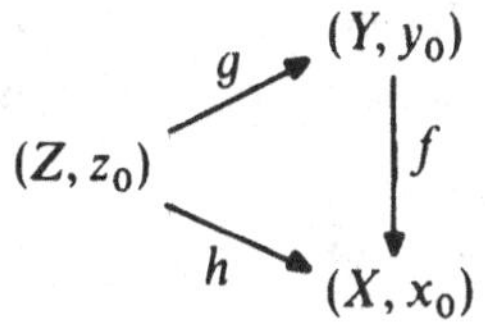

and define the natural homomorphisms

$$\delta_{n+1} : \pi_{n+1}(f, y_0) \to \pi_n(g, z_0)$$

by taking a representative $(b, a) : i^n \to f$ for each element of $\pi_{n+1}(f, y_0)$ and associating to it the class of the admissible pair $(a \circ b^n, c) : i^{n-1} \to g$, where $b^n: B^n \to S^n$ is the map introduced on page 6 and $c: S^{n-1} \to Z$ is the constant map. The fact that δ_{n+1} is a homomorphism comes out of the equation

$$\dot{p}^n \circ b^n = (b^n \vee b^n) \circ p^{n-1}.$$

The groups (pointed sets) together with the morphisms defined up to now can be arranged into a long exact sequence; more precisely:

Theorem A.8.9 *The sequence of groups (respectively pointed sets)*

$$\cdots \to \pi_{n+1}(f, y_0) \xrightarrow[\delta_{n+1}]{} \pi_n(g, z_0) \xrightarrow[(f,1)_n]{} \pi_n(h, z_0) \xrightarrow[(1,g)_n]{}$$

$$\pi_n(f, y_0) \xrightarrow[\delta_n]{} \pi_{n-1}(g, z_0) \to \cdots \to \pi_2(f, y_0) \xrightarrow[\delta_2]{}$$

$$\pi_1(g, z_0) \xrightarrow[(f,1)_1]{} \pi_1(h, z_0) \xrightarrow[(1,g)_1]{} \pi_1(f, y_0)$$

is exact.

Proof The proof is done by the standard arguments in diagram chasing. The only non-trivial step is to prove exactness at $\pi_n(g, z_0)$.

(1) $(f, 1)_n \circ \partial_{n+1} = 0$: Take a representative (b, a) for an element of $\pi_{n+1}(f, y_0)$. Its image is the pair $(f \circ a \circ b^n, c)$. Let $c' : (B^n, e_0) \to (Z, z_0)$ be the constant based map; then, $c' | S^{n-1} = c$, and, moreover, the composite map $b \circ c^n \circ (b^n \times 1)$ is a homotopy from $f \circ a \circ b^n$ to $h \circ c'$ rel. S^{n-1}.

(2) $ker(f, 1)_n \subset \operatorname{im} \partial_{n+1}$: Suppose that (b, a) represents an element of $\pi_n(g, z_0)$ which is mapped onto $0 \in \pi_n(h, z_0)$ by $(f, 1)_n$. Then, there is a based map $b' : B^n \to Z$ such that $b' | S^{n-1} = a$ and $f \circ b \simeq h \circ b'$ rel. S^{n-1}. Let

$$H : B^n \times I \to X$$

denote the corresponding homotopy; the map H factors through B^{n+1} via the map h^n defined on page 4, thus giving rise to a map $b'': B^{n+1} \to X$

with $b''\circ i_+ = h\circ b' = f\circ g\circ b'$ and $b''\circ i_- = f\circ b$. Therefore, there is a unique map $a'' : S^n \to Y$ with $a''\circ i_+ = g\circ b'$, $a''\circ i_- = b$ and $f\circ a'' = b''|S^n$. Hence, the pair (b'', a'') represents an element of $\pi_{n+1}(f, y_0)$. It remains to show that its image under ∂_{n+1} is the class of (b, a). This is done by considering the homotopy

$$H' : (S^{n-1}\times I)\times I \to Y$$

given by

$$H'(s,t,t') = \begin{cases} g\circ b'\circ c^{n-1}(s, t'-2t), & 0\leqslant t\leqslant t'/2, \\ b\circ c^{n-1}\left(s, \dfrac{2t-t'}{2-t'}\right), & t'/2\leqslant t\leqslant 1. \end{cases}$$

□

Some particular cases are of special interest:

Case 1 The maps f and g are inclusions of subspaces; then, the exact sequence of the theorem above becomes the exact sequence of the based triple (X, Y, Z, z_0) (see Spanier, 1966, Theorem 7.2.15).

Case 2 Suppose that X is a singleton space; then, one gets the exact homotopy sequence of the map g, namely:

$$\cdots \to \pi_n(Y, y_0) \to \pi_n(g, z_0) \to \pi_{n-1}(Z, z_0) \to \pi_{n-1}(Y, y_0) \to \pi_{n-1}(g, z_0) \to$$
$$\cdots \to \pi_1(Y, y_0) \to \pi_1(g, z_0) \to \pi_0(Z, z_0) \to \pi_0(Y, y_0).$$

Case 3 Moreover, if in the previous case, the map g is an inclusion, one obtains the usual homotopy sequence of the based pair (Y, Z, z_0) (see Spanier, 1966, Theorem 7.2.3). As an application, consider the case $Z = \{y_0\}$; then $\pi_n(Z, z_0) = 0$, for every $n\in\mathbf{N}$, and the homotopy sequence shows

$$\pi_n(Y, y_0) \cong \pi_n(Y, \{y_0\}, y_0),$$

again for all $n\in\mathbf{N}$.

The veracity of the following result stems from the exact sequence discussed in Case 2 above.

Proposition A.8.10 *A map $f : Y \to X$ is a weak homotopy equivalence iff $f_0 : \pi_0(Y, y_0) \to \pi_0(X, f(y_0))$ is onto and $\pi_n(f, y_0) = 0$, for every $y_0\in Y$ and every $n\in\mathbf{N}\setminus\{0\}$.* □

The condition given in the preceding proposition is equivalent to saying that every path-component of X meets $f(Y)$, and, for any admissible pair

$(b,a):i^{n-1}\to f(n\geqslant 1)$ there is an extension $b':B^n\to Y$ of a such that $f\circ b'$ is homotopic to b rel. S^{n-1}. A map $f:Y\to X$ is said to be *m-connected* ($m\in\mathbf{N}$) – or, f is said to be an *m-equivalence* – if this condition holds true for every n such that $1\leqslant n\leqslant m$; for $m=0$, this just means that every path-component of X meets $f(Y)$. In view of Proposition A.8.10, one can give the following characterization of m-connectivity:

Proposition A.8.11 *A map* $f:Y\to X$ *is m-connected* ($m\in\mathbf{N}$)

iff $f_0:\pi_0(Y,y_0)\to\pi_0(X,f(y_0))$ *is onto and* $\pi_n(f,y_0)=0$, *for every* $y_0\in Y$ *and every* $1\leqslant n\leqslant m$;

iff $f_m:\pi_m(Y,y_0)\to\pi_m(X,f(y_0))$ *is an epimorphism and* $f_n:\pi_n(Y,y_0)\to\pi_n(X,f(y_0))$ *is an isomorphism, for every* $y_0\in Y$ *and n such that* $0\leqslant n<m$. □

A pair (X,Y) is said to be *m-connected* ($m\in\mathbf{N}$) if the inclusion $Y\to X$ is m-connected; in this case one has the following version of the previous proposition.

Corollary A.8.12 (i) *The pair* (X,Y) *is* 0*-connected iff the function* $i_0:\pi_0(Y,y_0)\to\pi_0(X,y_0)$ *induced by the inclusion* $i:Y\to X$ *is onto, for any choice of the base point* $y_0\in Y$.

(ii) *The pair* (X,Y) *is m-connected* $(m\geqslant 1)$

iff $i_0:\pi_0(Y,y_0)\to\pi_0(X,y_0)$ *is onto and* $\pi_n(X,Y,y_0)=0$, *for every choice of base point* $y_0\in Y$ *and every n such that* $1\leqslant n\leqslant m$,

iff $i_m:\pi_m(Y,y_0)\to\pi_m(X,y_0)$ *is an epimorphism and* $i_n:\pi_n(Y,y_0)\to\pi_n(X,y_0)$ *is an isomorphism, for every* $y_0\in Y$ *and every n such that* $0\leqslant n<m$. □

The following two properties are easily derived from the concept of n-equivalence.

(1) If g is an m-equivalence, a composition $f\circ g$ is an m-equivalence iff f is an m-equivalence.

(2) Any map homotopic to an m-equivalence is an m-equivalence.

According to Proposition A.4.10 (iv), any map $f:Y\to X$ decomposes in the form $f=r_f\circ i_Y$, where $\operatorname{cod} i_Y=\operatorname{dom} r_f=M(f)$ is the mapping cylinder of f, i_Y is an inclusion and r_f is a homotopy equivalence. Selecting a point $y_0\in Y$ as a base point of both Y and $M(f)$ and taking $f(y_0)=x_0$ to be the base point of X, one may regard the maps f, r_f and i_Y as based. Then, using the exact sequence of Theorem A.8.9 and applying Proposition A.8.10 to r_f, one can conclude that the following result holds true.

Proposition A.8.13 *For every $n \geqslant 1$ and every choice of base point $y_0 \in Y$,*

$$(r_f, 1)_n : \pi_n(M(f), Y; y_0) \to \pi_n(f, y_0)$$

is an isomorphism. □

As in homology, homotopic maps of pairs induce the same homomorphisms of homotopy groups; more precisely, if $f_\lambda : (Y, D) \to (X, A)$, $\lambda = 0, 1$ are homotopic maps of pairs via homotopies moving the image of D only within A, then, for every $n \geqslant 1$, $(f_0)_n = (f_1)_n$. Consequently, a homotopy equivalence of pairs induces an isomorphism of the corresponding homotopy groups. This, together with Proposition A.8.13, allows one to state the following generalization of Theorem A.8.9:

Proposition A.8.14 *Given maps $g : (Z, z_0) \to (Y, y_0)$, $f : (Y, y_0) \to (X, x_0)$ and $h : (Z, z_0) \to (X, x_0)$ such that $f \circ g \simeq h$, there is a long exact sequence connecting the homotopy groups of (f, y_0), (g, z_0) and (h, z_0).* □

The action of the fundamental group of a space on its homotopy groups can be extended to an action of the fundamental group of the domain of a map on the homotopy groups of the map. More precisely, let $f : (Y, y_0) \to (X, x_0)$ be a based map; then, for every $(b, a): i^{n-1} \to f$ and every $\sigma: (I, 0) \to (Y, y_0)$ make the following construction: firstly, the maps a and σ induce a map $S^{n-1} \times \{0\} \cup \{e_0\} \times I \to Y$ which extends to a homotopy $a' : S^{n-1} \times I \to Y$; secondly, $f \circ a'$ and b induce a map $B^n \times \{0\} \cup S^{n-1} \times I \to X$ which extends to a map $b' : B^n \times I \to X$; thirdly, notice that the homotopies b' and a' end at an admissible pair $(b'_1, a'_1): i^{n-1} \to f$ (with respect to the base points $\sigma(1)$ and $f \circ \sigma(1)$) whose homotopy class depends only on the homotopy classes of (b, a) and σ. Thus, σ induces a well-defined function

$$\sigma_* : \pi_n(f, y_0) \to \pi_n(f, \sigma(1)),$$

which is an isomorphism. If one now assumes that σ is a loop and $n > 2$ (so that $\pi_n(f, y_0)$ is abelian), one obtains the announced operation of $\pi_1(Y, y_0)$ on $\pi_n(f, y_0)$, turning $\pi_n(f, y_0)$, $n > 2$ into a left Λ-module, where Λ is the integral group ring of $\pi_1(Y, y_0)$. If $n = 2$, one often considers only the case in which $f_1 : \pi_1(Y, y_0) \to \pi_1(X, f(y_0))$ is an isomorphism; then, $\pi_2(f, y_0)$ is abelian, as a quotient of the abelian group $\pi_2(X, f(y_0))$ and the above argument applies. The case in which f is the inclusion of a space A into a space X is of particular interest:

Proposition A.8.15 *Let (X, A) be a pair of spaces with A path-connected and suppose that $\pi_1(A, x_0)$ acts trivially on $\pi_n(X, A, x_0)$ for some base-point x_0. Then, there is a bijection between $\pi_n(X, A, x_0)$ and the set of free homotopy* *classes of maps* $(B^n, S^{n-1}) \to (X, A)$; *if* $\pi_1(A, x_0) = 0$, *this happens for every* $n \geqslant 1$.

Proof See Spanier, 1966, Section 7.3. □

Sometimes it is possible to decompose an element of a homotopy group of a map into a sum of elements. Let $(b, a) : i^{n-1} \to f$ be an admissible pair, with $f : Y \to X$; let $c : B^{n-1} \to Y$ be a map such that $c|S^{n-2} = a|S^{n-2}$ and $f \circ c = b|B^{n-1}$ (in particular, if $f : Y \subset X$ this means that b restricted to the equator factors through Y). Let $a_+ : S^{n-1} \to Y$ (resp. $a_- : S^{n-1} \to Y$) be defined by $a_+ \circ i_+ = a \circ i_+$, $a_+ \circ i_- = c$ (resp. $a_- \circ i_- = a \circ i_-$, $a_- \circ i_+ = c$). Then, under these assumptions:

Theorem A.8.16 (*The homotopy addition theorem*):

$$[b, a] = [b \circ i_+, a_+] + [b \circ i_-, a_-].$$

Proof Observe that

$$i_+ \vee i_- \circ p_n : (B^n, B^{n-1} \cup S^{n-1}) \to (B^n, B^{n-1} \cup S^{n-1})$$

is homotopic to $\mathbf{1}_{B^n}$, as maps of pairs. □

The homotopy sequence of a fibration

Another long, exact sequence of homotopy groups appears in connection with the theory of fibrations.

Proposition A.8.17 *Let* $p : Y \to X$ *be a fibration with fibre* $F = p^{-1}(x)$, $x \in X$. *Then, for any* $y \in F \subset Y$, *the following sequence of groups (and sets) is exact:*

$$\cdots \to \pi_n(F, y) \xrightarrow[i_n]{} \pi_n(Y, y) \xrightarrow[p_n]{} \pi_n(X, x) \to \pi_{n-1}(F, y) \to$$

$$\cdots \to \pi_1(X, x) \to \pi_0(F, y) \to \pi_0(Y, y) \to \pi_0(X, x)$$

(*here* i_n *is the morphism induced by the inclusion* $i : F \to Y$).

Proof See Spanier (1966, Theorem 7.2.10). □

Corollary A.8.18 *Let* $p : \tilde{X} \to X$ *be a covering projection with* X *path-connected and let* $F = p^{-1}(x_0)$. *If* $\pi_1(X, x_0)$ *acts fixed point free on* F, *then* $p : \tilde{X} \to X$ *is a universal covering projection.*

Proof The proposition implies that the sequence of groups and spaces

$$0 \to \pi_1(\tilde{X}, \tilde{x}_0) \to \pi_1(X, x_0) \to F \to 0$$

is exact; since $\pi_1(X, x_0)$ acts fixed point free on F, the map $\pi_1(X, x_0) \to F$ is injective and so, $\pi_1(\tilde{X}, \tilde{x}_0) = 0$. □

Fundamental groupoid

As seen before, among the functors π_n, the fundamental group functor π_1 plays an exceptionally important role. Sometimes it is necessary to study the fundamental group of a space in connection to several different base points at the same time. The abstract setting for this situation is given by the so-called *fundamental groupoid* ΠX of a space X. Recall that a *groupoid* is a small category with all its morphisms invertible. The objects of the groupoid ΠX are the points of the space X; the set $\Pi X(x, y)$ of its morphisms from x into y is the set of all homotopy classes (rel. end points) of paths from x into y, i.e., a morphism from x into y may be represented by a map

$$w: I \to X$$

with $w(0) = x$, $w(1) = y$ and two maps w_1, w_2 represent the same morphism iff $w_1 \simeq w_2$ rel. $\dot{I} = \{0, 1\}$. The identities in ΠX can be represented by constant maps; morphisms are composed in the obvious way. Moreover, for every $x_0 \in X$, there is a canonical isomorphism

$$\Pi X(x_0, x_0) \cong \pi_1(X, x_0).$$

Clearly, every continuous map $f : X \to Y$ gives rise to a functor $\Pi f : \Pi X \to \Pi Y$, and hence there is a functor Π from *Top* to the category of groupoids. Note the following useful properties of this functor.

(i) If $f : X \to Y$ is a homotopy equivalence, then Πf is an equivalence of groupoids (as categories; see Brown, 1988, 6.5.10 Corollary.).

(ii) If $f : Y \text{-/}\!\!\to A$ is a partial map, then the induced square

$$\begin{array}{ccc} \Pi Y & \longrightarrow & \Pi(A \sqcup_f Y) \\ \uparrow & & \uparrow \\ \Pi \operatorname{dom} f & \longrightarrow & \Pi A \end{array}$$

is a pushout of groupoids. (Because f is a partial map, $\operatorname{dom} f \subsetneq Y$ is a closed cofibration; see Brown, 1988, 8.4.2 Theorem.)

Next is a review of the statement of the celebrated Seifert–van Kampen theorem, and an interesting generalization of that result.

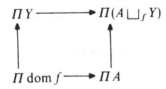

Theorem A.8.19 (*Seifert–van Kampen*) *If $f : Y -/\rightarrow A$ is a partial map such that A, Y and $D = \mathrm{dom}\, f$ are path-connected, then, for every $x \in D$, the induced square of groups*

$$\begin{array}{ccc} \pi_1(Y, x) & \longrightarrow & \pi_1(A \sqcup_f Y, f(x)) \\ \uparrow & & \uparrow \\ \pi_1(D, x) & \longrightarrow & \pi_1(A, f(x)) \end{array}$$

is a pushout in the category of groups.

Proof See Brown (1988, 8.4.2 Theorem).

Let (X, x_0) be a based space, with X path-connected. Let $\{U_\lambda, \lambda \in \Lambda\}$ be a covering of X by path-connected open sets such that:

(i) for every $\lambda \in \Lambda$, $x_0 \in U_\lambda$;
(ii) for any two indices λ_1, $\lambda_2 \in \Lambda$, there exists an index $\lambda \in \Lambda$ such that $U_{\lambda_1} \cap U_{\lambda_2} = U_\lambda$.

Let

$$\phi_{\lambda,\mu} : \pi_1(U_\lambda, x_0) \rightarrow \pi_1(U_\mu, x_0)$$

be the homomorphism induced by the inclusion $U_\lambda \subset U_\mu$. □

Under these conditions, the following result holds true.

Proposition A.8.20 *The group $\pi_1(X, x_0)$ is isomorphic to the colimit group of the diagram (i.e., small category) whose objects are all groups $\pi_1(U_\lambda, x_0)$ and whose morphisms are the identity homomorphisms and the homomorphisms $\phi_{\lambda,\mu}$.*

Proof See Massey (1984, Chapter 4). □

A.9 Dimension and embedding

In this book, the dimension of a space is always understood to be the covering dimension (see Pears, 1975, and Engelking, 1978). For normal spaces it is characterized by the following.

Theorem A.9.1 *A normal space X has dimension $\leqslant n$ iff for each closed subset $C \subset X$, each map $C \rightarrow S^n$ has a continuous extension over the entire space X.*

Proof See Engelking (1978, Theorem 3.2.10). □

Corollary A.9.2 *If Y is a closed subspace of a space X, then* $\dim Y \leqslant \dim X$. □

Example 1 The ball B^1 has dimension 1. By the intermediate value theorem, the identity map of S^0 cannot be extended over B^1 (maintaining the target S^0), implying that $\dim B^1 > 0$. On the other hand, consider a map $f : C \to S^1$ defined on a closed subset $C \subset B^1$. By continuity there are open subsets U, V of B^1 such that $e_0 \notin f(C \cap U)$, $-e_0 \notin f(C \cap V)$ and $B^1 = U \cup V \cup (B^1 \backslash C)$. Thus, $\{U, V, B^1 \backslash C\}$ is an open cover of B^1, and by compactness one may find finitely many numbers, say $a_0, a_1, \ldots, a_{k+1}$ such that $-1 = a_0 < a_1 < \cdots < a_{k+1} = 1$, and, for $i = 0, 1, \ldots, k$, $f(C \cap [a_i, a_{i+1}])$ does not contain both e_0 and $-e_0$. One may assume $a_i \in C$, for all i; otherwise, extend f by taking $f(a_i) = e_1$. Now, by Tietze's extension theorem, extend each restriction $f|(C \cap [a_i, a_{i+1}])$ over the whole interval $[a_i, a_{i+1}]$ and glue these extensions together to an extension of f over B^1. The possibility of this extension shows $\dim B^1 \leqslant 1$. □

In order to proceed to higher-dimensional balls, one needs two deeper theorems.

Theorem A.9.3 (*Product theorem*) *If X and Y are compact Hausdorff spaces, at least one of which is non-empty, then,*

$$\dim X \times Y \leqslant \dim X + \dim Y.$$

Proof See Engelking (1978, Theorem 3.2.13). □

Example 2 For every $n \in \mathbf{N}$, $\dim B^n \leqslant n$, since $B^n \cong (B^1)^n$ (see Proposition 1.0.2). □

Example 3 For every $n \in \mathbf{N}$, $\dim S^n \leqslant n$, since an n-sphere can be considered as a union of two n-balls (see Example 8 of Section A.4 and Proposition A.4.8 (v)). □

Theorem A.9.4 (*Brouwer theorem*) *The sphere S^n is not a retract of the ball B^{n+1}, for every $n \in \mathbf{N}$.*

Proof See Milnor (1965, §2.) For $n = 0$, this is just the intermediate value theorem. For $n > 0$, the following is a rough sketch of Milnor's argument. It is given in order to make clear that the proof can be done with analytical methods independent of the combinatorial ideas developed in the main text, and that the reader is not being misled by the use of the Brouwer theorem from the very beginning. Assume one is given a retraction $r: B^n \to S^{n-1}$. Then, the assignment $s \mapsto -r(s)$ describes a fixed point free map $B^{n+1} \to B^{n+1}$ which can be approximated by a fixed point free smooth map $f: B^{n+1} \to B^{n+1}$, by the Weierstrass approximation theorem. This f in turn induces a smooth retraction $\tilde{r}: B^{n+1} \to S^n$. By Sard's theorem, $\tilde{r}$ has at least one regular value, say $\tilde{s}$. The inverse image $\tilde{r}^{-1}(\tilde{s})$ is a compact smooth 1-manifold, and therefore it has an even number of boundary points. But $\tilde{r}^{-1}(\tilde{s})$ has a unique boundary point namely, the point $\tilde{s}$ itself. □

Example 2′ For every $n \in \mathbf{N}$, $\dim B^n = n$, since $\dim B^n \leqslant n-1$ would imply that the identity map of S^{n-1} could be extended over B^n (see Theorem A.9.1), contradicting the Brouwer theorem. □

Example 3′ For every $n \in \mathbf{N}$, $\dim S^n = n$, since every n-sphere contains n-balls as closed subsets (see Corollary A.9.2). □

Note: The question of the dimension of the Euclidean spaces is left open here because it is discussed in the main body of the book (see Section 2.2, Example 1).

To continue the development of this section, it is necessary to prove a rather technical lemma, which characterizes boundary points of compact sets in Euclidean spaces.

Lemma A.9.5 *Let X be a compact subset of $\mathbf{R}^n$ and x a point of X. Then, x is a boundary point of X iff x has arbitrarily small neighbourhoods U, open in X, with the property that any map $X \backslash U \to S^{n-1}$ can be extended over the entire X.*

Proof Cf. Hurewicz & Wallman (1948, Chapter VI, Section 6.)

$\Rightarrow$: Let B_ε denote the closed ε-neighbourhood of x in $\mathbf{R}^n$, for any positive real ε, and take $U = X \cap \mathring{B}_\varepsilon$. Given a map $f: X \backslash U \to S^{n-1}$, choose a continuous extension $f': \delta B_\varepsilon \to S^{n-1}$ of $f | X \cap \delta B_\varepsilon$; this is possible since δB_ε is a sphere of dimension $n-1$ (see Theorem A.9.1 and Example 3′). Since x is assumed to be a boundary point of X, one can find a point $y \in \mathring{B}_\varepsilon \backslash U$ and a retraction $r: B_\varepsilon \backslash \{y\} \to \delta B_\varepsilon$. The desired extension $g: X \to S^{n-1}$ of f can now be defined by taking $g | U = f' \circ (r | U)$.

$\Leftarrow$: Assume x to be an interior point of X. Choose a positive real number ε such that the closed ε-neighbourhood B_ε of x is completely contained in X. Identify its boundary δB_ε with S^{n-1} and choose a retraction $r : \mathbf{R}^n\backslash\{x\} \to S^{n-1}$. Now, if B_ε would contain an open neighbourhood U of x such that the map $f = r|(X\backslash U) : X\backslash U \to S^{n-1}$ could be extended to a map $g : X \to S^{n-1}$, then $g|B_\varepsilon$ would be a retraction from B_ε to S^{n-1}, which cannot exist according to the Brouwer Theorem (see Theorem A.9.4).

□

Theorem A.9.6 (*Theorem of the invariance of domain*) *if $X \subset \mathbf{R}^n$ is open and $f : X \to \mathbf{R}^n$ is an injective map, then $f(X)$ is an open subset of $\mathbf{R}^n$.*

Proof Take a point $x \in X$ and a closed ε-neighbourhood B_ε of x in $\mathbf{R}^n$ which is completely contained in X. The map f induces a homeomorphism $B_\varepsilon \to f(B_\varepsilon)$; thus, the fact that x is an interior point of B_ε implies that $f(x)$ cannot be a boundary point of $f(B_\varepsilon)$ (see Lemma A.9.5). Therefore, $f(x)$ is an interior point of $f(B_\varepsilon)$ with respect to R^n, and, consequently, an interior point of $f(X)$. □

The question of embedability is settled by the next two theorems.

Theorem A.9.7 (*Theorem of Menger–Nöbeling*) *A metrizable space of dimension n satisfying the second axiom of countability can be embedded into $\mathbf{R}^{2n+1}$.*

Proof See Engelking (1978, Theorem 1.11.4). □

Recall that the *Hilbert cube* is defined as the metric space consisting of the set

$$I^\infty = \left\{(x_0, x_1, \ldots,) : |x_i| \leqslant \frac{1}{i+1}, i \in \mathbf{N}\right\}$$

and the metric $d : I^\infty + I^\infty \to [0, \infty)$ given by

$$d(\mathbf{x}, \mathbf{y}) = \sqrt{\sum_{i=0}^{\infty} (x_i - y_i)^2}.$$

As a topological space, the Hilbert cube is homeomorphic to the cartesian product of countably many intervals I (see Dugundji, 1966, Chapter IX, Proposition 8.4).

Theorem A.9.8 *A metric space satisfies the second axiom of countability iff it is homeomorphic to a subspace of the Hilbert cube.*

Proof See Bourbaki (1966, Chapter IX, Section 2.8, Theorem 12). □

In this context, another important property of the Hilbert cube should be mentioned:

Proposition A.9.9 *If C is a compact metric space, then the function space $(I^\infty)^C$ satisfies the second axiom of countability.*

Proof Since C is compact metric, there is a sequence $\{f_n : n\in\mathbf{N}\}$ of maps $f_n: C\to\mathbf{R}$ that is separating, i.e., for every pair of points $x, y\in C$, with $x\neq y$, there is an $n\in\mathbf{N}$, such that $f_n(x)\neq f_n(y)$. Then, by the Weierstrass–Stone theorem (see Bourbaki, 1966, Chapter X, Section 4, Proposition 6), every map $f : C\to\mathbf{R}$ can be approximated, in the metric topology, by polynomials in the functions f_n with real coefficients. But, since the rational numbers form a dense subset of the reals, it is enough to consider only polynomials with rational coefficients. Thus, $\mathbf{R}^C$ has a countable dense subset in the metric topology and therefore, satisfies the second axiom of countability. Notice that the metric topology of $\mathbf{R}^C$ coincides with the compact-open topology (see Section A.1). The second axiom of countability property carries over to the subspace I^C of $\mathbf{R}^C$ and to the countable product $(I^C)^{\mathbf{N}} = (I^\infty)^C$ (cf. the exponential law, Section A.1). □

A.10 The adjoint functor generating principle

The following basic construction (see Kan, 1958a, b) is used in several places throughout the text. Let D be a small category and let *DSets* denote the category of contravariant functors $D\to Sets$. One might view the objects of *DSets* as sets graded by the objects of D with the morphisms of D operating on the right. (As a particular example of this situation, one can quote the definition of simplicial sets given in Section 4.2.) Let $\Phi : D\to Sets$ be an arbitrarily given covariant functor; associated to it, construct a pair of adjoint functors

$$\Gamma_\Phi : DSets \to Sets,$$
$$S_\Phi : Sets \to DSets,$$

as follows:

(1) The left adjoint functor Γ_Φ-called *realization functor*-associates to each object X of *DSets* the set of equivalence classes of pairs $(x, t)\in\bigsqcup X(d)\times\Phi(d)$ modulo the relation

$$(x\alpha, t)\sim(x, \Phi(\alpha)(t)).$$

Given a morphism $f : Y\to X$, i.e., a natural transformation, one has the function

$$\Gamma_\Phi f : \Gamma_\Phi Y \to \Gamma_\Phi X, [x,t] \to [f(x), t]$$

where $[x,t]$ denotes the equivalence class represented by the pair (x,t).

(2) The right adjoint functor S_Φ-called *singular functor*-associates to each set Z the object of *DSets* given by

$$(S_\Phi Z)(d) = Z^{\Phi(d)},$$

for each object d of D, and

$$\alpha^* : Z^{\Phi(d)} \to Z^{\Phi(d')}, \qquad x \to x \circ \Phi(\alpha),$$

for each morphism $\alpha : d' \to d$ of D.

The unit and the co-unit of the adjunction $\Gamma_\Phi \dashv S_\Phi$ are given as follows:

(1) Let X be an object of *DSets*. To each element $x \in X(d)$ associate the function

$$\bar{x} : \Phi(d) \to \Gamma_\Phi X, \qquad t \to [x,t];$$

then, the unit consists of the morphisms

$$\eta_X : X \to S_\Phi \Gamma_\Phi X, \qquad x \to \bar{x}.$$

(2) The co-unit consists of the functions

$$j_Z : \Gamma_\Phi S_\Phi Z \to Z, \qquad [x,t] \to x(t).$$

The situation can easily be generalized to functors Φ whose codomain is an arbitrary cocomplete category C instead of just sets. In this book, C is taken to be *Top*, *SiSets* or *PSiSets*. The construction of the singular functor $S_\Phi : C \to DSets$ remains almost unchanged: one takes

$$(S_\Phi Z)(d) = \{\Phi(d) \to Z\}$$

and

$$\alpha^* : \{\Phi(d) \to Z\} \to \{\Phi(d') \to Z\}, x \to x \circ \Phi(\alpha).$$

As for the realization functor, observe that if $C = Sets$, one can consider Γ_Φ as a coequalizer of an abstractly defined pair of functions; the definitions of these functions can be imitated in any cocomplete category C (see Gabriel & Zisman, 1967).

Exercises (to the whole Appendix)

1. Let $i : Y \to X$ be a closed cofibration. Prove that the mapping cone $C(i)$ has the same homotopy type as X/Y.
2. Let $i : A \to X$, $i' : A \to X'$ be closed cofibrations and let $f : X \to X'$ be a homotopy equivalence such that $f \circ i = i'$. Prove that f is a homotopy equivalence rel. A, i.e., there exists a homotopy inverse $g : X' \to X$ such that $g \circ i' = i$ and $g \circ f \simeq \mathbf{1}_X$ rel. A, $f \circ g \simeq \mathbf{1}_{X'}$ rel. A.
3. Let $p : Y \to X$ be a Hurewicz fibration and $f : Z \to X$ be a map; if X,

Y and Z are LEC spaces then, $Z \sqcap_f Y$ is LEC. (See Heath, 1986.)

4. Let $f : Y -/\rightarrow A$ be a partial map with Y and A paracompact. Prove that the resulting adjunction space X is also paracompact. (See Michael, 1953, 1956.)
5. Let $f : Y -/\rightarrow A$ be a partial map with Y and A stratifiable. Prove that $A \sqcap_f Y$ is stratifiable. (See Borges 1966.)
6. Show that the union space of an expanding sequence of stratifiable spaces is stratifiable. (See Borges, 1966.)
7. Let X be the union space of the expanding sequence $\{X_n : n \in \mathbf{N}\}$. Show that any compact subset of X is contained in some X_n.

Bibliography

Alder, M.D. (1974) Inverse limits of simplicial complexes, *Compositio Math.* **29**, *1–7.*

Alexander, James Waddell (1915) A proof of the invariance of certain constants of analysis situs, *Trans. Amer. Math. Soc.* **16**, *148–154.*

Alexandroff, Pavel Sergejewitsch (1925) Zur Begründung der *n*-dimensionalen Topologie, *Math. Ann.* **94**, *296–308.*

Allaud, Guy (1972) De-looping homotopy equivalences, *Arch. Math.* (*Basel*) **23**, *167–169.*

Barratt, Michael (1956) Simplicial and semisimplicial complexes, *Mimeographed lecture notes* (Princeton University).

Borges, Carlos R. (1966) On stratifiable spaces, *Pacific J. Math.* **17**- *1–16.*

Borovikov, V. (1952) On the intersection of a sequence of simplexes, *Uspehi Mat. Nauk* (*N.S.*) **7**, *179–180.*

Borsuk, Karol (1967) *Theory of Retracts*, Monografle Matematyczne 44, Warszawa (Państwowe wydawnictwo naukowe).

Bourbaki, Nicolas (1966) *General Topology*, Parts 1 and 2, Adiwes International Series in Mathematics, Don Mills (Addison–Wesley).

Brown, Ronald (1961) Some problems in algebraic topology: function spaces, function complexes and FD-complexes, *Thesis* (University of Oxford).

(1968) *Elements of Modern Topology*, European Mathematics Series, London–New York–Sydney–Toronto–Mexico–Johannesburg (McGraw-Hill). Revised, updated and expanded version:

(1988) *Topology, A geometric Account of General Topology, Homotopy Types and the Fundamental Groupoid*, Mathematics and its Applications, Chichester (Ellis Horwood)

Brown, Ronald & Heath, Philip Richard (1970) Coglueing homotopy equivalences, *Math. Z.* **113**, *313–325.*

Cauty, Robert (1976) Sur les espaces d'applications dans les CW-complexes, *Arch. Math.* (*Basel*) **27**, *306–311.*

Ceder, Jack G. (1961) Some generalizations of metric spaces, *Pacific J. Math.* **11**, *105–125.*

Curtis, Edward Baldwin (1971) Simplicial homotopy theory, *Advances in Math.* **6**, *107–209.*

tom Dieck, Tammo (1971) Partitions of unity in homotopy theory, *Compositio Math.* **23**, *159–167.*

tom Dieck, Tammo, Kamps, Klaus Heiner and Puppe, Dieter (1970)

Homotopietheorie, Lecture Notes in Mathematics 157, Berlin–Heidelberg–New York (Springer).

Dold, Albrecht (1963) Parition of unity and the theory of fibrations, *Ann. of Math.* **78**, *223–25.*

(1972) *Lectures on Algebraic Topology*, Die Grundlehren der Mathematischen Wissenschaften in Einzeldarstellungen mit besonderer Berücksichtigung der Anwendungsgebiete **200**, Berlin–Heidelberg–New York (Springer).

Dowker, Clifford Hugh (1952) Topology of metric complexes, *Amer. J. Math.* **74**, *555–577.*

Dugundji, James (1965) Locally equiconnected spaces and absolute neighborhood retracts, *Fund. Math.* **57**, *187–193*

(1966) *Topology*, Allyn and Bacon Series in Advanced Mathematics, Boston–London–Sydney (Allyn and Bacon).

Dydak, Jerzy & Geoghegan, Ross (1986) The singular cohomology of the inverse limit of a Postnikov tower is representable, *Proc. Amer. Math. Soc.* **98**, *649–654.*

Dyer, Eldon & Eilenberg, Samuel (1972) An adjunction theorem for locally equiconnected spaces, *Pacific J. Math.* **41**, *669–685.*

Eckmann, Beno & Hilton, Peter John (1958) Groupes d'homotopie et dualité, *C.R. Acad. Sci. Paris Sér.* **A-B 246**, Groupes absolus: *2444–2447*; suites exactes: *2555–2558*; coefficients: *2991–2993.*

Eilenberg, Samuel & MacLane, Saunders (1953) Acyclic models, *Amer. J. Math.* **75**, *189–199.*

Eilenberg, Samuel & Steenrod, Norman (1952) *Foundations of Algebraic Topology*, Princeton (Princeton University Press).

Eilenberg, Samuel & Zilber, J.A. (1950) Semi-simplicial complexes and singular homology, *Ann. of Math.* (2) **51**, *499–513.*

Engelking, Ryszard (1977) *General Topology*, Monografie Matematyczne **60**, Warszawa (Państwowe wydawnictwo naukowe).

(1978) *Dimension Theory*, North Holland Mathematical Library **19**, Amsterdam–Oxford–New York (North Holland).

Finney, Ross L. (1965) The insufficiency of barycentric subdivision, *Michigan Math. J.* **12**, *263–272.*

Fox, Ralph Hartzler (1943) On fibre spaces. II, *Bull. Amer. Math. Soc.* **49**, *733–735.*

Freudenthal, Hans (1938) Über die Klassen der Sphärenabbildungen I. Große Dimensionen, *Compositio Math.* **5**, *299–314.*

Freyd, Peter (1964) *Abelian Categories: An Introduction to the Theory of Functors*, Harper's Series in Modern Mathematics, New York–Evanston–London (Harper & Row).

Fritsch, Rudolf (1969) Zur Unterteilung semisimplizialer Mengen. I, *Math. Z.* **108**, *329–327*; II, *Math. Z.* **109**, *131–152.*

(1972) Simpliziale und semisimpliziale Mengen, *Bull. Acad. Polon. Sci. Ser. Sci. Math. Astronom. Phys.* **20**, *159–168.*

(1974) Relative semisimpliziale Approximation, *Arch. Math.* (*Basel*) **25**, 75–78.
(1976) An approximation theorem for maps into Kan fibrations, *Pacific J. Math.* **65**, *347–351.*
(1983) Remark on the simplicial-cosimplicial tensor product, *Proc. Amer. Math. Soc.* **87**, *200–202.*

Fritsch, Rudolf & Latch, Dana May (1981) Homotopy inverses for nerve, *Math. Z.* **177**, *147–179.*

Fritsch, Rudolf & Puppe, Dieter (1967) Die Homöomorphie der geometrischen Realisierungen einer semisimplizialen Menge und ihrer Normalunterteilung, *Arch. Math.* (*Basel*) **18**, *508–512.*

Gabriel, Peter & Zisman, Michel (1967) *Calculus of Fractions and Homotopy Theory,* Ergebnisse der Mathematik und ihrer Grenzgebiete **35**, Berlin–Heidelberg–New York (Springer).

Gale, David (1950) Compact sets of functions and function rings, *Proc. Amer. Math. Soc.* **1**, *303–308.*

Giever, John B. (1950) On the equivalence of two singular homology theories, *Ann. of Math.* (2) **51**, *178–190.*

Gugenheim, Victor K.A.M. (1968) Semisimplicial homotopy theory, *Studies in Modern Topology, 99–133,* Englewood Cliffs, New Jersey (Math. Assoc. Amer., distributed by Prentice–Hall).

Hanner, Olof (1951) Some theorems on absolute neighbourhood retracts, *Arkiv för Mathematik* **1**, *389–408.*

Heath, Philip Richard (1986) A pullback theorem for locally equiconnected spaces, *Manuscripta Math.* **55**, *233–237.*

Hilton, Peter John & Stammbach, Urs (1971) *A Course in Homological Algebra,* Graduate Texts in Mathematics **4**, New York–Heidelberg–Berlin (Springer).

Hilton, Peter John & Wylie, Samuel (1960) *Homology Theory*, Cambridge (Cambridge University Press).

Hu, Sze-Tsen (1964) *Elements of General Topology*, Holden–Day Series in Mathematics, San Francisco–London–Amsterdam (Holden–Day).

Hurewicz, Witold & Wallman, Henry (1948) *Dimension Theory* (revised edition),Princeton Mathematical Series **4**, Princeton (Princeton University Press).

Hyman, Daniel M. (1968) A category slightly larger than the metric and CW-categories, *Michigan Math. J.* **15**, *193–214.*

Kan, Daniel M. (1955) Abstract homotopy **I**, *Proc. Nat. Acad. Sci. U.S.A.* **41**, *1092–1096.*
(1956) Abstract homotopy **II**, *Proc. Nat. Acad. Sci. U.S.A.* **42**, *255–258*; **III**, ibid. *419–421*; **IV**, ibid. *542–544.*
(1957) On c.s.s. complexes, *Amer. J. Math.* **97**, *449–476.*
(1958a) Adjoint functors, *Trans. Amer. Math. Soc.* **87**, *294–329.*
(1958b) Functors involving c.s.s. complexes, *Trans. Amer. Math. Soc.* **87**, *330–346.*
(1958c) A combinatorial definition of homotopy groups, *Ann. of Math.* (2) **67**, *282–312.*

(1970) Is an ss complex a css complex?, *Advances in Math.* **4**, *170–171.*
Kaplan, S. (1947) Homology properties of arbitrary subsets of Euclidean spaces, *Trans. Amer. Math. Soc.* **62**, *248–271.*
Kelley, John L. (1955) *General Topology*, The University Series in Higher Mathematics, Princeton–New York–Toronto–London (Van Nostrand).
Kodama, Yukihiro (1957) A relation between two realizations of complete semisimplicial complexes, *Proc. Japan Acad.* **33**, *536–540.*
Lamotke, Klaus (1963) Beiträge zur Homotopietheorie simplizialer Mengen, *Bonner Math. Schriften* **17**.
(1968) *Semisimpliziale algebraische Topologie*, Die Grundlehren der mathematischen Wissenschaften in Einzeldarstellungen mit besonderer Berücksichtigung der Anwendungsgebiete **147**, Berlin–Heidelberg–New York (Springer).
Lefschetz, Solomon (1970) The early developments of algebraic topology, *Bol. Soc. Brasil. Mat.* **1**, *1–48.*
Lewis, L. Gaunce, jr. (1982) When is the natural map $X \to \Omega\Sigma X$ a cofibration?, *Trans. Amer. Math. Soc.* **273**, *147–155.*
Lillig, Joachim (1973) A union theorem for cofibrations, *Arch. Math.* (*Basel*) **24**, *410–415.*
Lundell, Albert T. and Weingram, Stephen (1969) *The Topology of CW-Complexes*, The University Series in Higher Mathematics, New York–Cincinnati–Toronto–London–Melbourne (Van Nostrand Reinhold).
MacLane, Saunders (1963) *Homology*, Die Grundlehren der mathematischen Wissenschaften in Einzeldarstellungen, **114**, Berlin–Heidelberg–New York (Springer).
(1971) *Categories for the working mathematician*, New York–Heidelberg–Berlin (Springer).
Mardešic, Sibe & Segal, Jack (1982) *Shape Theory*, North Holland Mathematical Library **26**, Amsterdam–New York–Oxford (North Holland).
Massey, William S. (1984) *Algebraic Topology: An Introduction*, 6th printing, Graduate Texts in Mathematics **56**, New York–Heidelberg–Berlin (Springer).
Mather, Michael R. (1964) Paracompactness and partitions of unity, *Mimeographed Notes* (Cambridge University).
May, John Peter (1967) *Simplicial Objects in Algebraic Topology*, Van Nostrand Mathematical Studies **11**, Princeton–Toronto–London–Melbourne (D. Van Nostrand Company).
McCord, Michael C. (1969) Classifying spaces and infinite symmetric products, *Trans. Amer. Math. Soc.* **146**, *273–298.*
Metzler, Wolfgang (1967) Beispiele zu Unterteilungsfragen bei CW- und Simplizialkomplexen, *Arch. Math.* (*Basel*) **18**, *513–519.*
Michael, Ernest (1953) Some extension theorems for continuous functions, *Pacific J. Math.* **3**, *789–806.*
(1956) Continuous selections I, *Ann. of Math.* **63**, *361–382.*
Milnor, John W. (1956) Construction of universal bundles. I, *Ann. of Math.* (2)**63**, *272–284.*

(1957) The geometric realization of a semi-simplicial complex, *Ann. of Math.* (2) **65**, *357–362.*
(1959) On spaces having the homotopy type of a CW-complex, *Trans. Amer. Math. Soc.* **90**, *272–280.*
(1961) Two complexes which are homeomorphic but combinatorially distinct, *Ann. of Math.* (2) **74**, *575–590.*
(1962) On axiomatic homology theory, *Pacific J. Math.* **12**, *337–341.*
(1965) *Topology from the Differentiable Viewpoint*, Charlottesville (The University Press of Virginia).

Miyazaki, Hiroshi (1952) The paracompactness of CW-complexes, *Tohoku Math. J.* (2) **4**, *309–313.*

Moore, John C. (1955) Le théorème de Freudenthal, la suite exacte de James et l'invariant de Hopf généralisé, *Algèbres d'Eilenberg–MacLane et homotopie*, Séminaire Henri Cartan de l'Ecole Normale Supérieure 1954/55, Paris (Secrétariat mathématique, 11 rue Pierre Curie).
(1958) Semi-simplicial complexes and Postnikov systems, *Symposium Internacional de Topologia Algebraica 232–247*, Mexico (Universidad Nacional Autonoma de Mexico y UNESCO).

Morita, Kiiti (1954) On spaces having the weak topology with respect to closed coverings **II**, *Proc. Japan Acad.* **30**, *711–717.*

Morita, Kiiti and Hanai, S. (1956) Closed mappings and mapping spaces, *Proc. Japan Acad.* **32**, *10–14.*

Northcott, Douglas Geoffrey (1960) *An Introduction to Homological Algebra*, (Cambridge University Press).

Pears, Alan R. (1975) *Dimension Theory of General Spaces*, (Cambridge University Press).

Piccinini, Renzo Angelo (1973) *CW-Complexes, Homology Theory*, Queen's Papers in Pure and Applied Mathematics **34**, Kingston Ontario (Queen's University).

Poincaré, Henri (1895) Analysis situs, *J. de l'École Polytech.* (2) **1**, *1–121.*

Puppe, Dieter (1958) Homotopie und Homologie in abelschen Gruppen–und Monoidkomplexen. **I**, *Math. Z.* **68**, *367–406.*
(1983) Homotopy cocomplete classes of spaces, *Topological Topics*, London Mathematical Society Lecture Notes Series **86**, *55–69*, Cambridge–London–New York–New Rochelle–Melbourne–Sydney (Cambridge University Press).

Quillen, Daniel G. (1968) The geometric realisation of a Kan fibration is a Serre fibration, *Proc. Amer. Math. Soc.* **19**, *1499–1500.*

Ringel, Claus Michael (1970) Eine Charakterisierung der Homotopiekategorie der CW-Komplexe, *Math. Z.* **115**, *359–365.*

Rourke, Colin P. and Sanderson, Brian J. (1971) Δ-sets, *Quart. J. Math. Oxford* (2) **22**, **I.** Homotopy theory: *321–338*; **II.** Block bundles and block fibrations: *465–485.*
(1972) *Introduction to Piecewise Linear Topology*, Ergebnisse der Mathematik und ihrer Grenzgebiete **69**, Berlin–Heidelberg–New York (Springer).

Ruiz Salguero, Carlos & Ruiz Salguero, Roberto (1978) Remarks about the Eilenberg-Zilber type decomposition in cosimplicial sets, *Rev. Colombiana Mat.* **12**, *61–82.*

Schön, Rolf (1977) Fibrations over a CWh-base, *Proc. Amer. Math. Soc.* **62**, *165–166.*

Schubert, Horst (1958) Semisimpliziale Komplexe, *Jahresber. Deutsch. Math.-Verein.* **61**, *126–138.*

(1964) *Topologie*, Mathematische Leitfäden, Stuttgart (B.G. Teubner). English translation:

(1968) *Topology*, Boston–London–Sydney (Allyn & Bacon).

Segal, Jack (1965) Isomorphic complexes, *Proc. Amer. Math. Soc.* **71**, *571–572.*

Seifert, Herbert & Threlfall, William Richard Maximilian Hugo (1934) *Lehrbuch der Topologie*, Leipzig–Berlin (B.G. Teubner).

Serre, Jean- Pierre (1951) Homologie singulière des espaces fibres, *Ann. of Math.* **54**, *425–505.*

Spanier, Edwin (1966) *Algebraic Topology*, McGraw-Hill-Series in Higher Mathematics, New York (McGraw-Hill).

Spanier, Edwin & Whitehead, John Henry Constantine (1957) Carriers and S-theory, *Algebraic Geometry and Topology (a symposium in honour of S. Lefschetz) 330–360*, Princeton (Princeton University Press).

Stasheff, James D. (1963) A classification theorem for fibre spaces, *Topology* **2**, *239–246.*

Stone, Arthur H. (1956). Metrizability of decomposition spaces, *Proc. Amer. Math. Soc.* **7**, *690–700.*

Strøm, Arne (1966) Note on cofibrations, *Math. Scand.* **19**, *11–14.*

(1968) Note on cofibrations **II**, *Math. Scand.* **22**, *130–142.*

(1972) The homotopy category is a homotopy category, *Arch. Math. (Basel)* **23**, *435–441.*

Varadarajan, Kalathoor (1966) Groups for which Moore spaces $M(\pi, 1)$ exist, *Ann. of Math.* **84**, *368–371.*

Wall, C. Terence (1965) Finiteness conditions for CW-complexes, *Ann. of Math.* *(2)***81**, *56–69.*

Whitehead, John Henry Constantine (1939) Simplicial spaces, nuclei, and m-groups, *Proc. London Math. Soc. (2)* **45**, *243–337.*

(1949a) Cominatorial Homotopy **I**, *Bull. Amer. Soc.* **55**, *213–245.*

(1949b) On the realizability of homotopy groups, *Ann. of Math.* **50**, *261–263.*

(1950) A certain exact sequence, *Ann. of Math.* **52**, *51–110.*

Winkler, Gerhard (1985) *Choquet Order and Simplices with Applications in Probabilistic Models*, Lecture Notes in Mathematics **1145**, Berlin–Heidelberg–New York–Tokyo (Springer).

Zeeman, Eric Christopher (1964).Relative simplicial approximation, *Proc. Camb. Phil. Soc.* **60**, *39–43.*

Symbols

Table 1. *Abbreviations*

Symbol	Explanation	Example	Introduced or first used on page
$\square$	indicates the end of a proof, remark or the description of an example		10
iff	introduces a necessary and sufficient condition		15
$\Rightarrow$	shows implication; in a proof: the given condition is necessary		111, 15
$\Leftarrow$	shows, in a proof, that the given condition is sufficient		15
$\Leftrightarrow$	indicates logical equivalence or a definition		111, 133
$\hookrightarrow$	denotes an embedding		94
$-/\rightarrow$	partial map	$f : Y -/\rightarrow A$	258

Table 2. *Operations*

Symbol	Meaning	Example	Introduced or first used on page
$\times$	product in the categorical sense	$B^n \times B^n$	2
$\times_c$	Cartesian product of spaces	$X \times_c Y$	59
$\vee$	binary wedge product	$B^n \vee B^n$	10
$\bigvee_\Gamma$	wedge product over objects indexed by Γ	$\bigvee_\Gamma S^n_\Gamma$	18
$\wedge$	binary smash product	$D \wedge A$	270, 4
$\sqcap$	pullback	$A \sqcap_f Y$	258
$\sqcup$	attaching	$A \sqcup_f Y$	259
$\bigsqcup\limits_{\lambda\in\Lambda} = \bigsqcup_{\lambda\in\Lambda}$	coproduct over objects indexed by Λ	$\bigsqcup_{\lambda\in\Lambda} X_\lambda$	57

Table 2. (*Cont.*)

Symbol	Meaning	Example	Introduced or first used on page
$\bigcup_{n=0}^{\infty}$	formation of a countable union, mostly of the union space of an expanding sequence (appears also in slightly different forms like $\bigcup\limits_{n=0}^{\infty}$ or $\bigcup_{n\in N}$)	$\bigcup_{n=0}^{\infty} X_n$	273
$\otimes$	tensor product	$X \otimes Y$	148
$\circ$	composition of maps	$b \circ c$	6
$\circ$	interior	$\mathring{B}^{n+1}$	1
δ	boundary	δB^{n+1}	1
$(-,-)_n$	induced homomorphism between the n-th homotopy groups	$(b,a)_n$	293
$[-,-]_*$	set of all based homotopy classes of based maps	$[Y,X]_*$	287
$C_{\bar{f}}(-)$	$\bar{f}$-collar	$C_{\bar{f}}(V)$	20
$C_{\infty}(-)$	infinite collar	$C_{\infty}(V_m)$	27
colim	colimit	colim ΔX	141
deg	degree	deg γ	29
dim	dimension of a cell	dim e	12
	a CW-complex	dim X	46
	of a Euclidean complex	dim K	100
	of a simplicial complex	dim K	121
	of an operator	dim α	132
	of a simplex in a simplicial set	dim x	139
	of a simplicial set	dim X	146
	of a presimplicial set	dim X	165
	of a space	dim X	300
dom	domain, source	*dom* f	258, 14
$-^n$	n-th power	$\mathbf{F}^n$	11
$-^n$	n-skeleton	X^n	22, 98
$-^{(n)}$	n-skeleton of a relative CW-complex	$X^{(n)}$	26
$s = s^{\#}s^{\flat}$	unique presentation of a point $\mathbf{s}\in\Delta^n$ by a face operator $\mathbf{s}^{\#}$ applied to an interior point $\mathbf{s}^{\flat}$		133
$x = x^{\#}x^{\flat}$	Eilenberg–Zilber decomposition of a simplex into a degeneracy operator		145

Table 2. (*Cont.*)

Symbol	Meaning	Example	Introduced or first used on page
	$x^\flat$ applied to a non-degenerate simplex $x^\#$		
$y = y^\# y^\flat$	Eilenberg–Zilber decomposition of a point in a cosimplicial set into a face operator $y\flat$ applied to an interior point $y^\#$		147

Table 3. *Objects*

Symbol	Explanation	Introduced or first used on page
B^0	0-ball	2
B^{n+1}	$(n+1)$-ball	1
B^∞	infinite ball	2
B_λ	indexed copy of a ball	12
BG	classifying set of the group G	192
$\mathbf{C}$	field of complex numbers	11
(C, B, p)	(Euclidean) cone with base B and peak p	92
$C(f)$	mapping cone of the map f	269, 63
CD	cone over the space D	269, 63
$C.(f)$	reduced mapping cone of the based map f	269
$C.D$	reduced cone over the pointed space D	269
$C(Y, X)$	set of maps $Y \to X$	241
$C_0(Y, X)$	space of maps $Y \to X$, provided with the compact-open topology	241
$\mathbf{C}P^n$	complex projective n-space	11, 25
e	cell	15
$\bar{e}$	closed cell	15
$\mathbf{F}$	$\mathbf{R}, \mathbf{C}$, or $\mathbf{H}$	11
$\mathbf{H}$	skewfield of quaternions	11
$\mathbf{H}P^n$	projective n-space over the quaternions	11
$K(\Lambda)$	simplicial complex obtained from a Λ-indexed family of sets	110
$L(p, q)$	lens space of type (p, q)	169
$M(f)$	mapping cylinder of the map f	264, 63
$M.(f)$	reduced mapping cylinder of the based map f	269
$M(\pi, n)$	Moore space of type (π, n)	18
$[n]$	ordered set of the numbers $0, 1, \ldots, n$	132

Table 3. (*Cont.*)

Symbol	Explanation	Introduced or first used on page
$\mathbf{N}$	set of natural numbers, including 0	11
pB	global set of a Euclidean cone with base B and peak p	93
pL	Euclidean complex describing a cone with base $\|L\|$ and peak p	102
$\mathbf{R}$	field of real numbers	11
$\mathbf{R}^{n+1}$	Euclidean $(n+1)$-space	1
$\mathbf{R}P^n$	real projective n-space	11
$S[p]$	simplicial p-sphere	145
S_λ	indexed copy of a sphere	12
S^n	n-sphere	1
S^∞	infinite sphere	2
$st_K p$	star of p in the Euclidean complex K	101
$St(L)$	star of the subset L in a CW-complex	36
$T(f)$	mapping track of the map f	270
$X(L)$	subcomplex of a CW-complex, generated by the subset L	36
$\tilde{X}$	covering space of X	256
$\mathbf{Z}$	additive group or ring of integers	78
$\mathbf{Z}_p = \mathbf{Z}/p\mathbf{Z}$	residue class group of $\mathbf{Z}$ mod p	169
$\mathbf{Z}\pi$	integral group ring over the group π	287
(Γ, R)	group presentation	80
Δ	geometric simplex	89
Δ^n	standard-n-simplex	93
$\Lambda^k[n]$	k-th horn of $\Delta[n]$	170
Ω'	set of all ordinals not greater than the first uncountable ordinal	245

Table 4. *Functions* (*maps, homomorphisms*)

Symbol	Explanation	Introduced or first used on page
b^n		6
c^n		4
d^n		2
f_*		284
$f_\#$		287
h^n		4
i^n		1
i_+, i_-		3
j^n		2
j_+, j_-	eggs of Columbus	2
k^n		5

Table 4. (*Cont.*)

Symbol	Explanation	Introduced or first used on page
$\dot{k}^n$		5
l^n		6
p^n		6
$\dot{p}^n$		7
$\hat{p}$		7
q^n		10
$\hat{q}^n$		10
$q\mathbf{R}^n, q\mathbf{C}^n, q\mathbf{K}^n$		11
r^{n+1}		7
R^{n+1}		8
v^n		9
$\dot{v}^n$		9
$\delta_i^n = \delta_i$	elementary face operator	133
$\varepsilon_i^n = \varepsilon_i$	vertex operator	134
μ, ν	face operators	134
ρ, τ	degeneracy operators	136
$\sigma_i^n = \sigma_i$	elementary degeneracy operator	135
$\omega_i^m = \omega_i$	preterminal operator	135
Φ		10
Ψ_Λ	$X \to \lvert K(\Lambda)\rvert$	114

Table 5. *Categories of ...*

Symbol	Meaning	Introduced or first used on page
CSiC	simplicial objects in *C*	138
CW	CW-complexes and maps	56
CW^c	CW-complexes and cellular maps	56
CW^r	CW-complexes and regular maps	56
DSets	contravariant functors $D \to Sets$	303
$hTop_*$	based spaces and based homotopy classes of maps	286
k(Top)	*k*-spaces	242
OSiCo	ordered simplicial complexes	111
PSiC	presimplicial objects over *C*	138
PSiSets	presimplicial sets	165
RCW^c	relative CW-complexes and cellular maps	56
Sets	Sets	139
SiC	simplicial objects over *C*	138
SiCo	simplicial complexes	111
SiSets	simplicial sets	140

Table 5. (*Cont.*)

Symbol	Meaning	Introduced or first used on page
$Top = wHk(Top)$	weak Hausdorff k-spaces (except Section A.1)	11
Top_*	based spaces and based maps	288
Δ	finite ordinals	132

Table 6. *Functors*

Symbol	Explanations	Introduced or first used on page
$\vert - \vert$	underlying polyhedron of a Euclidean complex	97
$\vert - \vert$	geometric realization of simplicial complexes	112
	of simplicial sets	139
$\wedge$	$M \to M$, $\mu \mapsto \hat{\mu}$	134
$-^{\perp}$	maximal section, retraction	136
$-_{\perp}$	minimal section, retraction	136
C	presimplicial cone functor	167
C	chain complex functor	284
C_n	n-chain functor	283, 284
C_-	category of simplices functor	140
D_x	$C_x \to \Delta$	141
E	$PSiSets \to SiSets$	165
Ex	$SiSets \to SiSets$	212
Ex^∞	$SiSets \to SiSets$	215
F	formation of the free group	80
FA	formation of the free abelian group	18
$H(-)$	homology functor	284
$H_n(-)$	n-th homology	284
k	k-ification	242
P	based path space functor	256
S	singular functor $Top \to SiSets$	156
S_Φ	(general) singular functor	303
Sd	barycentric subdivision of Euclidean complexes	102
SdX	normal subdivision of simplicial sets	200, 148
Sd^{op}	opnormal subdivision of simplicial sets	200
Sd^n	n times iterated normal subdivision	204
Γ_Φ	realization functor	303
Δ	standard simplices functor	141
$\Delta -$	Yoneda embedding	141
ΔX	$\Delta^- \circ D_X$	141
$\pi = \pi_0$	set of path components	14, 287
π_1	fundamental group	80

Table 6. (*Cont.*)

Symbol	Explanations	Introduced or first used on page
$\pi_n(-,-)$	n-th homotopy group of a based space	287, 69
	n-th homotopy group of a based map	292
$\pi_n(-,-,-)$	relative n-th homotopy group	292, 68
Π	fundamental groupoid functor	298
Σ	suspension	269, 63
Σ	reduced suspension	269, 4
Ω	loop space functor	256

Table 7. *Natural transformations*

Symbol	Explanation	Introduced or first used on page
d'	$\mathrm{Sd} \dot{\to} \mathbf{1}$	200
d''	$\mathrm{Sd}^{op} \dot{\to} \mathbf{1}$	200
e	$\mathbf{1}_{SiSets} \dot{\to} \mathrm{Ex}$	213
j	$\lvert S - \rvert \dot{\to} \mathbf{1}_{Top}$	156
∂_{n+1}	connecting homomorphism	293
η	$\mathbf{1}_{SiSets} \dot{\to} S\lvert - \rvert$	156

Index

In this index, no references are given to items in the bibliography.